ABU - 3090

D1450964

COMPUTER METHODS FOR PARTIAL DIFFERENTIAL EQUATIONS, VOLUME 1

Elliptic Equations and the Finite-Element Method

ROBERT VICHNEVETSKY
Rutgers University
New Brunswick, New Jersey

Prentice-Hall, Inc., Englewood Cliffs, New Jersey 07632

Library of Congress Cataloging in Publication Data

Vichnevetsky, Robert.
    Computer methods for partial differential
equations.

    (Prentice-Hall series in computational
mathematics)
    Bibliography: p. 336
    Includes index.
    Contents: v. 1. Elliptic equations and the finite-element method.
    1. Differential equations, Partial—Data
processing.   I. Title.   II. Series.
QA374.V53   515.3'53'02854   81-10673
ISBN 0-13-165233-8   AACR2

Prentice-Hall Series in Computational Mathematics

Cleve Moler, Advisor

**To Frans van den Dungen
and the kind of scholarliness
he represented
for those who knew him**

Editorial/production supervision
        and interior design by *Linda M. Paskiet*
Manufacturing buyers: *Gordon Osbourne* and *Joyce Levatino*

Printed in the United States of America

10 9 8 7 6 5 4 3 2 1

PRENTICE-HALL INTERNATIONAL, INC., *London*
PRENTICE-HALL OF AUSTRALIA PTY. LIMITED, *Sydney*
PRENTICE-HALL OF CANADA, LTD., *Toronto*
PRENTICE-HALL OF INDIA PRIVATE LIMITED, *New Delhi*
PRENTICE-HALL OF JAPAN, INC., *Tokyo*
PRENTICE-HALL OF SOUTHEAST ASIA PTE. LTD., *Singapore*
WHITEHALL BOOKS LIMITED, *Wellington, New Zealand*

Contents

PREFACE   *vii*

‖ I ‖   FUNDAMENTALS   *1*

‖ 1 ‖   INTRODUCTION   *3*

1.1 Introduction to Partial Differential Equations *3*
1.2 Finite-Difference Approximation of Elliptic Equations *16*
1.3 Rayleigh–Ritz and Galerkin Methods (Function-Expansion Methods) *24*
1.4 Introduction to the Finite-Element Method *29*

‖ 2 ‖   FOURIER SERIES   *35*

2.1 Complex Variables and Trigonometric Functions *35*
2.2 The Fourier Integral *39*   2.3 Fourier Series *43*
2.4 Discrete Fourier Series *47*   2.5 The Fast Fourier Transform *53*

||3|| FINITE-DIFFERENCE APPROXIMATIONS   *58*

3.1 Approximation of Differential Operators *58*   3.2 Operator Notations *68*
3.3 Truncation Error and Order of Accuracy *70*

||II|| ELLIPTIC EQUATIONS   *73*

||4|| TWO-POINT BOUNDARY-VALUE PROBLEMS I   *75*

4.1 Finite-Difference Approximation *75*
4.2 Truncation Error: Classical Analysis *79*
4.3 Truncation Error: Fourier Analysis *83*
4.4 A Fourier Method for Two-Point Boundary-Value Problems *89*
4.5 Initial-Value Methods *91*

||5|| FINITE-DIFFERENCE APPROXIMATION
OF ELLIPTIC EQUATIONS   *95*

5.1 Approximation and Convergence *95*   5.2 Higher-Order Approximations *102*
5.3 Approximation Near Boundaries *106*

||6|| NUMERICAL SOLUTION
OF ALGEBRAIC EQUATIONS   *109*

6.1 Direct Solution Methods *109*   6.2 Iterative Solution Methods *123*
6.3 Sparse Matrices *135*   6.4 Nonlinear Equations and Quasilinearization *141*

||7|| SPECIAL TOOLS   *147*

7.1 Special Poisson Solvers *147*   7.2 Networks and Analogs *158*

||III|| THE FINITE-ELEMENT METHOD   *163*

||8|| RAYLEIGH–RITZ AND GALERKIN METHODS   *165*

8.1 The Calculus of Variations *165*   8.2 The Rayleigh–Ritz Method *179*
8.3 The Galerkin Method *185*   8.4 Weighted Residual Methods *188*
8.5 Spectral Methods *191*   8.6 Introduction to the Finite-Element Method *194*

||9|| TWO-POINT BOUNDARY-VALUE PROBLEMS II   205

9.1 Linear Finite Elements for One-Dimensional Problems *205*
9.2 Quadratic Finite Elements *214*   9.3 Hermite Cubic Finite Elements *219*
9.4 The Engineers' Displacement Method *225*

||10|| POISSON'S EQUATION   235

10.1 Finite-Element Approximation *235*   10.2 Linear Elements on Triangles *240*
10.3 Special Cases *254*   10.4 Boundary Conditions *260*
10.5 The MERMAID Computer Code *269*
10.6 Bilinear Elements on Rectangles *278*

||11|| THEORY AND GEOMETRY   285

11.1 Finite Elements in Two Dimensions *285*
11.2 Families of Element Shapes on Rectangles *291*
11.3 Families of Element Shapes on Triangles *298*
11.4 Hermitian Elements in Rectangles *302*
11.5 Collocation on Finite Elements *305*

||12|| GEOMETRIC TRANSFORMATIONS AND ISOPARAMETRIC ELEMENTS   309

12.1 Change of Coordinates *309*
12.2 Finite Elements in Global Curvilinear Coordinates *314*
12.3 Isoparametric Elements *318*   12.4 Numerical Integration *327*

BIBLIOGRAPHY   336

INDEX   354

Preface

To compute the numerical solution of partial differential equations requires, with few exceptions, that some form of approximation be invoked: This book is about the theories on which those approximations are based, and about the corresponding methods that are used in practice, in particular, with the aid of computers. It is addressed to graduate students as a textbook, and to practicing scientists and engineers as a reference.

Partial differential equations were "invented" in the first half of the eighteenth century. For nearly two hundred years, they were used mostly as analytical tools to describe the physical world with which they are intimately related. While the concept of approximating their solutions by numerical calculations was reasonably common knowledge, the importance given to such calculations was slight, and they were not held in high esteem by the scientific community of the time.

A gradual change became perceivable toward the turn of the twentieth century. Descriptions of numerical approximation methods began to appear in the literature with an increased frequency. But a true revolution came with the appearance of electronic computers. While any numerical calculation can in principle be carried out by hand, the numerical solution of meaningful, and as a rule reasonably complex, scientific problems did not become a useful proposition until sufficiently large computers were available to automate the work. This happened during the 1950s.

The methods that were used at first were those that had been developed in precomputer days, intended for pencil and paper implementation. New tools were needed, and this precipitated a significant increase in the development of methods that were at first mostly adaptations of the existing ones to whatever new capabilities the emerging electronic computers were offering. The 1962 book by Forsythe and Wasow gave a good review of the state of the art that included this first generation of new methods.

The most important development in the 1950s and 60s that qualifies as "modern" was the finite-element method. While this point was unnoticed at first, it turned out that the finite-element method had the potential of utilizing the capabilities of electronic computers more than any of the other methods available at the time. Much of the drudgery of generating discretized equations prior to their solutions is now being delegated to finite-element computer codes, in particular for elliptic equations that describe structural problems of mechanical and civil engineering, and steady state field problems in many other applied sciences. While codes with a similar intent had been implemented with finite-difference methods, they were largely unable to adapt the discretization of space to the complex shapes and irregularities found in many real-life problems. With finite elements, complex shapes and irregularities create little complication, thus making the corresponding codes convenient to use and broad in their applicability.

Moreover, the finite-element method rests on more rigorous foundations than the finite-difference method, and this fact did not escape the attention of numerous mathematicians who have consequently devoted themselves to furthering the supporting theory.

Both mathematicians and engineers have had their share in the development of the finite-element method. However their efforts have, with few exceptions, been largely independent. While both may at times be concerned with the same problems, communication is inhibited by the fact that they do not speak the same language. Moreover, what interests mathematicians is not necessarily what concerns engineers, and vice versa. It is important that anyone entering the field realize that those two distinct schools exist, that they have different objectives, and that bringing the two together is not necessarily what they want or need.

This book is intended to be appealing to both audiences. It provides a much needed summary of the present state of the art in the computer solution of partial differential equations, serving as an introduction to more specialized study either in specific applications or in detailed mathematical analyses. It grew out of teaching materials from courses given by the author at Princeton University and Rutgers University and a wide variety of applied research projects, heavy on computing, in which he was engaged.

The material in Volume I mostly involves the approximation of elliptic equations. It is suitable for a sequence of two or three semesters of graduate courses. The first seven chapters are a self-contained introduction to the tools and techniques for the numerical approximation of elliptic equations with finite differences.

The finite-element method starts with Chapter 8. We begin with a review of the calculus of variations, followed by the Rayleigh–Ritz method, which is the mathematicians' approach. In Section 8.3 we present Galerkin's method, which is closer to the engineers' approach, and we also describe generalizations such as collocation and subdomain methods that engineers have introduced in applications.

Since the calculus of variations and Galerkin's method lead to essentially the same result by different routes, we often describe both of them side-by-side to emphasize their equivalence, and to give a choice to the reader as to which route he or she wants to follow.

Two-point boundary-value problems, while not technically partial differential equations, nevertheless offer the simplest device for the description of certain methods for partial differential equations of the boundary-value type. The corresponding finite-difference methods are given in Chapter 4, and finite-element methods in Chapter 9. Many of the components that contribute to the finite-element method come together in Chapter 10, where we approximate Poisson's equation with finite elements on triangles and rectangles. This is one of the cases where all the expressions can be evaluated analytically. The chapter is completed by a computer code that was developed at Rutgers University and has proved to be a very effective tool for learning as well as in applications. This chapter must be covered in detail in any course that intends to prepare students for applications, or by anyone who wants to learn how the finite-element method works.

Beyond theory, the finite-element method relies on a great deal of analytical geometry to provide the needed element shapes. The essential principles for the generation of families of elements are described in Chapter 11. Irregular elements, which are extremely important in practice, are introduced in Chapter 12. The mathematics of changes of coordinates and the elegant family of transformations known as "isoparametric" are described. Analytic evaluation of the expressions appearing in the corresponding implementation is no more feasible, and numerical integration (or quadrature) becomes an indispensable ingredient. This chapter, too, is a must for those who intend to apply the finite-element method to practical problems in two and three dimensions.

Volume II is about initial value (parabolic and hyperbolic) equations. While the finite-element method has had an impact there too, it has been much lighter than in the elliptic case: Finite-difference methods remain very important. Questions of stability, accuracy, and characterization of the errors for both finite differences and finite elements are covered as the necessary complement to the study of the methods for initial value problems that are described in that volume.

It is difficult to express acknowledgements to all those who have helped in the development of a book that has been growing out of courses and group projects over many years. I am grateful to my colleagues who have allowed me to benefit from their suggestions and criticisms in the course of many private discussions and informal seminars. I am also grateful to the many students who have helped both in the implementation of computer programs and in the debugging of earlier versions of the manuscript that were used as course notes in class.

Finally, I must express my gratitude to Linda M. Paskiet from Prentice-Hall's Editorial Production Department, who supervised the production of the manuscript to its final form with a degree of care and attention that must have been well beyond the call of duty.

Robert Vichnevetsky

I

Fundamentals

1

Introduction

1.1 INTRODUCTION TO PARTIAL DIFFERENTIAL EQUATIONS

1.1.1 PHYSICAL MODELS OF THE THREE CANONICAL TYPES OF EQUATIONS

Partial differential equations of the second order play a primordial role in mathematical physics. It is through attempts to describe the physical world that scientists of the eighteenth and nineteenth centuries were led to their formulation.

The model of hyperbolic equations is the wave equation

$$\frac{\partial^2 U}{\partial t^2} = c^2 \frac{\partial^2 U}{\partial x^2} \tag{1.1.1}$$

which describes (for instance) small-amplitude vibration of a taut string (see Figure 1.1.1). It was first expressed by Jean Le Rond d'Alembert in 1747, in what may amount to the "discovery" of partial differential equations. His own description of this discovery in the *Encyclopédie* reads as follows:

> Mr. d'Alembert[1] is the inventor of this branch of analysis, without which one could not solve in a rigorous and general manner

---

[1] Although written by d'Alembert, he uses the third person in referring to himself.

3

the problems where one deals with fluid or flexible bodies. This discovery, as important and maybe more difficult than that of integral calculus, has been less spectacular only because its author has expressed an entirely new thing by words and signs already known. [d'Alembert, 1785]

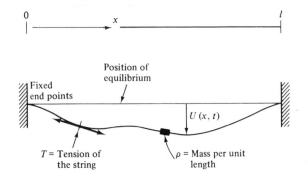

Vibrating string equation:

$$\frac{\partial^2 U}{\partial t^2} = c^2 \frac{\partial^2 U}{\partial x^2}$$

$$c^2 = \frac{T}{\rho}$$

$$U(0, t) = U(l, t) = 0$$

Figure 1.1.1

What is worth noting is that d'Alembert, as most mathematicians of his time, was primarily concerned with finding a mathematical explanation of the physical world, and that he points to the usefulness of partial differential equations in solving new problems of physics as the primary (and possibly only) reason for their interest.

This is more than a note of mere historical value; it is difficult, sometimes futile, to think of partial differential equations as abstract mathematical objects. There are many instances where an equation, in itself, fails to allow for a unique solution or even for a solution at all to exist, because of apparent ambiguities or contradictions. It is then only by returning to the physical world, of which the equation is an abstraction (sometimes imperfect), that those ambiguities and contradictions may be removed and the intended solution found.

The model of parabolic equations is the heat equation

$$\frac{\partial U}{\partial t} = \sigma \frac{\partial^2 U}{\partial x^2} \tag{1.1.2}$$

derived by Joseph Fourier[2] in 1807. As suggested by its name, it describes in its original derivation the important physical phenomenon of diffusion of heat in solid bodies (Figure 1.1.2).

A physical model of elliptic equations may be found in Laplace's equation (sometimes called the *potential equation*):

$$\frac{\partial^2 U}{\partial x^2} + \frac{\partial^2 U}{\partial y^2} = 0 \tag{1.1.3}$$

in two dimensions, or

$$\frac{\partial^2 U}{\partial x^2} + \frac{\partial^2 U}{\partial y^2} + \frac{\partial^2 U}{\partial z^2} = 0 \tag{1.1.4}$$

in three dimensions. An important difference between this equation and the two preceding ones is that none of the independent variables (x, y, and z) can be associated with Newtonian time as we perceive it through our own cognition of the physical world. By contrast, parabolic and hyperbolic equations are generally associated with phenomena that are *evolutionary* in their very physical nature, and the variable t in (1.1.1) and (1.1.2) is indeed intended to denote a time-like variable.

Elliptic equations appear in the description of several-dimensional *steady-state* (i.e., time-invariant) *fields* of different physical origin. For

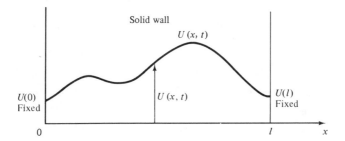

$U(x, t)$ = Temperature distribution

Heat equation:

$$\frac{\partial U}{\partial t} = \sigma \frac{\partial^2 U}{\partial x^2}$$

$$\sigma = \frac{k}{c}$$

k = Thermal conductivity

c = Thermal capacity

Figure 1.1.2

---

[2]French mathematician (1768–1830).

instance, Fourier's heat equation (1.1.2) in two spatial dimensions and time would be

$$\frac{\partial U}{\partial t} = \sigma\left(\frac{\partial^2 U}{\partial x^2} + \frac{\partial^2 U}{\partial y^2}\right) \tag{1.1.5}$$

from which one may conclude that a steady-state temperature distribution must satisfy (1.1.3) (from $\partial U/\partial t = 0$).

One of the ways in which elliptic equations came about was in the extension by eighteenth-century mathematicians of Newton's law of gravitation to cases where masses cannot be considered as point-concentrated. In modern notations, the original formulation of the inverse-square law of attraction states that the gravity vector in a point P external to a body B is given by the integral equation (Figure 1.1.3)

$$\mathbf{g}_P = -k \iiint_B \rho(x, y, z) \frac{\mathbf{P} - \mathbf{X}}{|\mathbf{P} - \mathbf{X}|^3} \, dx \, dy \, dz \tag{1.1.6}$$

where

$$\mathbf{P} \equiv \begin{bmatrix} x_P \\ y_P \\ z_P \end{bmatrix}; \quad \mathbf{X} \equiv \begin{bmatrix} x \\ y \\ z \end{bmatrix}$$

It is required that this integral be evaluated separately for each external point P.

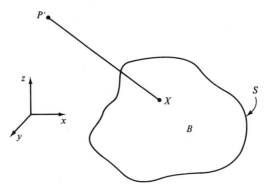

Gravity in P

$$\mathbf{g}_P = -k \iiint_B \rho(x, y, z) \frac{(\mathbf{P} - \mathbf{X})}{|\mathbf{P} - \mathbf{X}|^3} \, dx \, dy \, dz$$

Figure 1.1.3

It was discovered, however, that $\mathbf{g}_P$ could be defined as minus the gradient of a potential function,

$$\mathbf{g}_P = -\nabla V = - \begin{bmatrix} \dfrac{\partial V}{\partial x} \\[2mm] \dfrac{\partial V}{\partial y} \\[2mm] \dfrac{\partial V}{\partial z} \end{bmatrix} \qquad (1.1.7)$$

and, remarkably, that V satisfies the potential equation

$$\frac{\partial^2 V}{\partial x^2} + \frac{\partial^2 V}{\partial y^2} + \frac{\partial^2 V}{\partial z^2} = 0 \qquad (1.1.8)$$

This was formalized in the last quarter of the eighteenth century by Laplace, whence the name given to this equation. The important consequence of this is that if V can somehow be computed on the external surface S of B, then computing the gravity vector $\mathbf{g}_P$ in any point exterior to B can be obtained by merely solving Laplace's equation, subject to the given boundary condition on S. It was then shown that V_S may be computed by evaluation of the integral

$$V_S = k \iiint_B \frac{\rho(x, y, z)}{|\mathbf{P}_S - \mathbf{X}|} \, dx \, dy \, dz \qquad (1.1.9)$$

whence completing the formulation of the method.

As is often the case with names, Laplace's equation was not discovered by Laplace; it had probably been used at first by Euler in the 1750s. In 1752, Euler published a paper dealing with the motion of incompressible fluids, in which he defined a potential function from which velocity may be derived:

$$\text{velocity} = -\nabla V \qquad (1.1.10)$$

and in which he showed that V must satisfy the potential equation (1.1.8).

The next important step in the formulation of gravitation theory and elliptic equations came in the nineteenth century, when Poisson (1813) showed that for a point P located *inside* B, the gravity potential does not satisfy (1.1.8) but obeys

$$\frac{\partial^2 V}{\partial x^2} + \frac{\partial^2 V}{\partial y^2} + \frac{\partial^2 V}{\partial z^2} = -4\pi\rho \qquad (1.1.11)$$

which is known as Poisson's equation. This, in fact, completed the transformation of the expression of the law of Newtonian gravity of distributed bodies from its original integral formulation to the analytically much more convenient differential formulation. The magnitude of the impact of these new formulations upon electromagnetic theory which was developed shortly thereafter is only one example of the importance to modern applied science of the early work surrounding partial differential equations of physics.

1.1.2 CANONICAL CLASSIFICATION OF PARTIAL DIFFERENTIAL EQUATIONS

Not all partial differential equations can be classified into well-defined categories. The most comprehensive analysis relates to linear or quasilinear equations of the *second order* in *two independent variables*. Those can be classified into equations of the *elliptic, parabolic,* and *hyperbolic* type, with solutions of each type displaying specific properties. Generalizing the name of elliptic, hyperbolic, and parabolic to equations that are neither second order nor in two independent variables, but which possess similar properties, is often done as a matter of fact. The result is that the majority of problems of practical importance can be related to these three canonical classes.

Consider as a fundamental form the equation[3]

$$aU_{xx} + 2bU_{xy} + cU_{yy} + g = 0 \qquad (1.1.12)$$

with analytic coefficients $a, b, c,$ and g depending exclusively on $x, y, U_x,$ and U_y. Such an equation is called *quasi-linear* because it is linear in the derivatives of the highest order (i.e., it is linear in $U_{xx}, U_{xy},$ and U_{yy}). To gain an insight into the properties of that equation, we shall analyze the local behavior of its solution.

We begin this subject by recalling the process of finding the solution of an *ordinary* differential equation of the *initial-value* type by building up its *Taylor series* term by term. For

$$\frac{d^2Y}{dt^2} = F\left(Y, \frac{dY}{dt}, t\right) \qquad Y(0) \text{ and } \left(\frac{dY}{dt}\right)_{t=0} \text{ prescribed} \qquad (1.1.13)$$

we may write (if F is analytic)

$$\left(\frac{d^2Y}{dt^2}\right)^0 = F\left(Y^0, \left(\frac{dY}{dt}\right)^0, 0\right)$$

$$\left(\frac{d^3Y}{dt^3}\right)^0 = \left[\frac{\partial F}{\partial Y}\frac{dY}{dt} + \frac{\partial F}{\partial(dY/dt)}F + \frac{\partial F}{\partial t}\right]^0$$

$$\left(\frac{d^4Y}{dt^4}\right)^0 = \dots \quad \text{etc.}$$

---

[3] Conventional notations $U_x = \partial U/\partial x, \dots$ are used here.

and, finally,

$$Y(t) = Y^0 + t\left(\frac{dY}{dt}\right)^0 + \frac{t^2}{2!}\left(\frac{d^2Y}{dt^2}\right)^0 + \frac{t^3}{3!}\left(\frac{d^3Y}{dt^3}\right)^0 + \cdots \qquad (1.1.14)$$

which provides in the neighborhood of $t = 0$ a power (Taylor) series expression of Y.

Extending this process to the partial differential equation (1.1.12) requires that the initial point ($t = 0$) be replaced by an initial line [say, Γ in (x, y)], and that the initial data $[Y^0, (dY/dt)^0]$ be replaced by $U(x_\Gamma, y_\Gamma)$ and $(\partial U/\partial n)_\Gamma$, the derivative normal to Γ. Unless Γ is parallel to either the x or the y axis, this is equivalent to prescribing U, U_x, and U_y on Γ.

Solving equation (1.1.12) with boundary data prescribed on a single *initial* line is called the *Cauchy* or *Cauchy–Kowalewski problem*. The procedure consists in utilizing U, U_x, and U_y on Γ and equation (1.1.12) to compute successively the derivatives of second, third, and higher order, thereby obtaining the terms of a Taylor series analogous to (1.1.14) in the form of a two-dimensional power series which, in some neighborhood of Γ, is the solution $U(x, y)$ to (1.1.12).[4]

But there are certain lines in the (x, y) plane (called *characteristic lines* of the equation) along which this process does not work. Those are the lines along which U, U_x, U_y, and the equation do not permit the derivation of U_{xx}, U_{xy}, and U_{yy}. This can be clarified as follows. Let (dx, dy) be a small displacement along Γ. We have, by simple derivation,

$$\begin{aligned} d(U_x) &= U_{xx}\, dx + U_{xy}\, dy \\ d(U_y) &= \qquad\quad U_{xy}\, dx + U_{yy}\, dy \end{aligned} \qquad (1.1.15)$$

Equation (1.1.12), together with equations (1.1.15), can be expressed in matrix form as

$$\begin{bmatrix} a & 2b & c \\ dx & dy & 0 \\ 0 & dx & dy \end{bmatrix} \begin{bmatrix} U_{xx} \\ U_{xy} \\ U_{yy} \end{bmatrix} = \begin{bmatrix} -g \\ d(U_x) \\ d(U_y) \end{bmatrix} \qquad (1.1.16)$$

A solution U_{xx}, U_{xy}, U_{yy} exists and is unique in a given point of Γ if the determinant of the coefficient matrix

$$\det \begin{bmatrix} a & 2b & c \\ dx & dy & 0 \\ 0 & dx & dy \end{bmatrix} = a\, dy^2 - 2b\, dx\, dy + c\, dx^2 \qquad (1.1.17)$$

does not vanish. Thus, any displacement (dx, dy) defining locally the direction

---

[4]A general name given to this process is *analytic continuation*.

of Γ will be acceptable, except for those directions that make (1.1.17) equal to zero. Let

$$s = \frac{dy}{dx} = \text{slope of } \Gamma$$

For Γ to be a characteristic line, s must satisfy [from (1.1.17)]

$$as^2 - 2bs + c = 0 \qquad (1.1.18)$$

or

$$s = \frac{b \pm \sqrt{b^2 - ac}}{a} \qquad (1.1.19)$$

Three typical situations may thus occur:

1. $b^2 - ac > 0$ *in every point of* (x, y): there exist *two* real families of characteristic lines, with slopes given by (1.1.19). The equation (1.1.12) is then called *hyperbolic*.

2. $b^2 - ac = 0$: there exists *one* real family of characteristic lines with slope $s = b/a$. The equation is called *parabolic*.

3. $b^2 - ac < 0$: there are no *real* characteristics. The equation is called *elliptic*.

Although an equation may be of one type in one region of the plane (x, y) and of another type in another region, these cases are infrequent and can be ignored, in our discussion of the subject, with little loss of generality.

It may be shown that by means of a real change of variables,

$$\begin{aligned}\xi &= \xi(x, y) \\ \eta &= \eta(x, y)\end{aligned} \qquad (1.1.20)$$

(1.1.12) may be transformed into one of the three canonical forms:

$$\begin{aligned}U_{\xi\xi} - U_{\eta\eta} + \ldots &= 0 && \text{in the hyperbolic case} \\ U_{\xi\xi} + \ldots &= 0 && \text{in the parabolic case} \\ U_{\xi\xi} + U_{\eta\eta} + \ldots &= 0 && \text{in the elliptic case}\end{aligned}$$

Moreover, no real change of variables can alter the type of the equation.

1.1.3 MODELING PHYSICS AND WELL-POSED PROBLEMS

Partial differential equations that one is interested in are before all mathematical models of physical phenomena. For any problem describing a stable situation, one would expect that small variations in the data should result in correspondingly small variations in the solution. If this did not turn out to be true, we would be inclined to believe that the mathematical model of the physical problem has been badly formulated.

The proposition of finding in mathematical terms which problems are acceptable models of the physical world and which are not has led to the concept of a *well-posed problem* (Hadamard, 1932).

Definition. A partial differential equation, or a system of partial differential equations, is *well posed in the sense of Hadamard* if and only if its solution exists, is unique, and depends continuously on the data prescribed.

This definition shall be clarified by the following example devised by Hadamard. The elliptic equation

$$\nabla^2 U = \frac{\partial^2 U}{\partial x^2} + \frac{\partial^2 U}{\partial y^2} = 0 \qquad (1.1.21)$$

associated with the initial-value (Cauchy) data

$$U(x, 0) = 0; \qquad \frac{\partial U}{\partial y}(x, 0) = \frac{1}{k} \sin (kx) \qquad (1.1.22)$$

has the mathematically legitimate solution

$$U(x, y) = \frac{1}{k^2} \sin (kx) \sinh (ky) \qquad (1.1.23)$$

As we let $k \rightarrow \infty$, the initial data (1.1.22) approach zero uniformly, whereas for $y > 0$ the solution (1.1.23) oscillates between limits that increase indefinitely. Thus, this problem is not well posed by the preceding definition. We conclude that formulating an elliptic equation as a Cauchy (initial-value) problem is actually not the right question to pose, *at least if physically meaningful answers are to be expected.*

The proper way is to formulate elliptic equations as boundary-value problems, that is, prescribing boundary conditions in every point of the boundary ∂D of the domain of interest.

1.1.4 NATURAL PROBLEMS IN PARTIAL DIFFERENTIAL EQUATIONS

Most partial differential equations originate from *natural* situations in mechanics and physics. The character of additional boundary or initial conditions is similarly motivated by physical reality. We have just shown that formulating an elliptic equation as a boundary-value problem was proper, whereas formulating it as an initial-value (Cauchy) problem was not.

It would appear desirable to systematize what initial and boundary conditions are acceptable and natural for each type of equations. But one gets easily lost in attempting such a systematization from a strictly mathematical viewpoint. On the other hand, one may use typical examples from mathematical physics to establish natural classes of problems. We follow Courant and Hilbert (1962, vol. 2, p. 226) in stating a basic principle:

> Boundary-value problems are naturally associated with elliptic equations, while initial-value problems and mixed (initial/boundary-value) problems arise in connection with hyperbolic and parabolic differential equations.

Typical natural formulations of problems associated with the three canonical types of equations are given in Figures 1.1.4 to 1.1.9.

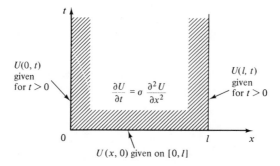

Figure 1.1.4   Example of a well-posed problem for Fourier's equation.

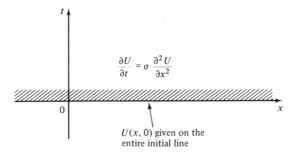

Figure 1.1.5   Example of a well-posed problem for Fourier's equation.

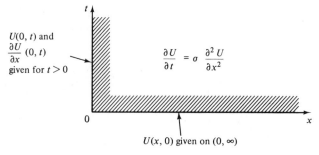

Figure 1.1.6 Example of an ill-posed problem for Fourier's equation.

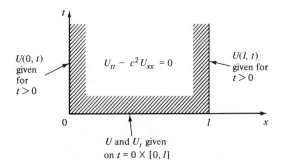

Figure 1.1.7 Example of a well-posed problem for the wave equation.

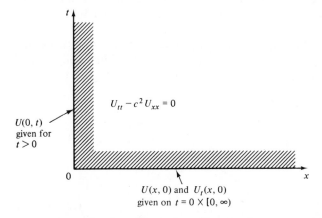

Figure 1.1.8 Example of a well-posed problem for the wave equation.

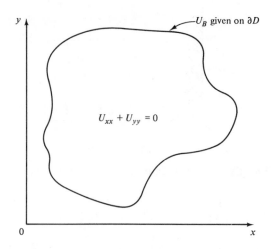

U_B given on ∂D

$$U_{xx} + U_{yy} = 0$$

Figure 1.1.9 Example of a well-posed problem for Laplace's equation.

1.1.5 DOMAIN OF DEPENDENCE

An important concept in partial differential equations is that of the *domain of dependence* of the solution on initial/boundary data. Referring to the typical problems illustrated above, we have the following.

Elliptic Equations. The solution in an interior point P depends on *all* the data given on the boundary line ∂D (Figure 1.1.10).

Parabolic Equations. The solution in an interior point P at time t depends on the data given on the entire initial line and on the data on the segments of the boundary lines for $0 < t < T$ (Figure 1.1.11).

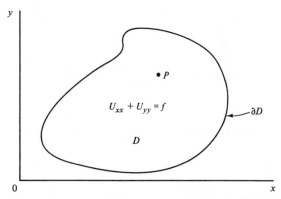

$\bullet\, P$

$$U_{xx} + U_{yy} = f$$

∂D

D

Figure 1.1.10 Domain of dependence for elliptic equations: the solution in P is dependent upon one boundary condition to be given on the entire boundary line ∂D.

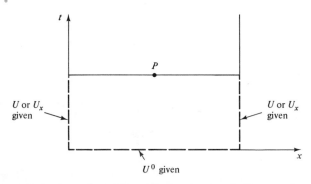

Figure 1.1.11 Domain of dependence for parabolic equations: the solution in P is dependent upon the initial/boundary conditions to be given where indicated by the dashed line.

Hyperbolic Equations. The solution in an interior point P depends on data given on those segments of the initial/boundary lines that are contained between the two characteristic lines passing through P (Figure 1.1.12).

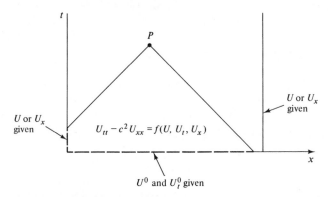

Figure 1.1.12 Domain of dependence for hyperbolic equations: the solution in P is dependent upon the initial/boundary conditions indicated by the dashed line.

1.1.6 FINDING SOLUTIONS

Thus far, we have discussed solutions of partial differential equations but have not said much about finding them. In the early days of this theory (starting with d'Alembert in the eighteenth century), mathematicians were interested primarily in analytical solutions. Elegant analytical methods were developed throughout the eighteenth and nineteenth centuries.

Finding particular solutions numerically, possibly by some more-or-less approximate set of calculations on numbers, was not recognized as a respectable pursuit until much later. To wit, Richardson found it necessary

15

to write an introduction to his 1910 paper, "The approximate arithmetical solution . . . of physical problems involving differential equations . . . ," in which, after having noted that there are many practical problems where analytical methods fail, he explains that

> there is a demand for rapid methods . . . applicable to unusual equations and irregular bodies. If they can be accurate, so much the better; but 1 percent would suffice for many purposes.

Many approximate numerical methods for partial differential equations were to be developed—"invented" around the turn of the century and shortly thereafter. The calculations were, of course, to be carried out by hand, or at best with the aid of desktop mechanical calculators. With the advent of the electronic computer in the 1940s, the scope of numerical methods became enlarged by orders of magnitude. It is about many of these methods, the theories on which they are based, and their applications to practical problems in science and engineering that much of this book is devoted.

1.2   FINITE-DIFFERENCE APPROXIMATION OF ELLIPTIC EQUATIONS

1.2.1   ELLIPTIC EQUATIONS

Boundary-value problems of potential theory are suggested by physical phenomena from such varied fields as gravity, electrostatics, steady heat conduction, and incompressible fluid flow. Their formulation leads in a natural way to elliptic equations. Laplace's equation in two dimensions,

$$\nabla^2 U = \frac{\partial^2 U}{\partial x^2} + \frac{\partial^2 U}{\partial y^2} = 0 \qquad (1.2.1)$$

is the most elementary example of an elliptic partial differential equation. Another simple example is Poisson's equation,

$$-\nabla^2 U = F(x, y) \qquad (1.2.2)$$

where F is a specified "source" function. In certain cases, elliptic equations may be interpreted as describing the steady-state (meaning $\partial/\partial t = 0$) solution of time-dependent problems.

1.2.2   WELL-POSED PROBLEMS

As we have seen in Section 1.1, for an elliptic problem to be well posed (i.e., small changes in the data should result in correspondingly small changes in the solution), it is required that a condition be imposed at every point of the boundary ∂D surrounding the domain of definition D (Figure 1.1.8). These (boundary) conditions may be of various kinds:

Boundary-Value Problem of the First Kind.   (Also called the *Dirichlet problem* when $F = 0$). The solution U is to satisfy (1.2.1) or (1.2.2) in D and take on prescribed values

$$U = U_B(x, y)$$

on the boundary ∂D of that region. A physical example is that of finding the steady temperature inside of a solid body whose outside surface temperature is maintained at the prescribed values $U_B(x, y)$ (a function of location on the boundary).

Boundary-Value Problem of Second Kind.   (Often called the *Neumann problem*). The solution is to possess prescribed normal derivatives

$$\frac{\partial U}{\partial n} = U'_B$$

on the surface ∂D bounding D.

Boundary-Value Problem of the Third Kind.   Boundary conditions of the mixed form

$$\frac{\partial U}{\partial n} = G + HU \qquad \text{on } \partial D$$

are called conditions of the *third kind.*

It is often the case that an elliptic problem is specified by boundary conditions that are of different kinds along different parts of ∂D.

1.2.3 DIFFERENCE APPROXIMATION OF ELLIPTIC EQUATIONS

A fundamental method for the numerical approximation of elliptic equations is the *finite-difference method*. The formal steps involved in the application of this method are:

1.   The domain of the problem is covered by a simple mesh.

2.   Values of the numerical solution are labeled at the intersections or *nodes* of the mesh.

3.   A *finite-difference approximation* to the differential equation is

formulated in each node that is not on a boundary of the first kind.[5] This results in a system of *algebraic* equations called *finite-difference equations*.

4.   The system of finite-difference equations approximating the problem is solved to produce the numerical solution. This process generally involves solving numerically large systems of linear algebraic equations. The corresponding computer algorithms are those of *numerical linear algebra* and are described in some detail in Chapter 6.

EXAMPLE 1

As an example, we seek a function $U(x, y)$ that satisfies Poisson's equation,

$$-\left(\frac{\partial^2 U}{\partial x^2} + \frac{\partial^2 U}{\partial y^2}\right) = F(x, y) \tag{1.2.3}$$

in a closed domain D, and which satisfies the condition

$$U(x, y) = G(x, y) \tag{1.2.4}$$

on the boundary ∂D. If $F \equiv 0$, then this becomes Laplace's equation,

$$\frac{\partial^2 U}{\partial x^2} + \frac{\partial^2 U}{\partial y^2} = 0 \tag{1.2.5}$$

A simple difference approximation to (1.2.3) is obtained by defining the rectangular mesh:

$$x_m = m\,\Delta x; \quad m = 0, 1, 2, \ldots$$
$$y_n = n\,\Delta y; \quad n = 0, 1, 2, \ldots$$

on which the solution is approximated by the array of discrete values

$$u_{m,n} \simeq U(x_m, y_n)$$

A distinction is to be made between *interior points* (where the solution is to be found) and *boundary points*, where the solution is given by (1.2.4). Simple central-difference approximations to derivatives expressed at the interior mesh points are

$$\left(\frac{\partial^2 U}{\partial x^2}\right)_{m,n} \simeq \frac{u_{m+1,n} + u_{m-1,n} - 2u_{m,n}}{\Delta x^2}$$

$$\left(\frac{\partial^2 U}{\partial y^2}\right)_{m,n} \simeq \frac{u_{m,n+1} + u_{m,n-1} - 2u_{m,n}}{\Delta y^2} \tag{1.2.6}$$

---

[5]The finite-difference approximation of boundary conditions of the second and third kinds is discussed in Chapter 5.

Substituting these in (1.2.3) results in the equation

$$-\left(\frac{u_{m+1,n} + u_{m-1,n} - 2u_{m,n}}{\Delta x^2} + \frac{u_{m,n+1} + u_{m,n-1} - 2u_{m,n}}{\Delta y^2}\right) = f_{m,n} \qquad (1.2.7)$$

for all interior points. This is a *five-point* approximation of the equation.

A simple mnemonic representation of this formula is obtained when Δx and Δy are chosen equal ($\Delta x = \Delta y = h$). Then the stencil notation,

$$\nabla^2 \simeq \frac{1}{h^2}\begin{bmatrix} & 1 & \\ 1 & -4 & 1 \\ & 1 & \end{bmatrix} \qquad (1.2.8)$$

is equivalent to the left-hand side of (1.2.7). It expresses the simple five-point approximation of the Laplacian operator

$$\nabla^2 \equiv \frac{\partial^2}{\partial x^2} + \frac{\partial^2}{\partial y^2}$$

That other simple approximations to the Laplacian operator can be derived is easily illustrated; for example, since this operator is invariant under coordinate rotation, it may be expressed in coordinates rotated 45° with respect to the mesh. The spacing between points becomes $\sqrt{2}h$ and the equivalent of (1.2.8) is

$$\nabla^2 \simeq \frac{1}{2h^2}\begin{bmatrix} 1 & & 1 \\ & -4 & \\ 1 & & 1 \end{bmatrix} \qquad (1.2.9)$$

Returning to (1.2.7), we see that this system of equations is linear, having as unknowns the values of $u_{m,n}$ in all *interior* points of the mesh. As an example, consider the case $F(x, y) = 0$ on a square domain D, with $M = N = 3$ (Figure 1.2.1). Relabeling the variables as indicated, we have

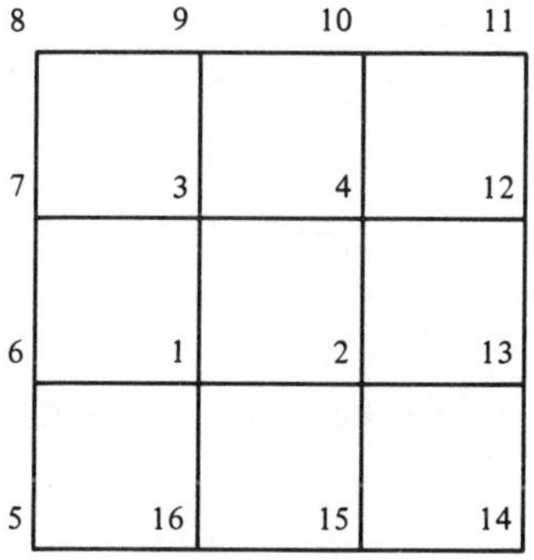

Figure 1.2.1

the equations

$$4u_1 - u_2 - u_3 - u_6 - u_{16} = 0$$
$$4u_2 - u_{13} - u_4 - u_1 - u_{15} = 0$$
$$4u_3 - u_4 - u_9 - u_7 - u_1 = 0 \qquad (1.2.10)$$
$$4u_4 - u_{12} - u_{10} - u_3 - u_2 = 0$$

Moving the known boundary values to the right, we obtain the linear system

$$\left.\begin{array}{c} 4u_1 - u_2 - u_3 \qquad = u_6 + u_{16} \\ -u_1 + 4u_2 \qquad - u_4 = u_{13} + u_{15} \\ -u_1 \qquad + 4u_3 - u_4 = u_7 + u_9 \\ -u_2 - u_3 + 4u_4 = u_{10} + u_{12} \end{array}\right\} \qquad (1.2.11)$$

or

$$\begin{pmatrix} 4 & -1 & -1 & 0 \\ -1 & 4 & 0 & -1 \\ -1 & 0 & 4 & -1 \\ 0 & -1 & -1 & 4 \end{pmatrix} \begin{pmatrix} u_1 \\ u_2 \\ u_3 \\ u_4 \end{pmatrix} = \begin{pmatrix} u_6 + u_{16} \\ u_{13} + u_{15} \\ u_7 + u_9 \\ u_{10} + u_{12} \end{pmatrix} \qquad (1.2.12)$$

Suppose now that $U(x, y) = 0$ on the north and south sides of the square and $= 10$ on the east and west sides. Then (1.2.12) becomes

$$\begin{pmatrix} 4 & -1 & -1 & 0 \\ -1 & 4 & 0 & -1 \\ -1 & 0 & 4 & -1 \\ 0 & -1 & -1 & 4 \end{pmatrix} \begin{pmatrix} u_1 \\ u_2 \\ u_3 \\ u_4 \end{pmatrix} = \begin{pmatrix} 10 \\ 10 \\ 10 \\ 10 \end{pmatrix}$$

which has the solution

$$u_1 = 5$$
$$u_2 = 5$$
$$u_3 = 5$$
$$u_4 = 5$$

EXAMPLE 2

As a second example, consider the simple problem

$$-\frac{d^2 U}{dx^2} = F(x) \qquad (1.2.13)$$

over $x \in [0, l]$, with boundary conditions

$$\begin{aligned} U(0) &= U_0 \quad \text{(given)} \\ U(l) &= U_l \quad \text{(given)} \end{aligned} \tag{1.2.14}$$

This is a two-point boundary-value problem (sometimes denoted TPBVP in the literature). Although not formally partial differential equations, two-point boundary-value problems possess most properties of elliptic problems (which, in fact, they are), and they therefore offer a convenient vehicle to illustrate and analyze computational methods.

To approximate equation (1.2.13) with finite differences, we first divide the domain $x \in [0, l]$ in discrete increments by the set of points

$$\{x_n = n\,\Delta x; \; n = 0, 1, 2, \ldots, N + 1\}; \quad \Delta x = \frac{l}{N + 1} \tag{1.2.15}$$

and let

$$\{u_n\} \simeq \{U(x_n)\} \tag{1.2.16}$$

be the set of discrete values intended to approximate U. At each point x_n, equation (1.2.13) may be written as

$$-\left(\frac{d^2U}{dx^2}\right)_n = F(x_n) \equiv f_n \tag{1.2.17}$$

If we expand U in a Taylor series around a point x_n [letting U_n stand for $U(x_n)$], we get

$$\begin{aligned} U_{n+1} &= U(x_n + \Delta x) = U_n + U_n'\,\Delta x + U_n''\frac{\Delta x^2}{2} + U_n'''\frac{\Delta x^3}{3!} + \cdots \\ U_{n-1} &= U(x_n - \Delta x) = U_n - U_n'\,\Delta x + U_n''\frac{\Delta x^2}{2} - U_n'''\frac{\Delta x^3}{3!} + \cdots \end{aligned} \tag{1.2.18}$$

whence, by simple addition and subtraction,

$$\frac{U_{n+1} - 2U_n + U_{n-1}}{\Delta x^2} = U_n'' + U^{(iv)}(\xi)\frac{\Delta x^2}{12} \tag{1.2.19}$$

where ξ is in $[x_n - \Delta x, x_n + \Delta x]$. The left-hand side of (1.2.19) is an *approximation* of the derivative d^2U/dx^2 in the point $x = x_n$ and provides an approximation of the equation (with $u \simeq U$)

$$-\frac{u_{n-1} - 2u_n + u_{n+1}}{\Delta x^2} = f_n \tag{1.2.20}$$

The *remainder*

$$T_n \equiv -\frac{U_{n-1} - 2U_n + U_{n+1}}{\Delta x^2} - f_n = -U^{(\text{iv})}(\xi)\frac{\Delta x^2}{12} \qquad (1.2.21)$$

obtained by inserting the exact solution U in place of the numerical solution u in the difference equation (1.2.20), is called the *truncation error* of the approximation.

A common way to describe a relation of the form (1.2.21) is to say that "T is of order Δx^2," also denoted as

$$T = O(\Delta x^2)$$

This expression means that there exists a constant K independent of Δx such that for Δx sufficiently small,

$$T \leq K \Delta x^2$$

Such measures of the rate at which functions vanish when one of their arguments approaches zero are frequently used in numerical analysis.

Expressing (1.2.20) in $n = 1, 2, \ldots, N$ and using the boundary conditions (1.2.14) results in

$$\left.\begin{aligned}
2u_1 - u_2 &= \Delta x^2 f_1 + U_0 \\
-u_1 + 2u_2 - u_3 &= \Delta x^2 f_2 \\
-u_2 + 2u_3 - u_4 &= \Delta x^2 f_3 \\
&\ \cdot \\
&\ \cdot \\
&\ \cdot \\
-u_{N-1} + 2u_N &= \Delta x^2 f_N + U_l
\end{aligned}\right\} \qquad (1.2.22)$$

This *system of linear equations* is to be solved to yield the numerical solution $\{u_n\}$.

1.2.4 CONVERGENCE

A formal theoretical foundation is necessary to establish the validity of numerical methods of approximation for partial differential equations. Although formulations such as those described above appear plausible in the sense that each term in the original partial differential equation seems reasonably approximated by finite differences, there is no assurance that the numerical solution will also be a reasonable approximation of the genuine solution. This problem became well recognized in the early days of the use of electronic computers (i.e., in the 1940s and early 1950s) for the solution of differential equations.

Central to this problem appears to be the concept of convergence, which we shall briefly define. Consider the domain D of an elliptic equation, divided by a mesh of size $\Delta x, \Delta y, \ldots$ on which a numerical approximation is expressed. A family or sequence of calculations is defined, in which $\Delta x, \Delta y,$ $\ldots$ are decreased, keeping the relation to one another specified. If the approximate solution $\{u_{m,n,\ldots}\}$ becomes equal to the exact solution in the limit, then the family of approximations so defined is called *convergent*. Formally, one must introduce a measure whereby the difference between $\{u_{m,n,\ldots}\}$ and $U(x, y)$ may be quantified. Such a measure is called a *norm*. Different kinds of norms may be used in practice. For example, a norm that is suggested by the classical expression for the length of a vector,

$$[\sum_{m,n,\ldots} |u_{m,n,\ldots} - U(x_m, y_n, \ldots)|^2]^{1/2} \tag{1.2.23}$$

is called the $\mathcal{L}_2$ norm of $u - U$, and

$$\max_{m,n,\ldots} |u_{m,n,\ldots} - U(x_m, y_n, \ldots)| \tag{1.2.24}$$

which expresses the maximum value (over all points) of the difference $u_{m,n,\ldots}$ $- U(x_m, y_n, \ldots)$, is another possible norm (sometimes called *sup-norm*). Convergence in one norm does not necessarily imply convergence in another (although this is often the case).

An important contribution to the analysis of the convergence of discrete approximation was made by Courant, Friedrichs, and Lewy (1928) in a paper which, interestingly enough, was not concerned with the *calculation of numerical solutions* but with proving the *existence of genuine solutions* of the partial differential equations of mathematical physics.

An important segment of the literature concerned with numerical methods for partial differential equations has been dominated by convergence studies. This is justified by the notion that if a family of algorithms for a partial differential equation is convergent, then the "programmer" (i.e., the person who computes numerical solutions rather than studies their properties) can always choose a value for $\Delta x, \Delta y, \ldots$ small enough that the error of the approximation will be as small as desired. This however is a weak argument, since what the programmer needs to know is what maximum values of $\Delta x, \Delta y, \ldots$ he or she can afford and still obtain sufficient accuracy. This is answered very incompletely by convergence theories, which are also not very specific when it comes to comparing the economical merits of several competing methods.

There is today, regrettably, somewhat of a schism in the published literature which goes by the name "numerical analysis." Studies of the convergence analysis type address themselves to points which, although well founded, are of little interest to practitioners. These practitioners are more concerned with efficient algorithms, with problems of stability, and with the qualitative and quantitative evaluation of errors in actual calculations, sometimes with no concern at all for convergence.

1.3 RAYLEIGH–RITZ AND GALERKIN METHODS (FUNCTION-EXPANSION METHODS)

1.3.1 INTRODUCTION

A function $U(x)$ with domain of definition D may be approximated by a linear combination of a finite number of prescribed functions $\varphi_k(x)$ (called *basis functions*)

$$U(x) \simeq u(x) = \sum_{k=1}^{N} \varphi_k(x)v_k \qquad (1.3.1)$$

Suppose now that $U(x)$ is to be the solution of the partial differential equation

$$LU = F \quad \text{in } D \qquad (1.3.2)$$

(where L is an operator that contains derivatives with respect to x), with boundary conditions of the type

$$BU - G = 0 \qquad (1.3.3)$$

to be satisfied (where B may be a scalar or a differential operator).

One may seek an approximation to the solution of this problem by expressing it in the form of (1.3.1), attempting to satisfy equation (1.3.2) and the boundary conditions (1.3.3) as well as possible.

That is, we are to find the coefficients v_k of (1.3.1) such that the *residuals*

$$\mathcal{R} = L[\sum_k \varphi_k(x)v_k] - F \quad \text{in } D \qquad (1.3.4)$$

and

$$\mathcal{B} = B[\sum_k \varphi_k(x)v_k] - G \quad \text{on } \partial D \qquad (1.3.5)$$

be made as small as possible, *in some sense to be defined.*

We may call *function-expansion methods* those methods that seek to approximate the solution of differential equations in this fashion. For reasons that will become clear later (see especially Chapter 8), the names *weighted residual method, error distribution method, variational method,*

Rayleigh or *Rayleigh–Ritz method*, and *Galerkin method* are also given (among others) to all or some of the methods that belong to this class.

EXAMPLE

As a motivating example, consider the boundary-value problem

$$\frac{d^2U}{dx^2} + F(x) = 0 \qquad \text{in } x \in (0, l) \tag{1.3.6}$$

$$U(0) = U(l) = 0 \tag{1.3.7}$$

We first note that this problem is easily solved *analytically* by the use of Fourier series (see Chapter 2). Let a set of basis functions be chosen as

$$\varphi_k(x) = \sin\left(\frac{k\pi x}{l}\right); \qquad k = 1, 2, \ldots \tag{1.3.8}$$

We may formally express $U(x)$ as the *infinite* series

$$U(x) = \sum_{k=1}^{\infty} \sin\left(\frac{k\pi x}{l}\right) v_k \tag{1.3.9}$$

With this particular choice of basis functions, the boundary conditions (1.3.7) are automatically satisfied. (Such a lucky choice is possible only because the boundary conditions are homogeneous, but is, in general, not possible when this circumstance is not present.)

The orthogonality property of Fourier series,

$$\int_0^l \sin\left(\frac{k\pi x}{l}\right) \sin\left(\frac{k'\pi x}{l}\right) dx = \begin{cases} 0 & \text{if } k \neq k' \\ \dfrac{l}{2} & \text{if } k = k' \neq 0 \end{cases} \tag{1.3.10}$$

may be used to derive the coefficients v_k. We first insert (1.3.9) into (1.3.6) to obtain

$$\sum_{k'} \frac{d^2}{dx^2}\left[\sin\left(\frac{k'\pi x}{l}\right)v_{k'}\right] + F(x) = \sum_{k'} -\left(\frac{k'\pi}{l}\right)^2 \sin\left(\frac{k'\pi x}{l}\right)v_{k'} + F(x) = 0$$

Then, multiplying by $\sin(k\pi x/l)$ and integrating over $(0, l)$, we find

$$-\frac{l}{2}\left(\frac{k\pi}{l}\right)^2 v_k + \int_0^l \sin\left(\frac{k\pi x}{l}\right) F(x) \, dx = 0$$

or

$$v_k = \frac{2}{l}\left(\frac{l}{k\pi}\right)^2 \int_0^l \sin\left(\frac{k\pi x}{l}\right) F(x) \, dx; \qquad k = 1, 2, \ldots \tag{1.3.11}$$

Moreover, theory tells us that the series is *convergent in the mean*. That is, with

$$u^N(x) = \sum_{k=1}^{N} \sin\left(\frac{k\pi x}{l}\right)v_k \qquad (1.3.12)$$

defined as the series (1.3.9) *truncated* after N terms, we have

$$\lim_{N\to\infty} \int_0^l |u^N(x) - U(x)|^2\, dx = 0 \qquad (1.3.13)$$

Approximation.   We have not raised the question of approximation so far, but we will now. *If we let N remain finite, then the resulting finite sum (1.3.12) is obviously an approximation to the solution $U(x)$ of the equation (1.3.6).*

This may be used as the starting point of a method of numerical approximation, computing the N first coefficients v_k by (1.3.11), with N finite. These may be evaluated analytically in special cases but, in general, are to be approximated by numerical quadrature. Point values of the solution may then be obtained by evaluating (1.3.12) numerically at fixed values of x:

$$U(x_n) \simeq u^N(x_n) = \sum_{k=1}^{N} \sin\left(\frac{k\pi x_m}{l}\right)v_k \qquad (1.3.14)$$

The approximation is best in the following sense: Of all the approximations of $U(x)$ which are expressed as the truncated Fourier series (1.3.12), that which minimizes the mean squared error,[6]

$$||e^N||_2 \equiv ||u^N - U(x)||_2 \equiv \left[\int_0^l |u_N - U|^2\, dx\right]^{1/2} \qquad (1.3.15)$$

is that for which the coefficients v_k are those given by (1.3.11).

We also have, by Bessel's inequality

$$||u^N||_2 \leq ||U(x)||_2 \qquad (1.3.16)$$

1.3.2   VARIATIONAL FORMULATION (RAYLEIGH–RITZ)

The branch of applied mathematics called "calculus of variations" tells us (see Chapter 8) that of all functions $U(x)$ which satisfy the boundary

---

[6]The notation $||(\cdot)||_2$ introduced here defines the $\mathcal{L}_2$ norm of $(\cdot)$.

conditions (1.3.7), that which minimizes the integral

$$J(U) \equiv \int_0^l \left[\frac{1}{2} \left(\frac{dU}{dx} \right)^2 - UF(x) \right] dx \qquad (1.3.17)$$

is precisely the solution of the differential equation (1.3.6). *We may thus seek an approximation of $U(x)$ in the form (1.3.12), by imposing that the parameters $\{v_k\}$ be those that minimize the analog of (1.3.17):*

$$J(u^N) = \int_0^l \left[\frac{1}{2} \left(\sum_k \frac{d\varphi_k}{dx} v_k \right)^2 - \left(\sum_k \varphi_k v_k \right) F(x) \right] dx \qquad (1.3.18)$$

The conditions for this minimum are

$$\frac{\partial J(u^N)}{\partial v_k} = \int_0^l \left[\frac{d\varphi_k}{dx} \left(\sum_{k'} \frac{d\varphi_{k'}}{dx} v_{k'} \right) - \varphi_k F(x) \right] dx$$
$$= 0; \qquad k = 1, 2, \ldots, N \qquad (1.3.19)$$

Integrating by parts allows us to rewrite (1.3.19) as

$$-\frac{\partial J(u^N)}{\partial v_k} = \int_0^l \varphi_k \left[\sum_{k'} \frac{d^2\varphi_{k'}}{dx^2} v_{k'} + F(x) \right] dx = 0 \qquad (1.3.20)$$

When this procedure is applied to the approximation (1.3.12) of U, the result is

$$-\left(\frac{k\pi}{l} \right)^2 \frac{l}{2} v_k + \int_0^l \sin \left(\frac{k\pi x}{l} \right) F(x) \, dx = 0 \qquad (1.3.21)$$

which, we may see, is identical to (1.3.11), which was arrived at by merely following Fourier's theory.

However, variational conditions of the form (1.3.19) may be used with any (reasonable) finite set of basis functions $\{\varphi_k(x)\}$ which *need not be orthogonal* [i.e., (1.3.10) need not hold], *and need not be eigenfunctions of the differential operator* **L**. Section 1.4 below provides powerful illustrations of this fact..

The use of a variational principle of the form (1.3.19) to approximate a partial differential equation is generally referred to as Rayleigh's method or the Rayleigh–Ritz method.

1.3.3 GALERKIN'S METHOD

The concepts contained in the preceding example have led, through a variety of considerations (separation into sinusoidal modes in the analytic Fourier series method, and variational formulation in the Rayleigh–Ritz method) to what may be summarized in the following steps:

1. Choose a set of N linearly independent basis functions

$$\varphi_k(x); \qquad k = 1, 2, \ldots, N \qquad (1.3.22)$$

2. Express an approximation of U in the form of a linear combination (or function expansion)

$$U(x) \simeq u(x) \equiv \sum_k \varphi_k(x) v_k \qquad (1.3.23)$$

3. Form the equation residual

$$\mathfrak{R}(x) = Lu - F = L[\sum_k \varphi_k(x) v_k] - F \qquad (1.3.24)$$

4. Find the coefficients v_k in (1.3.23) by asking that the residual $\mathfrak{R}$ be orthogonal to the N basis functions:

$$\langle \varphi_k(x), \mathfrak{R} \rangle \equiv \int_D \varphi_k(x) \mathfrak{R}(x)\, dx = 0; \qquad k = 1, 2, \ldots, N \qquad (1.3.25)$$

While in the preceding examples additional analytical properties were invoked, such as the exact verification of the equation by each mode v_k $\sin(k\pi x/l)$ in Fourier's method and the global satisfaction of a variational principle by $u(x)$ in the Rayleigh–Ritz method, one may take the pragmatic approach of accepting as an approximating precedure steps 1 through 4 above, without any constraint other than that the boundary conditions of the problem be satisfied by the approximation. This generalization was formally expressed by Galerkin (1915), and the method of approximation expressed by steps 1 through 4, without reference to other analytical principles, is referred to as *Galerkin's method.*

The power of this method relates principally to the fact that many problems of interest are not endowed with the nice analytic properties of the preceding example. Whereas Fourier's method and the Rayleigh–Ritz method do not apply in those cases, Galerkin's method does.

1.3.4 WEIGHTED RESIDUAL METHODS

Conditions (1.3.25) are identical to requiring that the integral of the residual, *weighted* by $\varphi_k(x)$, be equal to zero for $k = 1, 2, \ldots$.

A further generalization consists in choosing weights

$$w_k(x); \qquad k = 1, 2, \ldots, N$$

which may be different from the basis functions $\varphi_k(x)$, and replacing (1.3.25) with

$$\langle w_k, \Re \rangle \equiv \int_D w_k \Re \, dx = 0; \qquad k = 1, 2, \ldots, N \qquad (1.3.26)$$

Methods in this class are described as *weighted residual methods*, of which several variations are used in practice. Galerkin's method is just a special case of a weighted residual method. Other special cases are described in Chapter 8.

1.4   INTRODUCTION TO THE FINITE-ELEMENT METHOD

A newcomer to the field, the finite-element method may be considered as a crossbreed between the finite-difference methods described in Section 1.2 and the function-expansion methods described in Section 1.3.

In the finite-element method, the domain D is divided in subdomains or elements D_e with nodes $\{n\}$ located on interelement lines and/or inside the elements.

The numerical solution takes on discrete values $\{u_n\}$ in these nodes.

In each element D_e the solution is approximated by a simple geometrical shape, in general a polynomial function of the independent variables interpolating between the values $\{u_n\}$ in those nodes that belong to the element D_e.

One of the decisive features of the finite-element method is that the solution is expressed, as with function-expansion methods, by a weighted sum[7]:

$$U(x) \simeq u(x) = \sum_{n=1}^{N} \varphi_n(x)u_n \qquad (1.4.1)$$

Here the basis functions $\varphi_n(x)$ are nonzero only for those elements that contain or are adjacent to the node n. For example, the function $U(x)$ shown in Figure 1.4.1a may be approximated by pieces of straight line, and this approximation may be expressed as a linear combination of basis functions as in (1.4.1); the corresponding basis functions are those illustrated in Figure

---

[7]The subscripts $\{n\}$ used here are intended to indicate that to each n corresponds a (nodal) point in space, where $U \simeq u_n$. No such significance was attached to the subscripts $\{k\}$ in (1.3.1).

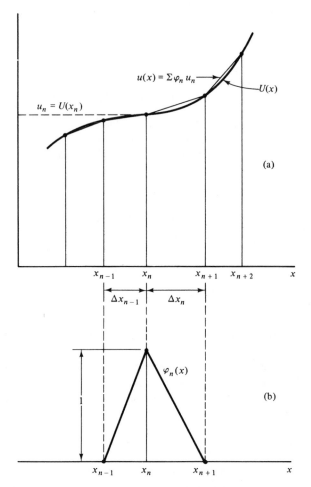

Figure 1.4.1 (*a*) Approximation of $U(x)$ by piecewise linear interpolation between nodal points (—●—); (*b*) linear basis functions in one dimension (sometimes called "chapeau" functions).

1.4.1b. Similarly, the function of two variables $U(x, y)$ shown in Figure 10.2.3b is approximated by pieces of planes on triangles, and the corresponding basis functions are the pyramidal functions of Figure 10.2.3a.

Having chosen this expression of a *function expansion* for the approximate solution, one may then use all the techniques of the Rayleigh–Ritz or Galerkin type.

EXAMPLE 1

Consider the following as an intentionally simple example:

$$-\frac{d^2U}{dx^2} = F(x) \tag{1.4.2}$$

$$U(0) = U(l) = 0 \tag{1.4.3}$$

A simple application of the finite-element method is obtained by dividing $D \equiv [0, l]$ in N discrete "elements" of length Δx_n,

$$0 = x_0 < x_1 < x_2 \ldots < x_n < x_{n+1} \ldots < x_{N+1} = l;$$

$$\Delta x_n = x_{n+1} - x_n \tag{1.4.4}$$

and choosing as basis function $\varphi_n(x)$ the piecewise-linear functions of Figure 1.4.1b. Upon inserting

$$U(x) \simeq u(x) = \sum_n \varphi_n(x)u_n \tag{1.4.5}$$

into (1.4.2), the Galerkin equations are found:

$$0 = \int \varphi_n \left(-\sum_m \frac{d^2\varphi_m}{dx^2}u_m - F \right) dx \tag{1.4.6}$$

It may be appropriate to approximate F in a similar fashion,

$$F(x) \simeq f(x) = \sum_n \varphi_n(x)f_n; \qquad f_n \equiv F(x_n) \tag{1.4.7}$$

whereupon (1.4.6) becomes, after evaluation of the integrals,

$$0 = \frac{u_{n-1} - u_n}{\Delta x_{n-1}} + \frac{u_{n+1} - u_n}{\Delta x_n}$$

$$+ \frac{\Delta x_{n-1}}{6}(f_{n-1} + 2f_n) + \frac{\Delta x_n}{6}(2f_n + f_{n+1}) \tag{1.4.8}$$

Here the first term in (1.4.6) has been integrated by parts as

$$+ \int \frac{d\varphi_n}{dx} \sum_m \frac{d\varphi_m}{dx}u_m \, dx \tag{1.4.9}$$

When the increments are chosen constant,

$$\Delta x_n = \Delta x = \text{constant}$$

then (1.4.8) may be rewritten as

$$0 = \frac{u_{n-1} - 2u_n + u_{n+1}}{\Delta x^2} + \tfrac{1}{6}(f_{n-1} + 4f_n + f_{n+1}) \qquad (1.4.10)$$

We observe that this expression greatly resembles the finite-differences approximation (1.2.20) of the same equation. The main difference between those two expressions is the replacement of f_n (finite-differences case) with

$$\tfrac{1}{6}(f_{n-1} + 4f_n + f_{n+1}) \qquad \text{(finite-element case)}$$

The latter expression may be observed to be an average value of $F(x)$ by the application of *Simpson's rule* of numerical quadrature:

$$F_{\text{average}} = \frac{1}{2\Delta x} \int_{x_n-\Delta x}^{x_n+\Delta x} F(x)\, dx$$

$$\simeq \tfrac{1}{6}(f_{n-1} + 4f_n + f_{n+1}) \qquad (1.4.11)$$

EXAMPLE 2

Consider as a second example the Dirichlet problem for Poisson's equation

$$-\nabla^2 U \equiv -\left(\frac{\partial^2 U}{\partial x^2} + \frac{\partial^2 U}{\partial y^2}\right) = F(x, y) \qquad \text{in } D$$
$$U = U_B(x, y) \qquad\qquad\qquad\qquad \text{on } \partial D \qquad (1.4.12)$$

One form of the application of the finite-element method consists in dividing D in triangular elements D_e as shown in Figure 1.4.2. Nodal values of a numerical solution $\{u_n\}$ are meant to be approximations of $U(x, y)$ in those nodes:

$$u_n \simeq U(x_n, y_n)$$

The finite-element numerical solution is assumed to consist of pieces of plane in each triangle (or "tiles") represented by (1.4.1) with the linear, pyramidal basis functions of Figure 10.2.3b. The Galerkin conditions (1.3.24) applied here read

$$\int \varphi_n \left(\sum_m \nabla^2 \varphi_m u_m + F\right) dD = 0, \qquad n = 1, 2, \ldots, N \qquad (1.4.13)$$

Since φ_n is nonzero only over those elements D_e that surround the node n, this integral need be evaluated over those few elements only. After evaluation, (1.4.13) becomes a system of linear algebraic equations of the form (details

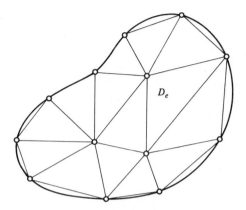

Figure 1.4.2 Division in irregular triangles of a domain D over which Poisson's equation is to be solved. The discrete equations for the unknown nodal values $\{u_n\}$ are derived by the mere evaluation of the integrals in (1.4.13). By contrast, deriving similar discrete equations over this irregular grid by standard finite-difference methods would be extremely difficult.

are given in Section 10.2)

$$A\{u_n\} = \{b_n\} \tag{1.4.14}$$

where A is a square, constant matrix (called the "stiffness matrix"). This system is then to be solved to produce the nodal values $\{u_n\}$ of the numerical solution. Because of the finite support property of the basis functions, the matrix A turns out to be sparse (i.e., many of its elements are zero). In fact, the matrix A bears a great resemblance to the matrices derived with classical finite-difference techniques, although the method used to derive A is quite different in the two approaches (see Example 1).

Of great practical importance is the fact that going from (1.4.12) to (1.4.14) can be accomplished by simple calculations. The system of linear equations (1.4.14) is of a type similar to the finite-difference approximation to the same problem. But one may choose in the finite-element method an irregular distribution of nodal points in (x, y) and corresponding irregular elements. This adds little complication to the evaluation of (1.4.13). By contrast, deriving classical finite-difference equations over a similarly irregular mesh would be extremely complicated, if not impossible.

This relative simplicity of derivation of the discrete equations in applications of the finite-element method to the odd-shaped domains that one finds in real-life problems has permitted the implementation of *general-purpose computer codes*, in which not only are discrete equations of the type

(1.4.14) solved, but the calculations implied in deriving those equations [i.e., in going from (1.4.12) to (1.4.14)] are also taken care of. Attempting to implement similar codes with finite-difference methods had proven to be a very difficult task. Several general-purpose computer codes intended to solve partial differential equations with finite-difference methods do exist. But they are less flexible and convenient to use than are comparable (and often more recent) finite-element codes.

2

Fourier Series

2.1 COMPLEX VARIABLES
AND TRIGONOMETRIC FUNCTIONS

> The imaginary number is a fine and wonderful recourse of the divine spirit, almost an amphibian between being and not being.
>
> G. W. Leibnitz (1646–1716)

Before discussing the details of Fourier series and integrals, we shall briefly survey some basic notions and notations in complex variables and functions which are a fundamental tool in this theory.

A complex number

$$z = a + ib \qquad (i = \sqrt{-1}) \qquad (2.1.1)$$

consists of a real part a and an imaginary part b. The notations

$$a = \text{Re}\,(z) = \text{real part of } z$$
$$b = \text{Im}\,(z) = \text{imaginary part of } z$$

are also used.

A convenient way to visualize a complex number is by using its representations as a *vector* in the plane (called the *complex plane*), as shown in Figure 2.1.1.

The *absolute value* or *modulus* of a complex number z is the length of the vector representing it: by the Pythagorean theorem,

$$|z| = \sqrt{a^2 + b^2} \qquad (2.1.2)$$

Its *phase angle* or *argument* is the angle of this vector with respect to the Re (z) axis:

$$\angle z = \arg(z) = \tan^{-1}\left(\frac{b}{a}\right) \qquad (2.1.3)$$

There are strong connections between complex numbers and trigonometric functions. This is probably most apparent when considering the relation

$$e^{i\alpha} = \cos\alpha + i\sin\alpha \qquad (2.1.4)$$

Proof. Express the three Taylor series,

$$e^{i\alpha} = 1 + i\alpha - \frac{\alpha^2}{2!} - \frac{i\alpha^3}{3!} + \cdots$$

$$\sin\alpha = \alpha - \frac{\alpha^3}{3!} + \frac{\alpha^5}{5!} - \cdots$$

$$\cos\alpha = 1 - \frac{\alpha^2}{2!} + \frac{\alpha^4}{4!} - \cdots$$

which, inserted in (2.1.4), indeed result in an identity.   □

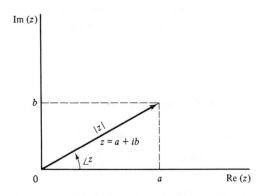

Figure 2.1.1 Representation of a complex number in a plane.

This remarkable expression was known early in the eighteenth century. Its discovery is generally attributed to Euler and is often referred to as *Euler's formula*. However, a more complete statement of this formula,

$$(\cos \alpha + i \sin \alpha)^n = \cos (n\alpha) + i \sin (n\alpha) \qquad (2.1.5)$$

known as the *de Moivre's theorem*, was published as early as 1707 (see Boyer, 1968, p. 466).

Euler's (or de Moivre's) theorem allows the expression of a complex number to be given in the form

$$z = |z| e^{iLz} \qquad (2.1.6)$$

Applying previous definitions, we also find that

$$\begin{aligned} |e^{i\alpha}| &= 1 \\ \angle e^{i\alpha} &= \alpha \end{aligned} \qquad (2.1.7)$$

Thus, the graph of the function $e^{i\alpha}$ in the complex plane is a circle of unit radius and center at the origin (Figure 2.1.2). From (2.1.4) and

$$e^{-i\alpha} = \cos \alpha - i \sin \alpha$$

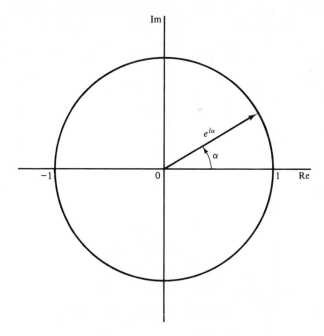

Figure 2.1.2 Graph of the function $e^{i\alpha}$.

we also get, by addition and subtraction,

$$\cos \alpha = \frac{e^{i\alpha} + e^{-i\alpha}}{2} \tag{2.1.8}$$

$$\sin \alpha = \frac{e^{i\alpha} - e^{-i\alpha}}{2i} \tag{2.1.9}$$

2.1.1 ALGEBRAIC OPERATIONS ON COMPLEX NUMBERS

The operations of addition and multiplication of two complex numbers are expressed algebraically as follows [the only rule we need to remember is that $(i)^2 = -1$]:

$$\begin{aligned} z_1 + z_2 &\equiv (a_1 + ib_1) + (a_2 + ib_2) \\ &= (a_1 + a_2) + i(b_1 + b_2) \end{aligned} \tag{2.1.10}$$

and

$$\begin{aligned} z_1 z_2 &= (a_1 + ib_1)(a_2 + ib_2) \\ &= (a_1 a_2 - b_1 b_2) + i(a_1 b_2 + a_2 b_1) \end{aligned} \tag{2.1.11}$$

A simple interpretation of these relations is given by considering the vector representations of z_1 and z_2 in the complex plane. *Addition* consists in adding vectors in the usual geometrical sense (Figure 2.1.3). The *product* of two complex numbers is obtained by *multiplying* their modulus, and *adding* their argument, which becomes evident when we consider the polar representation

$$z_1 = |z_1| e^{i\angle z_1}$$
$$z_2 = |z_2| e^{i\angle z_2}$$

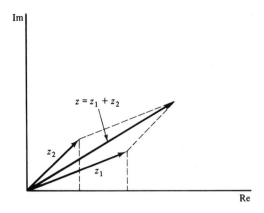

Figure 2.1.3   Addition of two complex numbers.

Then, clearly,

$$z = z_1 z_2 = |z_1| \cdot |z_2| \, e^{i(\angle z_1 + \angle z_2)}$$

2.2 THE FOURIER INTEGRAL

Fourier's Theorem is not only one of the most beautiful results of modern analysis but it may be said to furnish an indispensable instrument in the treatment of nearly every recondite question in modern physics.

LORD KELVIN and PETER GUTHRIE TAIT,
Treatise on Natural Philosophy

2.2.1 THE FOURIER INTEGRAL
AND SOME OF ITS PROPERTIES

Given a function of the real variable x, which is square-integrable,

$$\int_{-\infty}^{\infty} |U(x)|^2 \, dx \neq \infty$$

the integral (called the Fourier integral)

$$\hat{U}(\omega) = \int_{-\infty}^{\infty} U(x)e^{-i\omega x} \, dx \tag{2.2.1}$$

defines $\hat{U}(\omega)$ as the *Fourier transform* of $U(x)$. The notation

$$\hat{U}(\omega) = \mathfrak{F}(U(x))$$

is also used occasionally.

The inverse relation, known as the *inverse Fourier transform*, is

$$U(x) = \int_{-\infty}^{\infty} \hat{U}(\omega)e^{i\omega x} \frac{d\omega}{2\pi} \tag{2.2.2}$$

and is also denoted as

$$U(x) = \mathfrak{F}^{-1}(\hat{U}(\omega))$$

Although $U(x)$ is often assumed to be real, its Fourier transform is, in general, complex. Two useful rules, which are easily verifiable by simple substitution, are:

1. The *derivation rule:*

$$\mathfrak{F}\left(\frac{d^r U}{dx^r}\right) = (i\omega)^r \mathfrak{F}(U) \tag{2.2.3}$$

2.   The *shifting rule:*

$$\mathcal{F}(U(x + h)) = e^{ih}\mathcal{F}(U) \qquad (2.2.4)$$

Some common Fourier transform pairs are given in Figure 2.2.1.

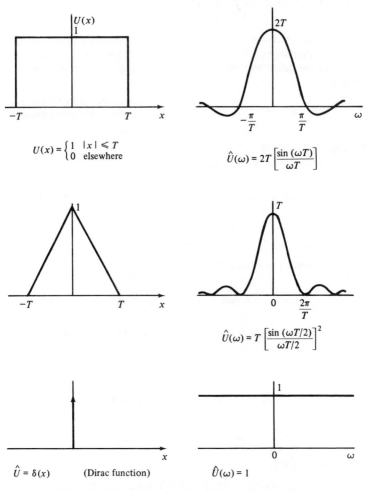

$$U(x) = \begin{cases} 1 & |x| \leqslant T \\ 0 & \text{elsewhere} \end{cases} \qquad \hat{U}(\omega) = 2T\left[\frac{\sin{(\omega T)}}{\omega T}\right]$$

$$\hat{U}(\omega) = T\left[\frac{\sin{(\omega T/2)}}{\omega T/2}\right]^2$$

$$\hat{U} = \delta(x) \qquad \text{(Dirac function)} \qquad \hat{U}(\omega) = 1$$

Figure 2.2.1   Common Fourier transform pairs.

2.2.2   *CONVOLUTION THEOREMS*

The convolution of two functions $U_1(x)$ and $U_2(x)$, denoted $U_1(x) \otimes U_2(x)$, is defined as the function

$$V(x) = U_1(x) \otimes U_2(x) \equiv \int_{-\infty}^{\infty} U_1(\xi)U_2(x - \xi)\,d\xi \qquad (2.2.5)$$

We seek the Fourier transform of $V(x)$:

$$\mathfrak{F}(V(x)) = \int_{-\infty}^{\infty} \left[\int_{-\infty}^{\infty} U_1(\xi) U_2(x - \xi) \, d\xi \right] e^{-i\omega x} \, dx \qquad (2.2.6)$$

Changing the order of integration, we may write

$$\mathfrak{F}(V(x)) = \int_{-\infty}^{\infty} \left[\int_{-\infty}^{\infty} U_2(x - \xi) e^{-i\omega x} \, dx \right] U_1(\xi) \, d\xi$$

$$= \int_{-\infty}^{\infty} \left[\int_{-\infty}^{\infty} U_2(x - \xi) e^{-i\omega(x-\xi)} \, d(x - \xi) \right] U_1(\xi) e^{-i\omega\xi} \, d\xi$$

$$= \hat{U}_1(\omega) \hat{U}_2(\omega) \qquad (2.2.7)$$

That is, *the Fourier transform of the convolution of two functions is the product of their respective Fourier transforms.*

The converse can be established by a strictly symmetrical proof. We need give the result only: The *convolution* of two Fourier transforms is defined as

$$\hat{U}_1(\omega) \otimes \hat{U}_2(\omega) \equiv \int_{-\infty}^{\infty} \hat{U}_1(\beta) \hat{U}_2(\omega - \beta) \frac{d\beta}{2\pi} \qquad (2.2.8)$$

Its inverse is the simple product of the original functions:

$$\mathfrak{F}^{-1}(\hat{U}_1(\omega) \otimes \hat{U}_2(\omega)) = U_1(x) U_2(x) \qquad (2.2.9)$$

That is,

$$\mathfrak{F}(U_1(x) U_2(x)) = \hat{U}_1(\omega) \otimes \hat{U}_2(\omega) \qquad (2.2.10)$$

2.2.3   PARSEVAL'S RELATION: FORM I

We form the integrated product of two functions, denoted by angular brackets:

$$\langle U_1, U_2 \rangle \equiv \int_{-\infty}^{\infty} U_1(x) U_2(x) \, dx \qquad (2.2.11)$$

The result is a scalar number and $\langle U_1, U_2 \rangle$ is sometimes called the *scalar product* of U_1 and U_2. We shall seek an expression of (2.2.11) in terms of the respective Fourier transforms. By the convolution theorem (2.2.10), we have

$$U_1(x) U_2(x) = \int_{-\infty}^{\infty} \left[\int_{-\infty}^{\infty} \hat{U}_1(\omega) \hat{U}_2(\beta - \omega) \frac{d\omega}{2\pi} \right] e^{i\beta x} \frac{d\beta}{2\pi} \qquad (2.2.12)$$

and

$$\langle U_1, U_2 \rangle = \int_{-\infty}^{\infty} dx \int_{-\infty}^{\infty} \left[\int_{-\infty}^{\infty} \hat{U}_1(\omega)\hat{U}_2(\beta - \omega) \frac{d\omega}{2\pi} \right] e^{i\beta x} \frac{d\beta}{2\pi} \qquad (2.2.13)$$

But

$$\int_{-\infty}^{\infty} e^{i\beta x} \, dx = 2\pi \, \delta(\beta) \qquad (2.2.14)$$

(δ is the Dirac delta function), whence (2.2.13) becomes

$$\langle U_1, U_2 \rangle = \int_{-\infty}^{\infty} \hat{U}_1(\omega)\overline{\hat{U}}_2(\omega) \frac{d\omega}{2\pi} \qquad (2.2.15)$$

where $\overline{\hat{U}}_2(\omega) = \hat{U}_2(-\omega)$ is the complex conjugate of $\hat{U}_2(\omega)$. Also, defining

$$\langle \hat{U}_1, \hat{U}_2 \rangle \equiv \int_{-\infty}^{\infty} \hat{U}_1(\omega)\overline{\hat{U}}_2(\omega) \frac{d\omega}{2\pi} \qquad (2.2.16)$$

we have simply

$$\langle U_1, U_2 \rangle = \langle \hat{U}_1, \hat{U}_2 \rangle \qquad (2.2.17)$$

This is known as (one of the forms of) Parseval's relation.[1] Note that the scalar product is related to the convolution by the identities

$$\langle U_1, U_2 \rangle = [U_1 \otimes U_2]_{x=0}$$
$$\langle \hat{U}_1, \hat{U}_2 \rangle = [\hat{U}_1 \otimes \hat{U}_2]_{\omega=0} \qquad (2.2.18)$$

2.2.4 $\mathcal{L}_2$ NORM OF A FUNCTION (A SECOND FORM OF PARSEVAL'S RELATION)

The $\mathcal{L}_2$ or Euclidian norm of a function $U(x)$, defined everywhere on the real axis x, is defined as

$$\| U(x) \|_2 \equiv \left[\int_{-\infty}^{\infty} U(x)^2 \, dx \right]^{1/2} \qquad (2.2.19)$$

Functions $U(x)$ of a linear function space for which this norm is finite are said to belong to the class $\mathcal{L}_2$ or to be in $\mathcal{L}_2$. They are also said to be square-integrable or to belong to a Hilbert space H. The analogy of (2.2.19) with

---

[1] Sometimes known as Plancherel's relation. It is given Parseval's name more precisely when $U_1 = U_2$.

the length of an infinite-dimensional vector in Euclidian space (via the Pythagorean theorem) should be self-evident.

We now turn to the equivalent representation of $U(x)$ in the frequency domain. We note that the space of $\hat{U}(\omega)$ is also a linear function space of domain $\omega \in (-\infty, \infty)$.

The $\mathcal{L}_2$ norm of $\hat{U}(\omega) = \mathcal{F}(U)$ in that space is defined as

$$\| \hat{U}(\omega) \|_2 \equiv \left[\int_{-\infty}^{\infty} | \hat{U}(\omega) |^2 \frac{d\omega}{2\pi} \right]^{1/2} \tag{2.2.20}$$

To obtain a relation between $\| U(x) \|_2$ and $\| \hat{U}(\omega) \|_2$, we turn to Parseval's relation (2.2.17). We readily find (Parseval's relation, form II)

$$\| \hat{U}(\omega) \|_2^2 = \int_{-\infty}^{\infty} | \hat{U}(\omega) |^2 \frac{d\omega}{2\pi} = \int_{-\infty}^{\infty} \hat{U}(\omega) \overline{U}(\omega) \frac{d\omega}{2\pi}$$

$$= \int_{-\infty}^{\infty} | U(x) |^2 \, dx = \| U(x) \|_2^2$$

or

$$\| \hat{U}(\omega) \|_2 = \| U(x) \|_2 \tag{2.2.21}$$

We thus have the important property that the Fourier transformation of a function is $\mathcal{L}_2$-norm-preserving.

In several branches of engineering, $\| U \|_2^2$ is called the energy of the function $U(x)$ and $| \hat{U}(\omega) |^2 / \pi$ is called its *spectral energy density* (see, e.g., Laning and Battin, 1956):

$$\text{energy} = \int_0^{\infty} \frac{| \hat{U}(\omega) |^2}{\pi} \, d\omega \tag{2.2.22}$$

2.3 FOURIER SERIES

A function $U(x)$ defined over a *finite* domain $x \in (0, l)$ may be formally expressed as the (Fourier) series

$$U(x) = \frac{A_0}{2} + \sum_{k=1}^{\infty} \left[A_k \cos \left(\frac{2k\pi x}{l} \right) + B_k \sin \left(\frac{2k\pi x}{l} \right) \right]$$

$$= \frac{A_0}{2} + \sum_{k=1}^{\infty} [A_k \cos (k\omega_0 x) + B_k \sin (k\omega_0 x)]; \quad (\omega_0 = 2\pi/l) \tag{2.3.1}$$

This expresses $U(x)$ as a linear combination of the basis functions

$$1, \cos(k\omega_0 x), \cos(2k\omega_0 x), \ldots$$
$$\sin(k\omega_0 x), \sin(2k\omega_0 x), \ldots \tag{2.3.2}$$

The following relations, derivable through elementary calculus,

$$
\left.
\begin{aligned}
\int_0^l \cos(k_1\omega_0 x)\cos(k_2\omega_0 x)\, dx &= \int_0^l \sin(k_1\omega_0 x)\sin(k_2\omega_0 x)\, dx \\
&= \begin{cases} 0 & \text{if } k_1 \neq k_2 \\ \dfrac{l}{2} & \text{if } k_1 = k_2 \neq 0 \end{cases} \\
\int_0^l \cos(k_1\omega_0 x)\sin(k_2\omega_0 x)\, dx &= 0 \\
\text{and}\qquad\quad \int_0^l \cos(k\omega_0 x)\, dx &= 0
\end{aligned}
\right\} \tag{2.3.3}
$$

express the fact that these basis functions form an *orthogonal set*. From this fact results the property that the coefficients A_k and B_k may be expressed simply as

$$A_k = \frac{2}{l}\int_0^l U(x)\cos(k\omega_0 x)\, dx; \qquad k = 0, 1, 2, \ldots$$
$$B_k = \frac{2}{l}\int_0^l U(x)\sin(k\omega_0 x)\, dx; \qquad k = 1, 2, \ldots \tag{2.3.4}$$

2.3.1 VALUE OF THE SERIES OUTSIDE $x \in (0, l)$

The sine and cosine functions in (2.3.1) are defined for all x. Outside $x \in (0, l)$, the function defined by the series (2.3.1) repeats itself periodically by simple translation, with period l (Figure 2.3.1). The Fourier series (2.3.1) may also be visualized as representing a function that is wrapped around a cylinder of perimeter l, as shown in Figure 2.3.2.

Figure 2.3.1 Value of the Fourier series (2.3.1) for all x.

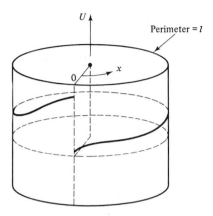

Figure 2.3.2 Circular representation of the Fourier series (2.3.1).

2.3.2 $\mathcal{L}_2$ NORM AND PARSEVAL'S THEOREM

The $\mathcal{L}_2$ norm of the function $U(x)$ defined in $x \in (0, l)$ is, by definition,

$$\| U(x) \|_2 = \left[\int_0^l |U(x)|^2 \, dx \right]^{1/2} \tag{2.3.5}$$

The square of this quantity is called the energy of $U(x)$. Note that a *necessary* condition for $U(x)$ to be expressible in Fourier series form is that $\| U(x) \|_2$ be finite. An important relation between $\| U(x) \|_2^2$ and the Fourier coefficients of U can be derived; the expression

$$\int_0^l |U(x)|^2 \, dx = \int_0^l \left\{ \frac{A_0}{2} + \sum_k [A_k \cos (k\omega_0 x) + B_k \sin (k\omega_0 x)] \right\}^2 dx \tag{2.3.6}$$

becomes, *taking the orthogonality relations.* (2.3.3) *into account,*

$$\| U(x) \|_2^2 = \frac{l}{2} \left[\frac{A_0^2}{2} + \sum_{k=1}^{\infty} (A_k^2 + B_k^2) \right] \tag{2.3.7}$$

This expression is known as (one of the forms of) *Parseval's relation* and is the equivalent of (2.2.21) for functions defined on a finite domain $(0, l)$ instead of $(-\infty, \infty)$.

2.3.3 SPECTRAL DECOMPOSITION OF THE ENERGY

The important interpretation of Parseval's relation lies in the observation that the energy $\| U \|_2^2$ of U can be decomposed as a sum of independent contributions of the energy at the various frequencies $\omega_k = 2k\pi/l$. With

$$\| U_k \|_2^2 = \begin{cases} \dfrac{l}{4} A_0^2 & \text{for } k = 0 \\[2ex] \dfrac{l}{2}(A_k^2 + B_k^2) & \text{for } k > 0 \end{cases} \tag{2.3.8}$$

defined as the spectral components of the energy of U, we may write simply

$$\| U \|_2^2 = \sum_k \| U_k \|_2^2 \tag{2.3.9}$$

This relation is equivalent to (2.2.22).

2.3.4 FOURIER SINE SERIES

When dealing in boundary-value problems where the solution $U(x)$ is to satisfy the boundary conditions $U(0) = U(l) = 0$, it is convenient to use the following artifice: Instead of considering the interval $x \in (0, l)$, we consider the extended interval $x \in (0, l^*)$, $l^* = 2l$. Over $x \in (l, 2l)$, we let

$$U(x) = -U(2l - x) \tag{2.3.10}$$

(See Figure 2.3.3.) The function $U(x)$ (for all x) becomes antisymmetric with respect to the points

$$-l, 0, l, 2l, \ldots$$

over the whole x axis, and must be equal to zero in those points. *We thus have found a way to satisfy the imposed zero boundary conditions by a mere doubling of the interval of the problem.* Moreover, antisymmetry means that all cosine terms vanish

$$A_k = 0; \quad k = 0, 1, 2, \ldots \tag{2.3.11}$$

and we may thus write

$$\begin{aligned} U(x) &= \sum_{k=1}^{\infty} B_k \sin\left(\frac{2k\pi x}{l^*}\right) \\ &= \sum_{k=1}^{\infty} B_k \sin\left(\frac{k\pi x}{l}\right) \end{aligned} \tag{2.3.12}$$

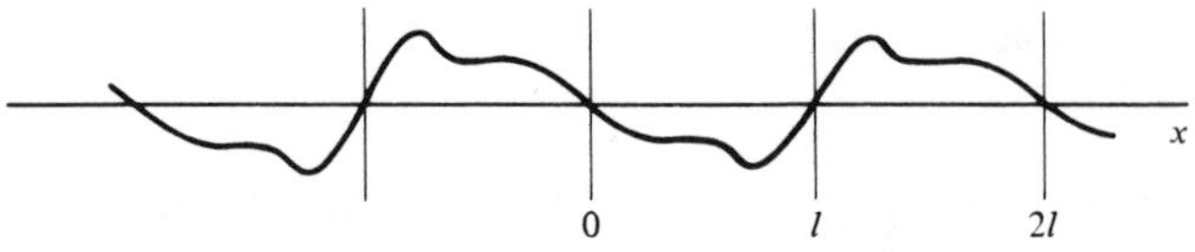

Figure 2.3.3   Periodicity of a Fourier sine series outside of $x \in [0, l]$.

The integrals defining the B_k can (because of symmetry) be limited to $x \in (0, l)$:

$$B_k = \frac{2}{l^*} \int_0^{l^*} U(x) \sin \left(\frac{2k\pi x}{l^*} \right) dx = \frac{2}{l} \int_0^l U(x) \sin \left(\frac{k\pi x}{l} \right) dx \qquad (2.3.13)$$

In fact, we may observe that the basis of functions which is implicit in (2.3.12) is

$$\sin \left(\frac{\pi x}{l} \right), \sin \left(\frac{2\pi x}{l} \right), \ldots$$

It forms an orthogonal set over $(0, l)$:

$$\int_0^l \sin \left(\frac{k_1 \pi x}{l} \right) \sin \left(\frac{k_2 \pi x}{l} \right) dx = \begin{cases} 0 & \text{if } k_1 \neq k_2 \\ \dfrac{l}{2} & \text{if } k_1 = k_2 \neq 0 \end{cases} \qquad (2.3.14)$$

We may consider (2.3.12) as a different type of Fourier series, called the *Fourier sine series*, of which (2.3.13) is the expression of the coefficients.

2.3.5 PARSEVAL'S RELATION

Parseval's relation (2.3.7) becomes for Fourier sine series

$$\| U(x) \|_2^2 = \int_0^l | U(x) |^2 \, dx = \frac{l}{2} \sum_{k=1}^\infty B_k^2 \qquad (2.3.15)$$

2.4 DISCRETE FOURIER SERIES

2.4.1 INTRODUCTION

In dealing with the approximation of functions, it is generally the case that they are represented (incompletely) by a finite set of discrete values taken at a finite number of points in the domain. It is relatively easy (as we shall see) to extend to those cases the preceding concepts, and to develop an *exact, self-contained theory of discrete Fourier analysis*.

Let $\{u_n\}$ be the set of values of a function $u(x)$, defined (only) in $2N$ discrete, equidistant points $x_n = n \, \Delta x$ on the segment of the x axis $x \in [0, l)$; see Figure 2.4.1.

$$u_n = U(n \, \Delta x), \qquad n = 0, 1, 2, \ldots, (2N - 1); \qquad \Delta x = \frac{l}{2N} \qquad (2.4.1)$$

We may express $\{u_n\}$ formally as a discrete Fourier series:

$$u_n = \frac{a_0}{2} + \sum_{k=1}^{N-1} [a_k \cos (k\omega_0 x_n) + b_k \sin (k\omega_0 x_n)]$$

$$+ \frac{a_N}{2} \cos (N\omega_0 x_n); \qquad n = 0, 1, 2, \ldots, 2N - 1 \qquad (2.4.2)$$

where the $\{a_k, b_k\}$ are called the discrete Fourier coefficients of $\{u_n\}$, and, as in (2.3.1),

$$\omega_0 = \frac{2\pi}{l} \qquad (2.4.3)$$

Note that the last term in (2.4.2) is equal to

$$\cos (N\omega_0 x_n) = (-1)^n \qquad (2.4.4)$$

The $2N$ coefficients

$$\{a_k, b_k\} \equiv a_0, a_1, a_2, \ldots, a_{N-1}, a_N$$
$$b_1, b_2, \ldots, b_{N-1}$$

shall be determined by the conditions that (2.4.2) must hold in the $2N$ collocation points $\{x_n, u_n\}$. The corresponding transformation from the $2N$ numbers $\{u_n\}$ to the $2N$ numbers $\{a_k, b_k\}$ is linear. It can be written in matrix form as

$$
\begin{bmatrix}
\frac{1}{2} & 1 & 0 & 1 & \cdots & 0 & \frac{1}{2} \\
\frac{1}{2} & \cos (\omega_0 \Delta x) & \sin (\omega_0 \Delta x) & & & \sin [(N-1)\omega_0 \Delta x] & -\frac{1}{2} \\
\frac{1}{2} & \cos (2\omega_0 \Delta x) & \sin (2\omega_0 \Delta x) & & & \sin [2(N-1)\omega_0 \Delta x] & +\frac{1}{2} \\
& & & \bullet & & & -\frac{1}{2} \\
& & & & \bullet & & +\frac{1}{2} \\
& & & & & \bullet & \vdots \\
& & & \bullet & & \bullet & \ddots \\
\frac{1}{2} & \cos [(2N-1)\omega_0 \Delta x] & \sin [(2N-1)\omega_0 \Delta x] & \cdots & & & -\frac{1}{2}
\end{bmatrix}
$$

$$
\cdot
\begin{bmatrix}
a_0 \\ a_1 \\ b_1 \\ a_2 \\ \vdots \\ \vdots \\ b_{N-1} \\ a_N
\end{bmatrix}
=
\begin{bmatrix}
u_0 \\ u_1 \\ u_2 \\ \vdots \\ \vdots \\ \vdots \\ \vdots \\ u_{2N-1}
\end{bmatrix}
\qquad (2.4.5)
$$

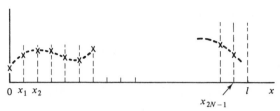

Figure 2.4.1

That the inverse exists results from the fact that the determinant of this system is not zero, which in turn is a consequence of the fact that the sets

$$\{\sin (k\omega_0 x_n)\}; \quad \{\cos (k\omega_0 x_n)\}$$

are linearly independent. From the discrete orthogonality relations

$$\sum_{n=0}^{2N-1} \cos (k_1\omega_0 x_n)\cos (k_2\omega_0 x_n) = \sum_{n=0}^{2N-1} \sin (k_1\omega_0 x_n) \sin (k_2\omega_0 x_n)$$

$$= \begin{cases} 0 & \text{if } k_1 \neq k_2 \\ N & \text{if } k_1 = k_2 \neq 0 \end{cases} \tag{2.4.6a}$$

$$\sum_{n=0}^{2N-1} \cos (k_1\omega_0 x_n) \sin (k_2\omega_0 x_n) = 0 \tag{2.4.6b}$$

and

$$\sum_{n=0}^{2N-1} (1)^2 = \sum_{n=0}^{2N-1} [\cos (N\omega_0 x_n)]^2 = 2N \tag{2.4.6c}$$

we may derive simply

$$\sum_{n=0}^{2N-1} u_n \cos (k\omega_0 x_n) = Na_k; \quad k = 0, 1, 2, \ldots, N$$
$$\sum_{n=0}^{2N-1} u_n \sin (k\omega_0 x_n) = Nb_k; \quad k = 1, 2, \ldots, N-1 \tag{2.4.7}$$

These are the expressions that give the $\{a_k, b_k\}$ in relation to the $\{u_n\}$. That is, we may also use (2.4.7) to write the inverse of the linear system (2.4.5) in matrix form:

$$\begin{bmatrix} a_0 \\ a_1 \\ b_1 \\ a_2 \\ \cdot \\ \cdot \\ \cdot \\ b_{N-1} \\ a_N \end{bmatrix} = \frac{1}{N} \begin{bmatrix} 1 & 1 & 1 & \cdots & 1 \\ 1 & \cos (\omega_0 \Delta x) & \cos (2\omega_0 \Delta x) & & \cos [(2N-1)\omega_0 \Delta x] \\ 0 & \sin (\omega_0 \Delta x) & \sin (2\omega_0 \Delta x) & & \sin [(2N-1)\omega_0 \Delta x] \\ 1 & \cos (2\omega_0 \Delta x) & \cos (4\omega_0 \Delta x) & & \\ 0 & & & \ddots & \\ \cdot & & & & \\ \cdot & & & & \\ 1 & -1 & 1 & \cdots & -1 \end{bmatrix} \begin{bmatrix} u_0 \\ u_1 \\ \cdot \\ \cdot \\ \cdot \\ \cdot \\ u_{2N-1} \end{bmatrix}$$

$$\tag{2.4.8}$$

These relations may also be written as

$$a_k = \frac{2}{l}\left[\Delta x \sum_{n=0}^{2N-1} u_n \cos(k\omega_0 x_n)\right]$$

$$b_k = \frac{2}{l}\left[\Delta x \sum_{n=0}^{2N-1} u_n \sin(k\omega_0 x_n)\right]$$

(2.4.9)

They are called discrete Fourier transformations, sometimes denoted as

$$\{a_k, b_k\} = \mathcal{F}(\{u_n\})$$

The similarity of (2.4.9) with the continuous-case transform (2.3.4) appears clearly. As N is increased, Δx decreases and the brackets in (2.4.9) tend toward the integrals in (2.3.4).

The discrete analog of Fourier *sine* series is also easily established. For the set of $N - 1$ interior values,

$$\{u_n; n = 1, 2, \ldots, N - 1\}$$

(2.4.10)

(where $u_0 = u_N = 0$ is implicit), we seek a Fourier sine series of the form

$$u_n = \sum_{k=1}^{N-1} b_k \sin\left(\frac{k\pi x_n}{l}\right)$$

(2.4.11)

From the discrete orthogonality relations

$$\sum_{n=1}^{N-1} \sin\left(\frac{k_1 \pi x_n}{l}\right) \sin\left(\frac{k_2 \pi x_n}{l}\right) = \begin{cases} 0 & \text{if } k_1 \neq k_2 \\ \dfrac{N}{2} & \text{if } k_1 = k_2 \neq 0 \end{cases}$$

(2.4.12)

we derive

$$\sum_{n=1}^{N-1} u_n \sin\left(\frac{k\pi x_n}{l}\right) = \frac{N}{2} b_k$$

whence the discrete analog of (2.3.13):

$$b_k = \frac{2}{N} \sum_{n=1}^{N-1} u_n \sin\left(\frac{k\pi x_n}{l}\right)$$

$$= \frac{2}{l}\left[\Delta x \sum_{n=1}^{N-1} u_n \sin\left(\frac{k\pi x_n}{l}\right)\right]; \quad k = 1, 2, \ldots, N - 1$$

(2.4.13)

referred to as the *discrete Fourier (sine) transform* of $\{u_n\}$:

$$\{b_k\} = \mathcal{F}(\{u_n\})$$

2.4.2 DISCRETE $\mathfrak{L}_2$ NORM AND THE DISCRETE FORM OF PARSEVAL'S RELATION

The discrete $\mathfrak{L}_2$ norm of the set

$$\{u_n; n = 0, 1, 2, \ldots, 2N - 1\} \tag{2.4.14}$$

is defined as

$$\|u\|_2 = \|\{u_n\}\|_2 \equiv \left(\Delta x \sum_{n=0}^{2N-1} |u_n|^2\right)^{1/2}$$

[which is the discrete approximation of a continuous integral of the form (2.3.5) obtained by application of the rectangular rule].

A discrete form of Parseval's relation (2.3.7) may be established. We may write

$$\|u\|_2^2 = \Delta x \left(\sum_{n=0}^{2N-1} \left\{\frac{a_0}{2} + \sum_{k=1}^{N-1} [a_k \cos (k\omega_0 x_n) + b_k \sin (k\omega_0 x_n)] + (-1)^n \frac{a_N}{2}\right\}\right)^2 \tag{2.4.15}$$

which becomes, *taking the orthogonality relations (2.4.6) into account,*

$$\|u\|_2^2 = 2 \Delta x N\left[\frac{a_0^2 + a_N^2}{4} + \frac{1}{2} \sum_{k=1}^{N-1} (a_k^2 + b_k^2)\right]$$

$$= \frac{l}{2}\left[\frac{a_0^2 + a_N^2}{2} + \sum_{k=1}^{N-1} (a_k^2 + b_k^2)\right] \tag{2.4.16}$$

This is a discrete version of *Parseval's relation*. It expresses the decomposition of the energy $\|u\|_2^2$ in its spectral components.

2.4.3 SAMPLING AND ALIASING

Let $U(x)$ be a function defined over $x \in [0, l)$ and $\{U_n\}$ be the discrete set of $2N$ points obtained by *sampling* $U(x)$ at the discrete points:

$$\begin{cases} x_n = n \Delta x; \quad \Delta x = \dfrac{l}{2N}, \quad n = 0, 1, 2, \ldots, 2N - 1 \\ U_n = U(x_n) \end{cases} \tag{2.4.17}$$

We wish to obtain an expression of the Fourier coefficients $\{a_k, b_k\}$ of $\{U_n\}$ in relation to those of the continuous function $U(x)$. We may note at the onset that:

1. The Fourier series of $U(x)$ has, in general, an infinite number of coefficients, whereas that of $\{U_n\}$ has $2N$ coefficients only.

2. *If the Fourier coefficients $\{A_k, B_k\}$ of $U(x)$ were zero for $k > N$, then we* would simply have

$$a_k = A_k; \qquad b_k = B_k \qquad (2.4.18)$$

Returning to our objective of finding a general relation between the $\{a_k, b_k\}$ of the "sampled" set $\{U_n\}$ and the $\{A_k, B_k\}$ of the original function $U(x)$, we write an equality between two expansions (which, by definition, holds at the collocation or "sampling" points):

$$U_n = \frac{a_0}{2} + \sum_{k=1}^{N-1} [a_k \cos (k\omega_0 x_n) + b_k \sin (k\omega_0 x_n)] + \frac{a_N}{2}(-1)^n$$

$$= \frac{A_0}{2} + \sum_{k=1}^{\infty} [A_k \cos (k\omega_0 x_n) + B_k \sin (k\omega_0 x_n)];$$

$$n = 0, 1, 2, \ldots, 2N - 1 \qquad (2.4.19)$$

With $\omega_0 = 2\pi/l$, the following identities hold for any k, j, and n:

$$\left.\begin{aligned}
\cos (2jN\omega_0 x_n) &= 1 \\
\cos [(2jN + k)\omega_0 x_n] &= \cos (k\omega_0 x_n) \\
\cos [(2jN - k)\omega_0 x_n] &= \cos (k\omega_0 x_n) \\
\sin [(2jN + k)\omega_0 x_n] &= \sin (k\omega_0 x_n) \\
\sin [(2jN - k)\omega_0 x_n] &= -\sin (k\omega_0 x_n) \\
\cos [(2j + 1)N\omega_0 x_n] &= \cos (N\omega_0 x_n) = (-1)^n
\end{aligned}\right\} \qquad (2.4.20)$$

Using those identities allows us to write, from (2.4.19),

$$\left.\begin{aligned}
a_0 &= A_0 + 2 \sum_{j=1}^{\infty} A_{2jN} \\
a_k &= A_k + \sum_{j=1}^{\infty} (A_{2jN+k} + A_{2jN-k}) \\
b_k &= B_k + \sum_{j=1}^{\infty} (B_{2jN+k} - B_{2jN-k}) \\
a_N &= 2A_N + 2 \sum_{j=1}^{\infty} A_{(2j+1)N}
\end{aligned}\right\} \qquad (2.4.21)$$

An interesting phenomenon displayed by these relations is called "aliasing": components of frequency number greater than N in the Fourier series of $U(x)$ are "folded" into components of lower frequency in the discrete Fourier series $\{a_k, b_k\}$. After sampling, the components $A_{2jN+\beta}$ and $B_{2jN+\beta}$ in the original function become indistinct from their low-frequency alias.

The equivalent of (2.4.21) for a Fourier sine series is

$$b_k = B_k + \sum_{j=1}^{\infty} (B_{2jN+k} - B_{2jN-k}) \qquad (2.4.22)$$

2.5 THE FAST FOURIER TRANSFORM

One of the algorithms that acquired great visibility soon after the development of electronic computers is the fast Fourier transform (FFT) algorithm. There are many problems (in the approximate solution of partial differential equations and elsewhere) which require, or may be formulated so as to require, the numerical calculation of discrete Fourier transforms.

> The *fast Fourier transform* uses the symmetry of trigonometric functions to regroup the equations in calculating discrete Fourier transforms so as to minimize the computational effort.

Whereas computing the discrete Fourier transform of a function represented by N discrete data points requires on the order of N^2 operations, when implemented in the straightforward way, organizing the same calculation as prescribed by the FFT requires only of the order $N \log_2 N$ operations (for $N = 1024$, this represents a saving of more than 100 to 1).

2.5.1 HISTORICAL NOTE

The original description of the fast Fourier transform was given in a paper by Cooley and Tukey published in *Mathematics of Computation* in 1965, although the technique was not unknown in the field. For example, the basic idea of the method may be found on page 239 of C. Lanczos' *Applied Analysis*, published in 1956, and earlier results are described in Danielson and Lanczos (1942). Danielson and Lanczos, in turn, refer to Runge (1903, 1905) for the source of their method. Other pre-1965 users of "special techniques" which were in the same vein as the FFT are given in the book *The Fast Fourier Transform* by Brigham (1974).

"Computational tricks" were not held in high esteem in the precomputer days, and attempts to give them visibility under a unified heading were few. The name "numerical analysis" was coined soon after the advent of electronic computers. The number of people involved in doing scientific calculations grew dramatically, and many techniques that had not been well known came to the surface or were reinvented at that time. The "invention" of the fast Fourier transform in the 1960s is an example.

2.5.2 HOW THE FFT WORKS

Consider $2N$ discrete values $\{u_n\}$ of a function, intended to represent this function in $2N$ equidistant points $\{x_n\}$ on the x axis in $[0, l)$:

$$u_n = u(x_n); \quad x_n = n\,\Delta x; \quad \Delta x = \frac{l}{2N};$$

$$n = 0, 1, 2, \ldots, 2N - 1 \qquad (2.5.1)$$

The discrete Fourier transform of $\{u_n\}$ is defined (see Section 2.4) (with $\omega_0 = 2\pi/l$) as

$$\left. \begin{aligned} a_k &= \frac{1}{N} \sum_{n=0}^{2N-1} u_n \cos\left(k\omega_0 x_n\right) \\ b_k &= \frac{1}{N} \sum_{n=0}^{2N-1} u_n \sin\left(k\omega_0 x_n\right) \end{aligned} \right\} \qquad k = 0, 1, 2, \ldots, N \qquad (2.5.2)$$

It is convenient to rewrite (2.5.2) in complex notations as

$$\hat{u}_k \equiv a_k - ib_k = \frac{1}{N} \sum_{n=0}^{2N-1} u_n e^{-ik\omega_0 x_n} \qquad (2.5.3)$$

Suppose now that we group the sample points in their odd- and even-numbered subsets of N values each:

$$\{v_n\} = \{u_n; n \text{ even}\}; \qquad \{w_n\} = \{u_n; n \text{ odd}\}$$

This process is illustrated in Figure 2.5.1. Let $\{\hat{v}_k\}$ and $\{\hat{w}_k\}$ be the discrete Fourier transforms of the points $\{v_n\}$ and $\{w_n\}$, respectively, defined term by term by the usual relations:

$$\hat{v}_k = \frac{2}{N} \sum_{\substack{n=0 \\ n \text{ even}}}^{2N-2} v_n e^{-ik\omega_0 x_n} \qquad (2.5.4)$$

$$\hat{w}_k = \frac{2}{N} \sum_{\substack{n=0 \\ n \text{ odd}}}^{2N-2} w_n e^{-ik\omega_0 x_{n-1}} \qquad (2.5.5)$$

Note that the origin has been shifted by Δx in (2.5.5) to place w_1 at the origin as prescribed in the standard formulation of discrete Fourier transforms. Since the number of samples in $\{v_n\}$ and $\{w_n\}$ is N each, the values taken by k in the transforms (2.5.4)–(2.5.5) are

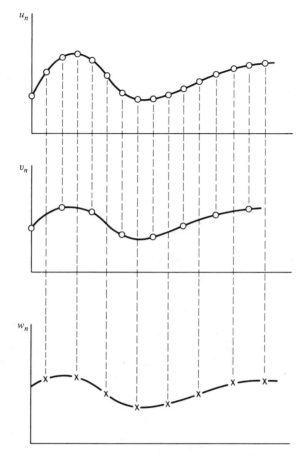

Figure 2.5.1 Odd-even re-grouping of the sample points.

$$k = 0, 1, 2, \ldots, \frac{N}{2}$$

[i.e., half the number of values in (2.5.3)]. The Fourier transform that we want is (2.5.3), which may be rewritten in terms of the odd- and even-numbered points as

$$\hat{u}_k = \frac{1}{N} \sum_{n \text{ even}} v_n e^{-ik\omega_0 x_n} + \frac{1}{N} \sum_{n \text{ odd}} w_n e^{-ik\omega_0 x_n} \qquad (2.5.6)$$

Comparing with (2.5.4)–(2.5.5), we observe that for $k \leq N/2$, we have simply

$$\hat{u}_k = \tfrac{1}{2}(\hat{v}_k + e^{-ik\omega_0 \, \Delta x} \hat{w}_k); \qquad k \leq \frac{N}{2} \qquad (2.5.7)$$

The following relation is essential to the FFT: Because of the symmetry of the function $e^{i\alpha}$, we have

$$e^{-i(k'-N)\omega_0 n \, \Delta x} = \begin{cases} e^{-ik'\omega_0 n \, \Delta x} & \text{for } n \text{ even} \\ -e^{-ik'\omega_0 n \, \Delta x} & \text{for } n \text{ odd} \end{cases} \tag{2.5.8}$$

This allows us to write for $N/2 < k' \le N$:

$$\hat{u}_{k'} = \tfrac{1}{2}(\overline{\hat{v}}_k - e^{+ik\omega_0 \, \Delta x}\overline{\hat{w}}_k); \qquad \frac{N}{2} < k' \le N \tag{2.5.9}$$

where $(\overline{\cdot})$ stands for the complex conjugate of $(\cdot)$ and $k = N - k' < N/2$ (thus $\hat{v}_k$ and $\hat{w}_k$ have already been computed). Comparing the complexity of (2.5.3) with that of (2.5.7)–(2.5.9), we see that:

> computing $\hat{u}_k$ for each k by (2.5.3) requires on the order of $4N$ additions and multiplications to be repeated N times for $k = 0, 1, 2, \ldots, N$, thus a total of $4N^2$ additions and multiplications

while

> computing $\hat{v}_k$ and $\hat{w}_k$ for each k by (2.5.4)–(2.5.5) requires on the order of $4N$ additions and multiplications, to be repeated $N/2$ times for $k = 0, 1, 2, \ldots, N/2$, followed by the $2N$ additions and multiplications (2.5.7)–(2.5.9), for a total of $(2N^2 + 2N)$ additions and multiplications.

Assume now that N had been chosen as a power of 2:

$$2N = 2^m \tag{2.5.10}$$

Then the discrete Fourier transforms $\{\hat{v}_k\}$ and $\{\hat{w}_k\}$ may themselves be computed by the same process of odd/even separation, and be obtained in a total of

$$(N^2 + 2N) \text{ additions plus multiplications}$$

The process may be repeated $m = \log_2(2N)$ times. The only Fourier transforms that need be computed in the end are $2N$ transforms of one sample each, which are equal to those samples themselves. To reconstruct $\{\hat{u}_k\}$ implies *only the $2N$ (additions and multiplications)* which appear at each step [the equivalent of (2.5.7)–(2.5.9)] for a total of

$2Nm$ (additions + multiplications)

$$= 2N \log_2(2N) \text{ (additions + multiplications)}$$

The ratio of computing effort with the FFT and without is thus

$$\frac{N' \log_2 (N')}{(N')^2} = \frac{\log_2 (N')}{N'}$$

(where $N' = 2N$ is the number of sample points). With $N' = 1024$, this ratio is about 1 to 100 in favor of the FFT method.

Finite-Difference
Approximations

3.1 APPROXIMATION
OF DIFFERENTIAL OPERATORS

Brooks Taylor, in his book, "The Method of Increments" (1715–1717), was the first to consider equations of finite differences. As a matter of fact, the relations between the terms of arithmetic and geometric progressions, which were known for a long time, represent the simplest examples of difference equations. But they were not considered from the point of view of their relation to a general theory. This viewpoint was a veritable discovery.

LE MARQUIS DE LAPLACE,
Théorie Analytique des Probabilités (1820)

3.1.1 DISCRETE APPROXIMATION
OF DIFFERENTIAL OPERATORS I:
POLYNOMIAL INTERPOLATION

In general terms, the synthesis of numerical methods for the solution of partial differential equations consists of two steps:

1. The replacement of the equation(s) by a discrete approximation.

2. The selection of some algorithm whereby this discrete approximation may be solved numerically.

We shall, in this chapter, go over some of the general techniques that are relevant to step 1 (i.e., *the approximation of differential operators by discrete forms*).

There is more than one possible approach to this problem: the one used here is that which is generally referred to as *finite-difference approximation*. Its historical origin lies in the "calculus of differences," which was a well-developed branch of mathematics in the precomputer days (see, e.g., Boole, 1860).

A modern approach to the development of finite-difference approximations consists in assuming that *functions which are known at discrete points only are to be "filled in" by polynomials between these points*. This approach reproduces all the results of the classical calculus of differences when grids of equally spaced points are used, but is not restricted to that case as the classical theory was.

Other methods that are used to derive discrete approximations of differential operators and *which are not considered as being of the finite-difference family* are those obtained from more general theories in function space, of which the finite-element method is an example that has received a considerable amount of attention in recent years.

Numerical Differentiation.   A typical situation is as follows:

1. A function $U(x)$ is known only by its values at discrete points along the x axis (called grid points; see Figure 3.1.1):

$$U_n = U(x_n); \qquad n = \ldots, -2, -1, 0, 1, 2, \ldots \qquad (3.1.1.)$$

2. We wish to find an approximation of a derivative at one of those grid points:

$$\left(\frac{\partial^r U}{\partial x^r}\right)_n$$

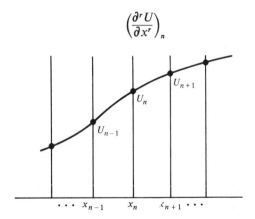

Figure 3.1.1

To derive such an approximation, we have to assume more-or-less arbitrarily what the characteristics of the function $U(x)$ between the points x_n are. To each such assumption will correspond a different approximation.

A standard approach to this problem is to assume that $U(x)$ may be approximated in the neighborhood of x_n by a *polynomial* $P_n(x)$ of order m, at least equal to r, with $(m + 1)$ undermined coefficients, and to determine these coefficients by the condition that $P_n(x)$ must be equal to U_n in $(m + 1)$ points generally taken adjacent to x_n, called the *collocation* points of P_n. Then an approximation of the rth derivative of U is provided by the rth derivative of that interpolating polynomial:

$$\left(\frac{\partial^r U}{\partial x^r}\right)_n \simeq \left(\frac{\partial^r P_n}{\partial x^r}\right)_n \tag{3.1.2}$$

Note that although we use this approximation here in $x = x_n$ only, it applies everywhere in the neighborhood of the collocation points defining P_n. We may write $P_n(x)$ in the general form

$$P_n(x) = \alpha_0 + \alpha_1(x - x_n) + \alpha_2(x - x_n)^2 + \ldots + \alpha_m(x - x_n)^m \tag{3.1.3}$$

The coefficients $\alpha_0, \alpha_1, \alpha_2, \ldots, \alpha_m$ are determined by the $(m + 1)$ equations derived from the collocation conditions:

$$\left.\begin{aligned}
\alpha_0 + \alpha_1(x_{n+\beta_0} - x_n) + \alpha_2(x_{n+\beta_0} - x_n)^2 + \ldots &= U_{n+\beta_0} \\
\alpha_0 + \alpha_1(x_{n+\beta_1} - x_n) + \alpha_2(x_{n+\beta_1} - x_n)^2 + \ldots &= U_{n+\beta_1} \\
&\vdots \\
\alpha_0 + \alpha_1(x_{n+\beta_m} - x_n) + \alpha_2(x_{n+\beta_m} - x_n)^2 + \ldots &= U_{n+\beta_m}
\end{aligned}\right\} \tag{3.1.4}$$

where $\{\beta\} \equiv \{\beta_0, \beta_1, \beta_2, \ldots, \beta_m\}$ are integers which may take on negative as well as positive values.

In the regular grid case (i.e., $x_{n+1} - x_n = \Delta x = $ constant), these equations become

$$\left.\begin{aligned}
\alpha_0 + \alpha_1\beta_0 \Delta x + \alpha_2(\beta_0 \Delta x)^2 + \ldots &= U_{n+\beta_0} \\
\alpha_0 + \alpha_1\beta_1 \Delta x + \alpha_2(\beta_1 \Delta x)^2 + \ldots &= U_{n+\beta_1} \\
&\vdots \\
\alpha_0 + \alpha_1\beta_m \Delta x + \alpha_2(\beta_m \Delta x)^2 + \ldots &= U_{n+\beta_m}
\end{aligned}\right\} \tag{3.1.5}$$

(3.1.4) and (3.1.5) are systems of $(m + 1)$ linear equations in $(m + 1)$ unknowns (the α_i). The approximation of $(\partial^r U/\partial x^r)_n$ is then given by

$$\left(\frac{\partial^r U}{\partial x^r}\right)_n \simeq \left(\frac{\partial^r P_n}{\partial x^r}\right)_n = r! \, \alpha_r \qquad (3.1.6)$$

Because the α's are obtained by solving a system of *linear* equations, it follows that they are always expressed as *linear combinations* of the $U_{n+\beta_i}$. These may be written as

$$\left(\frac{\partial^r U}{\partial x^r}\right)_n \simeq r! \, \alpha_r = \sum_\beta a_\beta U_{n+\beta} \qquad (3.1.7)$$

EXAMPLE

Over a regular grid $x_n = n \, \Delta x$, we seek an approximation of $\partial U/\partial x$ and $\partial^2 U/\partial x^2$. We do this by approximating U in the neighborhood of x_n by the polynomial

$$U(x) \simeq P_n(x) = \alpha_0 + \alpha_1(x - x_n) + \alpha_2(x - x_n)^2 \qquad (3.1.8)$$

where we request that $P_n(x_{n+\beta})$ equal $U_{n+\beta}$ in the three collocation points

$$\{x_{n-1}, x_n, x_{n+1}\} \qquad (3.1.9)$$

The α_i are determined by the collocation conditions (3.1.5):

$$\left. \begin{array}{l} \alpha_0 - \alpha_1 \, \Delta x + \alpha_2 \, \Delta x^2 = U_{n-1} \\ \alpha_0 \qquad\qquad\qquad\quad = U_n \\ \alpha_0 + \alpha_1 \, \Delta x + \alpha_2 \, \Delta x^2 = U_{n+1} \end{array} \right\} \qquad (3.1.10)$$

The approximations are then simply found to be

$$\left(\frac{\partial U}{\partial x}\right)_n \simeq \left(\frac{\partial P_n}{\partial x}\right)_n = \alpha_1 = \frac{U_{n+1} - U_{n-1}}{2 \, \Delta x} \qquad (3.1.11)$$

and

$$\left(\frac{\partial^2 U}{\partial x^2}\right)_n \simeq \left(\frac{\partial^2 P_n}{\partial x^2}\right)_n = 2\alpha_2 = \frac{U_{n+1} + U_{n-1} - 2U_n}{\Delta x^2} \qquad (3.1.12)$$

(*Note:* We could have obtained an approximation of $(\partial U/\partial x)_n$ based on $\{\beta = -1, 0\}$. We would have found

$$\left(\frac{\partial U}{\partial x}\right)_n = \frac{U_n - U_{n-1}}{\Delta x} \qquad (3.1.13)$$

This is called a *backward-difference* approximation[1] of $\partial U/\partial x$ [whereas (3.1.11) is a *central-difference* approximation].)

One may similarly derive higher-order central-difference approximations of $\partial U/\partial x$ and $\partial^2 U/\partial x^2$, by choosing $m = 2k + 1$ collocation points centered on x_n:

$$\{\beta = -k, -k + 1, -, 0, -, k - 1, k\}$$

The resulting central-finite-difference approximations are given in Table 3.1.1 for $k = 1, 2, 3,$ and 4.

3.1.2 DERIVATION OF DIFFERENCE APPROXIMATION FORMULAS BY THE USE OF THE LAGRANGE FORM OF POLYNOMIALS

Identical results to those above may be obtained by using interpolation polynomials written in the Lagrange form. Consider again the situation depicted in Figure 3.1.1. That is, we have a function $U(x)$ which is known only by the values $\{U_n\}$ it takes at the discrete points $\{x_n\}$. The polynomial of degree m which passes through any $(m + 1)$ such points, say $n_1, n_2, \ldots, n_m$, may be written as

$$P(x) = \left[\frac{(x - x_{n_2})(x - x_{n_3}) \ldots (x - x_{n_m})}{(x_{n_1} - x_{n_2})(x_{n_1} - x_{n_3}) \ldots (x_{n_1} - x_{n_m})}\right]U_{n_1}$$

$$+ \left[\frac{(x - x_{n_1})(x - x_{n_3}) \ldots (x - x_{n_m})}{(x_{n_2} - x_{n_1})(x_{n_2} - x_{n_3}) \ldots (x_{n_2} - x_{n_m})}\right]U_{n_2}$$

$$+ \ldots$$

$$+ \left[\frac{(x - x_{n_1})(x - x_{n_2}) \ldots (x - x_{n_{m-1}})}{(x_{n_m} - x_{n_1})(x_{n_m} - x_{n_2}) \ldots (x_{n_m} - x_{n_{m-1}})}\right]U_{n_m} \qquad (3.1.14)$$

called the Lagrange form of this polynomial. That this is the case is easily proved by verifying that:

---

[1] Also sometimes called an "upwind"-difference approximation in the context of fluid dynamics problems when flow is in the $+x$ direction.

Table 3.1.1a Coefficients of central-difference approximations of the first derivative:

$$\left(\frac{\partial U}{\partial x}\right)_n \simeq \sum_\beta a_\beta U_{n+\beta}$$

Coefficients a_β

Number of Points	a_{-4}	a_{-3}	a_{-2}	a_{-1}	a_0	a_1	a_2	a_3	a_4
3				$-\dfrac{1}{2\,\Delta x}$	0	$\dfrac{1}{2\,\Delta x}$			
5			$\dfrac{1}{12\,\Delta x}$	$\dfrac{-8}{12\,\Delta x}$	0	$\dfrac{8}{12\,\Delta x}$	$\dfrac{-1}{12\,\Delta x}$		
7		$\dfrac{-1}{60\,\Delta x}$	$\dfrac{9}{60\,\Delta x}$	$\dfrac{-45}{60\,\Delta x}$	0	$\dfrac{45}{60\,\Delta x}$	$\dfrac{-9}{60\,\Delta x}$	$\dfrac{1}{60\,\Delta x}$	
9	$\dfrac{3}{840\,\Delta x}$	$\dfrac{-32}{840\,\Delta x}$	$\dfrac{168}{840\,\Delta x}$	$\dfrac{-672}{840\,\Delta x}$	0	$\dfrac{672}{840\,\Delta x}$	$\dfrac{-168}{840\,\Delta x}$	$\dfrac{32}{840\,\Delta x}$	$\dfrac{-3}{840\,\Delta x}$

(*Continued*)

Table 3.1.1b Coefficients of central-difference approximations of the second derivative:

$$\left(\frac{\partial^2 U}{\partial x^2}\right)_n = \sum_\beta a_\beta U_{n+\beta}$$

Coefficients a_β

Number of Points	a_{-4}	a_{-3}	a_{-2}	a_{-1}	a_0	a_1	a_2	a_3	a_4
3				$\dfrac{1}{\Delta x^2}$	$\dfrac{-2}{\Delta x^2}$	$\dfrac{1}{\Delta x^2}$			
5			$\dfrac{-1}{12\,\Delta x^2}$	$\dfrac{16}{12\,\Delta x^2}$	$\dfrac{-30}{12\,\Delta x^2}$	$\dfrac{16}{12\,\Delta x^2}$	$\dfrac{-1}{12\,\Delta x^2}$		
7		$\dfrac{2}{180\,\Delta x^2}$	$\dfrac{-27}{180\,\Delta x^2}$	$\dfrac{270}{180\,\Delta x^2}$	$\dfrac{-490}{180\,\Delta x^2}$	$\dfrac{270}{180\,\Delta x^2}$	$\dfrac{-27}{180\,\Delta x^2}$	$\dfrac{2}{180\,\Delta x^2}$	
9	$\dfrac{-9}{5040\,\Delta x^2}$	$\dfrac{128}{5040\,\Delta x^2}$	$\dfrac{-1008}{5040\,\Delta x^2}$	$\dfrac{8064}{5040\,\Delta x^2}$	$\dfrac{-14{,}350}{5040\,\Delta x^2}$	$\dfrac{8064}{5040\,\Delta x^2}$	$\dfrac{-1008}{5040\,\Delta x^2}$	$\dfrac{128}{5040\,\Delta x^2}$	$\dfrac{-9}{5040\,\Delta x^2}$

1. $P(x)$ is a polynomial of degree m (each bracketed term [·] is a polynomial of degree m).

2. $P(x_{n_i}) = U_{n_i}$; $i = 1, 2, \ldots, m$ (observe that the term inside the ith bracket [·] in (3.1.14) is equal to 1 in $x = x_{n_i}$ and zero in $x = x_{n_j}$, $j \neq i$).

Thus, (3.1.14) and (3.1.3) are the same polynomial [from the known theorem: two polynomials of degree m having $(m + 1)$ distinct common points are identical]. An approximation of $\partial^r U/\partial x^r$, $r \leq m$, for any value of x' of x may be obtained by first computing analytically $\partial^r P/\partial x^r$, and then substituting the value of x' in the resulting expression.

EXAMPLE

Consider three nonequidistant points in which $U(x)$ is given (Figure 3.1.2). The Lagrange polynomial of degree 2 passing through these points is

$$P(x) = \frac{(x - x_2)(x - x_3)}{(x_1 - x_2)(x_1 - x_3)} U_1 + \frac{(x - x_1)(x - x_3)}{(x_2 - x_1)(x_2 - x_3)} U_2$$

$$+ \frac{(x - x_1)(x - x_2)}{(x_3 - x_1)(x_3 - x_2)} U_3 \tag{3.1.15}$$

Its first and second derivatives are, respectively,

$$\frac{\partial P}{\partial x} = \frac{(x - x_2) + (x - x_3)}{(x_1 - x_2)(x_1 - x_3)} U_1 + \frac{(x - x_1) + (x - x_3)}{(x_2 - x_1)(x_2 - x_3)} U_2$$

$$+ \frac{(x - x_1) + (x - x_2)}{(x_3 - x_1)(x_3 - x_2)} U_3 \tag{3.1.16}$$

and

$$\frac{\partial^2 P}{\partial x^2} = \frac{2}{(x_1 - x_2)(x_1 - x_3)} U_1 + \frac{2}{(x_2 - x_1)(x_2 - x_3)} U_2$$

$$+ \frac{2}{(x_3 - x_1)(x_3 - x_2)} U_3 \tag{3.1.17}$$

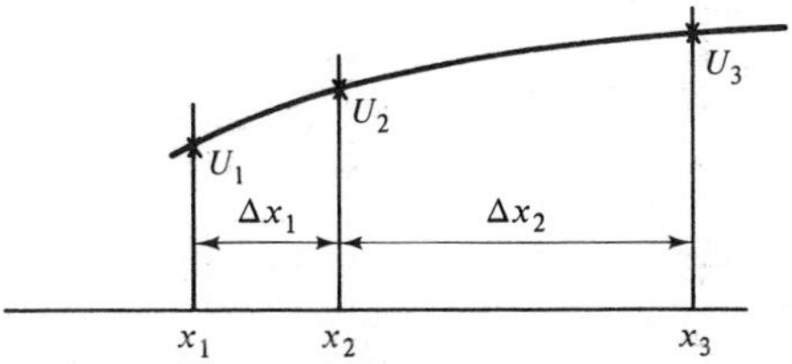

Figure 3.1.2

From the above, we may derive expressions for $\partial P/\partial x$ at several values of x; for example,

$$\left(\frac{\partial U}{\partial x}\right)_{x_1} \simeq \left(\frac{\partial P}{\partial x}\right)_{x_1} = \frac{-\Delta x_1 - (\Delta x_1 + \Delta x_2)}{\Delta x_1(\Delta x_1 + \Delta x_2)}U_1 + \frac{-(\Delta x_1 + \Delta x_2)}{-\Delta x_1 \Delta x_2}U_2$$

$$+ \frac{-\Delta x_1}{(\Delta x_1 + \Delta x_2)\Delta x_2}U_3 + \ldots \qquad (3.1.18)$$

But only one value for $\partial^2 P/\partial x^2$ can be obtained (the same everywhere), which is the best three points afford:

$$\frac{\partial^2 U}{\partial x^2} \simeq \frac{\partial^2 P}{\partial x^2} = \frac{2U_1}{(\Delta x_1 + \Delta x_2)\Delta x_1} - \frac{2U_2}{\Delta x_1 \Delta x_2} + \frac{2U_3}{\Delta x_2(\Delta x_1 + \Delta x_2)}$$

$$= \frac{2}{\Delta x_1 \Delta x_2}\left(-\frac{U_2 - U_1}{\Delta x_1} + \frac{U_3 - U_2}{\Delta x_2}\right) \qquad (3.1.19)$$

3.1.3 SYNTHESIS OF DIFFERENCE APPROXIMATIONS BY USE OF THE TAYLOR SERIES (METHOD OF UNDETERMINED COEFFICIENTS)

A third formalism that may be used to derive the same difference approximations is that called the "method of undetermined coefficients." We may expand $U(x)$ around the point x_n in a Taylor series as

$$U_{n+\beta} = U_n + \left(\frac{\partial U}{\partial x}\right)_n \beta \, \Delta x + \left(\frac{\partial^2 U}{\partial x^2}\right)_n \frac{(\beta \, \Delta x)^2}{2!} + \ldots \qquad (3.1.20)$$

Inserting this in the desired form of the approximation

$$\left(\frac{\partial^r U}{\partial x^r}\right)_n \simeq \sum_\beta a_\beta U_{n+\beta} \qquad (3.1.21)$$

results in

$$\left(\frac{\partial^r U}{\partial x^r}\right)_n \simeq U_n\left(\sum_\beta a_\beta\right) + \left(\frac{\partial U}{\partial x}\right)_n \Delta x\left(\sum_\beta \beta a_\beta\right)$$

$$+ \left(\frac{\partial^2 U}{\partial x^2}\right)_n \frac{\Delta x^2}{2!}\left(\sum_\beta \beta^2 a_\beta\right) + \ldots \qquad (3.1.22)$$

With $(m + 1)$ free coefficients a_β, we may impose as $(m + 1)$ conditions to make left- and right-hand sides conform as well as possible. For example, we may *impose that the first $(m + 1)$ terms of the right-hand side be exactly equal to the left-hand side, resulting in*

$$\sum_{\beta} a_{\beta} = 0$$

$$\sum_{\beta} \beta a_{\beta} = 0$$

$$\cdot$$
$$\cdot$$
$$\cdot$$

$$\sum_{\beta} \beta^r a_{\beta} = \frac{1}{(\Delta x)^r \, r!}$$

$$\sum_{\beta} \beta^{r+1} a_{\beta} = 0$$

$$\cdot$$
$$\cdot$$
$$\cdot$$

$$\sum_{\beta} \beta^m a_{\beta} = 0$$

$$(3.1.23)$$

The solution of these equations in the $\{a_{\beta}\}$ will give the same results as those obtained by assuming polynomial interpolation. By way of proof, it is easily verified that the approximation so obtained is *exact* if $U(x)$ is a polynomial of degree no greater than m (i.e., if all its derivatives of degree greater than m are identically zero).

Table 3.1.2 gives some simple finite-difference approximations of derivatives.

Table 3.1.2 Simple finite-difference approximations of derivatives

First Derivative: Central Differences

$$\left(\frac{\partial U}{\partial x}\right)_n \simeq \frac{U_{n+1} - U_{n-1}}{2\,\Delta x} \qquad \text{(a)}$$

Second Derivative: Central Differences

$$\left(\frac{\partial^2 U}{\partial x^2}\right)_n \simeq \frac{U_{n-1} - 2U_n + U_{n+1}}{\Delta x^2} \qquad \text{(b)}$$

Third Derivative: Central Differences

$$\left(\frac{\partial^3 U}{\partial x^3}\right)_n \simeq -\frac{U_{n-2} + 2U_{n-1} - 2U_{n+1} + U_{n+2}}{2\,\Delta x^3} \qquad \text{(c)}$$

Fourth Derivative: Central Differences

$$\left(\frac{\partial^4 U}{\partial x^4}\right)_n \simeq \frac{U_{n-2} - 4U_{n-1} + 6U_n - 4U_{n+1} + U_{n+2}}{\Delta x^4} \qquad \text{(d)}$$

3.2 OPERATOR NOTATIONS

It is convenient to use operator notations in the description of discrete approximations of differential operators. As before, let $\{U_n\}$ be the set of discrete values of a function in a grid of equally spaced points $\{x_n = n \, \Delta x\}$.
The *shift* or *displacement operator*, E, is defined by

$$EU_n = U_{n+1} \tag{3.2.1}$$

Similarly,

$$E^\beta U_n = U_{n+\beta} \tag{3.2.2}$$

(read "the operator E to the power β"). An interesting relation exists between Taylor series and the displacement operator E: With $D = \partial/\partial x$, the classical expression of a Taylor series of the function $U(x)$ is

$$U(x + \Delta x) = U(x) + (\Delta x \, D)U(x) + \frac{(\Delta x \, D)^2}{2!}U(x)$$
$$+ \frac{(\Delta x \, D)^3}{3!}U(x) + \dots \tag{3.2.3}$$

which, we may observe, may also be written as

$$U(x + \Delta x) = e^{\Delta x \, D} U(x) \tag{3.2.4}$$

That is,

$$\boxed{E = e^{\Delta x \, D} \qquad\qquad (3.2.5)}$$

Other useful discrete operators are given in Table 3.2.1.

Table 3.2.1 Discrete operators

$EU_n = U_{n+1}$	shift or displacement (3.2.6)
$\Delta = E - 1$	forward difference (3.2.7)
$E^{-1}\Delta = 1 - E^{-1}$	backward difference (3.2.8)
$DU = \dfrac{\partial U}{\partial x}$	differentiation (3.2.9)
$\delta = E^{1/2} - E^{-1/2}$	central difference (3.2.10)
$\delta^2 = E^{-1} - 2 + E$	second difference (3.2.11)
$\quad = \Delta \, E^{-1}\Delta$	

3.2.1 OPERATORS FOR FINITE-DIFFERENCE APPROXIMATIONS

The following operators, derived from those of Table 3.2.1, are sometimes used in the context of the analysis of finite-difference approximations:

Forward divided difference.

$$D_+ = \frac{E - 1}{\Delta x} = \frac{\Delta}{\Delta x} \tag{3.2.12}$$

Backward divided difference.

$$D_- = \frac{1 - E^{-1}}{\Delta x} = E^{-1}D_+ \tag{3.2.13}$$

Central divided difference.

$$D_0 = \frac{D_+ + D_-}{2} = \frac{E - E^{-1}}{2\,\Delta x} \tag{3.2.14}$$

Second-order divided difference.

$$D_+D_- = \frac{E^{-1} - 2 + E}{\Delta x^2} \equiv \frac{\delta^2}{\Delta x^2} \tag{3.2.15}$$

These operators are the simplest finite-difference approximations to the first and second derivatives. As will be seen later, higher-order approximations to derivatives may also be expressed in terms of those simple operators. A general form in which linear constant-coefficient finite-difference approximations may be expressed is

$$(LU)_n \simeq \tilde{L}U_n = \sum_\beta a_\beta U_{n+\beta}$$
$$= (\sum_\beta a_\beta E^\beta)U_n \tag{3.2.16}$$

Thus, the operator $\tilde{L}$ may be written as[2]:

$$\tilde{L} = \sum_\beta a_\beta E^\beta \tag{3.2.17}$$

For example, from Table 3.1.2, equation (c):

$$L \equiv \frac{\partial^3}{\partial x^3} \simeq \frac{-E^{-2} + 2E^{-1} - 2E + 2E^2}{2\,\Delta x^3} \tag{3.2.18}$$

---

[2] $\sim$ is the Spanish "tilde."

which is of the form (3.2.17) with

$$a_2 = -a_{-2} = \frac{1}{2\,\Delta x^3}; \qquad a_1 = -a_{-1} = -\frac{1}{\Delta x^3}; \qquad a_0 = 0 \qquad (3.2.19)$$

3.3 TRUNCATION ERROR AND ORDER OF ACCURACY

Consider the linear differential operator L and its discrete (finite-difference) approximation written as

$$(LU)_n \simeq \tilde{L}U_n = \sum_\beta a_\beta U_{n+\beta} \qquad (3.3.1)$$

[or $L \simeq \tilde{L} = \sum_\beta a_\beta E^\beta$]. The notation L_h (with $h = \Delta x$) instead of $\tilde{L}$ also used.

As before, it is assumed here that the $\{U_n\}$ are discrete values of a function $U(x)$ taken on a grid of equally spaced points:

$$\left.\begin{array}{l} U_n = U(x_n) \\ x_n = n\,\Delta x; \quad n = \ldots, -2, -1, 0, 1, 2, \ldots \end{array}\right\} \qquad (3.3.2)$$

With $U(x)$ assumed to be a sufficiently smooth function of x, the difference

$$\begin{aligned} T &= (\tilde{L} - L)U \\ &= (\sum_\beta a_\beta E^\beta - L)U \end{aligned} \qquad (3.3.3)$$

is called the *truncation error* of the approximation of L by $\tilde{L}$. Since $\tilde{L}$ is a discrete operator, this expression has meaning at the grid points only; that is, (3.3.3) is, more precisely,

$$T_n = \tilde{L}U_n - (LU)_n \qquad (3.3.4)$$

for every n.

As an example, consider the simple differential operator

$$L = \frac{\partial^r}{\partial x^r} \qquad (3.3.5)$$

approximated by (3.3.1). One of the ways to express the truncation error is by using remainders of Taylor series. We write

$$U_{n+\beta} = U_n + \beta\,\Delta x\left(\frac{\partial U}{\partial x}\right)_n + \frac{\beta\,\Delta x^2}{2}\left(\frac{\partial^2 U}{\partial x^2}\right)_n + \cdots \qquad (3.3.6)$$

and insert this expression in (3.3.1) and (3.3.3).

If $\tilde{L}$ is to be a consistent approximation of L, then the linear combination obtained by inserting (3.3.6) in (3.3.1) must have a leading nonzero term which matches exactly $(\partial^r U/\partial x^r)_n$:

$$\tilde{L}U_n = \left(\frac{\partial^r U}{\partial x^r}\right)_n + R \tag{3.3.7}$$

Because of the fact that $(\partial^r U/\partial x^r)_n$ has the common factor Δx^r in (3.3.6), it follows that:

1. All coefficients a_β in (3.3.1) have a factor Δx^{-k}, where $k \geq r$.

2. The remainder R (which is precisely the truncation error) must be of the form

$$R \equiv T = K \Delta x^q \left(\frac{\partial^{r+q} U}{\partial x^{r+q}}\right) + \cdots \tag{3.3.8}$$

where K is a constant which is independent of Δx and U.

The leading term of this expression is called the *principal part* of the truncation error, and the number q is called the *order of accuracy* of the approximation of L by $\tilde{L}$. This number plays an important role in analyzing how fast the error of a given approximation decreases when $\Delta x \longrightarrow 0$ (analysis of convergence). The common expression for this is

$$T = O(\Delta x^q) \tag{3.3.9}$$

II

Elliptic Equations

Two-Point
Boundary-Value Problems I

4.1 FINITE-DIFFERENCE APPROXIMATION

4.1.1 TWO-POINT BOUNDARY-VALUE PROBLEMS

Consider the differential equation

$$-\frac{d^2U}{dx^2} = F(x) \tag{4.1.1}$$

over the domain $0 \le x \le l$, associated with the boundary conditions

$$
\begin{aligned}
U(0) &= U_0 \quad \text{(given)} \\
U(l) &= U_l \quad \text{(given)}
\end{aligned} \tag{4.1.2}
$$

This constitutes a simple case of a *two-point boundary-value problem*. Although not technically partial differential equations, *two-point boundary-value problems may from most viewpoints be assimilated to elliptic equations*, with which they share both mathematical properties and computational methods. In particular, two-point boundary-value problems and elliptic equations share the property that their solution in an interior point depends on all boundary conditions, precluding the possibility of *directly*[1] constructing the solution by marching away from one boundary.

---

[1] As opposed to *iteratively*.

With minor modifications, (4.1.1) may describe:

The steady-state displacement of a taut string subjected to a distributed load.

The distribution of temperature in an internally heated homogeneous wall, with surfaces kept at given temperatures.

The steady-state concentration of a pollutant in a porous soil.

The distribution of electrical potential between two flat electrodes.

While they are representative of many problems in the sciences, two-point boundary-value problems are also often used in the literature to illustrate concepts and methods for the approximation of elliptic equations. We shall find it convenient to follow this approach here. The remainder of this chapter deals with solution techniques for two-point boundary-value problems. In Chapter 5 these concepts are generalized to elliptic equations in several dimensions.

A more general two-point boundary-value problem than (4.1.1) is given by the equation

$$LU(x) \equiv -\frac{d}{dx}\left(P\frac{dU}{dx}\right) + QU = F \qquad (4.1.3)$$

associated to the same boundary conditions (4.1.2). Here P, Q, and F are given functions of x. This problem is sometimes called a linear source problem.

4.1.2 FINITE-DIFFERENCE APPROXIMATION

A simple technique for the numerical approximation of (4.1.1) is the method of finite differences. Let, with $h = l/(N + 1)$,

$$\{x_n = nh; \, n = 0, 1, 2, \dots, N + 1\}$$

be a regular discretization of the x axis. Also let

$$\mathbf{u} \equiv \{u_n\} \simeq \{U(x_n)\}$$

be the set of values intended to approximate $U(x)$ in those discrete points.[2]
The strict identity

$$(LU)_n \equiv -\left(\frac{d^2U}{dx^2}\right)_n = F(x_n) \equiv f_n \qquad (4.1.4)$$

_____

[2]As a rule, upper case letters are used to denote known functions or exact solutions of equations, and lower case letters are used to denote their approximation.

may be approximated by using the standard finite-difference approximation (3.1.12) to second derivatives, giving

$$L_h u_n \equiv -\frac{u_{n-1} - 2u_n + u_{n+1}}{h^2} = f_n; \qquad n = 1, 2, \ldots, N \qquad (4.1.5)$$

This system of equations, together with the boundary conditions (4.1.2), gives the system of linear equations:

$$\left. \begin{array}{l} 2u_1 - u_2 = h^2 f_1 + U_0 \\ -u_1 + 2u_2 - u_3 = h^2 f_2 \\ -u_2 + 2u_3 - u_4 = h^2 f_3 \\ \qquad \vdots \\ -u_{N-1} + 2u_N = h^2 f_N + U_l \end{array} \right\} \qquad (4.1.6)$$

which is to be solved numerically to produce the solution

$$\mathbf{u} \equiv \left\{ \begin{array}{c} u_1 \\ u_2 \\ \cdot \\ \cdot \\ \cdot \\ u_N \end{array} \right\}$$

4.1.3 SOLUTION OF LINEAR ALGEBRAIC EQUATIONS

It is appropriate to say a few words here about the *solution* of systems of linear equations such as (4.1.6). Although this topic is covered in some detail in Chapter 6, we describe here a computational method that may be applied when the system of equations is *tridiagonal*. This method, with minor variations, is to be found in many places of the technical literature of the late 1950s and early 1960s. It is called the double-sweep method by Russian authors [e.g., in Godunoff and Ryabenkii (1964), p. 239 of the English translation] and Thomas's method by engineers [e.g., in Lapidus (1962), p. 255].

Solution of a Tridiagonal System of Equations. Consider the system of equations (4.1.6), rewritten simply as

$$\mathbf{A}\mathbf{u} = \mathbf{r} \qquad (4.1.7)$$

where A is a tridiagonal matrix; that is, in general:

$$
A \equiv
\begin{bmatrix}
a_1 & c_1 & & & & & \\
b_2 & a_2 & c_2 & & & 0 & \\
 & \cdot & \cdot & \cdot & & & \\
 & & \cdot & \cdot & \cdot & & \\
 & & & \cdot & \cdot & \cdot & \\
 & & & & \cdot & \cdot & c_{N-1} \\
 & 0 & & & & \cdot & \cdot \\
 & & & & & b_N & a_N
\end{bmatrix}
\tag{4.1.8}
$$

We assume that A is regular and that its diagonal elements are nonzero. As we will show, such a matrix is easily transformed into a product of two matrices of the form:

$$
A =
\begin{bmatrix}
\alpha_1 & \cdot & & & & \\
b_2 & \cdot & & & 0 & \\
 & \cdot & \cdot & & & \\
 & & \cdot & \cdot & & \\
 & 0 & & \cdot & \cdot & \\
 & & & b_N & \alpha_N
\end{bmatrix}
\begin{bmatrix}
1 & \beta_1 & & & & \\
 & 1 & \beta_2 & & 0 & \\
 & & \cdot & \cdot & & \\
 & & & \cdot & \cdot & \\
 & 0 & & & 1 & \beta_{N-1} \\
 & & & & & 1
\end{bmatrix}
$$

$$
= LU \tag{4.1.9}
$$

where the letters L and U stand for *lower* and *upper* (matrices), respectively. We obtain by identification line by line of (4.1.8) and (4.1.9):

$$
\alpha_1 = a_1; \qquad \beta_1 = \frac{c_1}{\alpha_1}
$$

$$
\left.
\begin{aligned}
\alpha_n &= a_n - b_n \beta_{n-1} \\
\beta_n &= \frac{c_n}{\alpha_n}
\end{aligned}
\right\}
\qquad n = 2, 3, \ldots, N
\tag{4.1.10}
$$

Define now the intermediate vector

$$
y \equiv
\begin{bmatrix}
y_1 \\
y_2 \\
\cdot \\
\cdot \\
\cdot \\
y_n
\end{bmatrix}
$$

as the solution of:

$$
Ly = r \tag{4.1.11}
$$

which may be computed by the *forward sweep*:

$$y_1 = \frac{r_1}{\alpha_1}$$

$$y_n = \frac{r_n - b_n y_{n-1}}{\alpha_n}; \qquad n = 2, 3, \ldots, N \tag{4.1.12}$$

Comparison of (4.1.7) and (4.1.11) shows that

$$\mathbf{Ly} = \mathbf{LUu}$$

whence

$$\mathbf{Uu} = \mathbf{y} \tag{4.1.13}$$

Owing to the particular bidiagonal form of the matrix $\mathbf{U}$, this system may be solved for $\mathbf{u}$ by a simple *backward sweep*:

$$u_N = y_N$$

$$u_n = y_n - \beta_n u_{n+1}; \qquad n = N - 1, N - 2, \ldots, 1 \tag{4.1.14}$$

The solution of (4.1.7) may thus be obtained by implementing the simple "sweeps" (4.1.10), (4.1.12), and (4.1.14). If the solution is to be obtained for a different right-hand side vector $\mathbf{r}$, then only (4.1.12) and (4.1.14) need be repeated.

This method is generalized in Section 6.1 to produce the $\mathbf{LU}$ decomposition of any regular matrix.

4.2 TRUNCATION ERROR: CLASSICAL ANALYSIS

4.2.1 ERROR ANALYSIS: CLASSICAL THEORY

The differences between the solution $\{u_n\}$ of the equations (4.1.5) given by a computer and the exact $\{U_n\}$ are *global errors*. There are two sources that contribute to those errors:

1. *Discretization or truncation errors:* These are discrepancies that result from the approximation of the differential operator L by its difference approximation L_h.

2. *Roundoff errors:* These are errors that result from the fact that computers carry numbers with a finite number of digits. The computer calculations for the solution of the system of equations (4.1.5) yield not its

exact solution, but its exact solution plus *roundoff errors* which would disappear if computer arithmetic were carried out with infinite word length.

The Truncation Error.   We ignore roundoff errors in the remainder. In each point of the net (4.1.3), the *local truncation error* (or *truncation error* for short) is defined as the difference

$$T_{h,n} \equiv (L_h U)_n - (LU)_n$$
$$= -\frac{U(x_n - h) - 2U(x_n) + U(x_n + h)}{h^2} + \left(\frac{d^2 U}{dx^2}\right)_n \qquad (4.2.1)$$

Here, $U(x)$ is any sufficiently smooth function. In particular, we may take $U(x)$ to be the exact solution of the problem (4.1.1). The truncation error T_h measures the error in approximating the differential operator L by its discrete approximation L_h. An equivalent notation for (4.2.1) is

$$T_h = L_h U - LU$$

which is to be interpreted as in Section 3.3.

Order of Accuracy.   A measure of T_h is obtained by using Taylor's theorem; we write

$$U(x_n \pm h) = U_n \pm h U'_n + \frac{h^2}{2} U''_n \pm \frac{h^3}{3!} U'''_n + \frac{h^4}{4!} U_n^{(iv)}$$
$$+ \text{ higher-order terms} \qquad (4.2.2)$$

where we have assumed that $U^{(iv)}$ is bounded near x_n. Upon substituting these results into (4.2.1), we find that

$$T_{h,n} = -\frac{h^2}{12} U_n^{(iv)} + \text{ higher-order terms} \qquad (4.2.3)$$

which may also be expressed as

$$T_{h,n} = -\frac{h^2}{12} U_\xi^{(iv)}; \qquad n - 1 \leq \xi \leq n + 1 \qquad (4.2.4)$$

Both expressions have the same meaning that, for h sufficiently small, $|T_h|$ goes to zero as Kh^2, where

$$K \leq \tfrac{1}{12} \max U^{(iv)} \qquad (4.2.5)$$

is assumed to be a finite number. This is also written as

$$T_h = O(h^2) \qquad (4.2.6)$$

The exponent of h in these expressions of the truncation error (2 in the present case) is called the *order of accuracy* of the approximation L_h to the differential operator L.

Global Errors and Convergence.   The *global error* or difference between approximate and exact solutions of the equation

$$LU = F \tag{4.2.7}$$

is the set of values

$$\mathbf{e} \equiv \{e_n\} = \{u_n - U_n\} \tag{4.2.8}$$

where U_n denotes the local value of an exact solution. The property

$$\lim_{h \to 0} \|\mathbf{e}\| = 0 \tag{4.2.9}$$

(where $\| \cdot \|$ is some chosen norm) is called (when it holds true) the property of *convergence* of the approximation. To establish that the approximation (4.1.5) is convergent, we may proceed as follows. We note that, for each n,

$$L_h U_n - f_n = T_{h,n}$$

(U_n = exact solution), and

$$L_h u_n - f_n = 0$$

(u_n = finite difference solution), whence, by subtraction,

$$L_h \mathbf{e} = -T_h \tag{4.2.10}$$

Or, conversely

$$\mathbf{e} = -L_h^{-1} T_h \tag{4.2.11}$$

where L_h^{-1} is the inverse of the operator or matrix L_h.

Various proofs of convergence may then be established. Most proofs are similar in their structure. Typical steps are:

1. It is established that the operator L_h^{-1} is bounded in some sense when $h \to 0$ (called the *stability* of the operator L_h).

2. For a stable operator L_h, the global error (4.2.8) goes to zero at least as T_h. The property of the truncation error

$$\lim_{h \to 0} \|T_h\| = 0 \tag{4.2.12}$$

(according to some norm $\| \cdot \|$) is called the *consistency* of the approximation.

From steps 1 and 2, we conclude that an approximation is *convergent* if it is both *consistent* and *stable*.

As an example, see Keller (1968), p. 76.

4.2.2 DISCRETIZATIONS WITH A HIGHER ORDER OF ACCURACY

The order of accuracy (i.e., the exponent of h in the truncation error) is often taken in the classical theory as the measure of the quality of a semi-discretization. With this in mind, we express a more general three-point semidiscretization of (4.1.1) as

$$-\frac{u_{n-1} - 2u_n + u_{n+1}}{h^2} = w_0 f_{n-1} + w_1 f_n + w_2 f_{n+1} \qquad (4.2.13)$$

where (w_0, w_1, w_2) are undetermined weights. One way to rationalize this expression is to observe that its left-hand side is a value of $-d^2 U/dx^2$ somewhere in the interval $x \in [x_{n-1}, x_{n+1}]$. One may consider the right-hand side as a mean value of the source function $F(x)$ in the same interval. For example, if we were to approximate F by a quadratic polynomial over that interval with collocation points

$$\{(x_m, f_m); m = n - 1, n, n + 1\} \qquad (4.2.14)$$

then the w_i would be the coefficients of Simpson's rule of quadrature:

$$w_0 = w_2 = \tfrac{1}{6}; \qquad w_2 = \tfrac{2}{3} \qquad (4.2.15)$$

This formula is called "optimal" in Babuska, Prager, and Vitasek (1966, p. 181).[3] As we shall now see, this formula *does not* have the highest possible order of accuracy for all formulas of the form (4.2.13).

Truncation Error.   We want an expression of the truncation error for the approximations (4.2.13) which will lead to an expression identical to (4.2.11) for the global error. To this end, we rewrite (4.2.13) as

$$L_h u_n \equiv -\frac{u_{n-1} - 2u_n + u_{n+1}}{h^2} - (w_0 u''_{n-1} + w_1 u''_n + w_2 u''_{n+1}) \qquad (4.2.16)$$

where by equation (4.1.1),

$$-u''_n = f_n$$

---

[3]See also Section 1.4, in particular equation (1.4.10).

The truncation error is then defined as before:

$$T_h = L_h U \qquad (4.2.17)$$

where $U(x)$ is any sufficiently smooth function.

It is easily verified that as desired, (4.2.11) still holds.

Equations (4.1.5) and (4.2.13) are special cases and have truncation errors

w_0	w_1	w_2	Truncation error
0	1	0	$-\dfrac{h^2}{12}U_\xi^{(\text{iv})}$
$\dfrac{1}{6}$	$\dfrac{4}{6}$	$\dfrac{1}{6}$	$+\dfrac{h^2}{12}U_\xi^{(\text{iv})}$

Interestingly, the $(\frac{1}{6}, \frac{4}{6}, \frac{1}{6})$ formula is no better (according to the *order of accuracy* criterion) than is the finite-difference case (0, 1, 0). Since the truncation error for both has the same principal part with opposite signs, one may expect the formula with coefficients half-the-sum-of-the-two to be of higher order of accuracy. This, indeed, is found to be the case. The operator

$$L_h u_n \equiv -\frac{u_{n-1} - 2u_n + u_{n+1}}{h^2} - \left(\tfrac{1}{12}u''_{n-1} + \tfrac{10}{12}u''_n + \tfrac{1}{12}u''_{n+1}\right) \qquad (4.2.18)$$

has truncation error

$$T_h \equiv L_h U = O(h^4) \qquad (4.2.19)$$

It is known as the *Störmer* or *Störmer–Numerov formula* (see e.g., Birkhoff and Gulati, 1974).

4.3 TRUNCATION ERROR: FOURIER ANALYSIS

What we shall do in this section may be simply stated as follows:

Given the two-point boundary-value problem

$$-\frac{d^2 U}{dx^2} = F(x); \qquad x \in (0, l); \qquad U(0) = U(l) = 0 \qquad (4.3.1)$$

approximated by the finite-difference formula

$$-\frac{u_{n-1} - 2u_n + u_{n+1}}{h^2} = f_n; \qquad n = 1, 2, \ldots, N \qquad (4.3.2)$$

where

$$u_n \simeq U(x_n); \qquad f_n = F(x_n); \qquad x_n = nh; \qquad h = \frac{l}{N+1}$$

we shall compute the discrete $\mathcal{L}_2$ norm of the error of the approximation

$$\|\{e_n\}\|_2 = \|\mathbf{e}\|_2 = \left\{ h \sum_{n=1}^{N} [u_n - U(x_n)]^2 \right\}^{1/2} \qquad (4.3.3)$$

The calculation will be carried out in the frequency domain, taking advantage of the discrete expression of Parseval's relation to express $\|\mathbf{e}\|_2$.

4.3.1 $\mathcal{L}_2$ NORM OF THE ERROR

Before analyzing the error, we recall the expression of the exact solution in the form of a Fourier sine series. We may write formally

$$U(x) = \sum_{k=1}^{\infty} B_k \sin\left(\frac{k\pi x}{l}\right) \qquad (4.3.4)$$

and, similarly,

$$F(x) = \sum_{k=1}^{\infty} \hat{F}_k \sin\left(\frac{k\pi x}{l}\right) \qquad (4.3.5)$$

Inserting this in equation (4.3.1) gives

$$\sum_{k=1}^{\infty} B_k \left(\frac{k\pi}{l}\right)^2 \sin\left(\frac{k\pi x}{l}\right) = \sum_{k=1}^{\infty} \hat{F}_k \sin\left(\frac{k\pi x}{l}\right) \qquad (4.3.6)$$

We now multiply both sides by $\sin(k'\pi x/l)$ and integrate over $(0, l)$ to obtain

$$B_{k'} \left(\frac{k'\pi}{l}\right)^2 = \hat{F}_{k'}$$

or

$$B_k = \hat{F}_k \left(\frac{l}{k\pi}\right)^2 \qquad (4.3.7)$$

An expression of the error of the approximation will be obtained by seeking the corresponding expression for the numerical solution. Let

$$u_n = \sum_{k=1}^{N} b_k \sin\left(\frac{k\pi x_n}{l}\right) \tag{4.3.8}$$

and

$$f_n = \sum_{k=1}^{N} \hat{f}_k \sin\left(\frac{k\pi x_n}{l}\right) \tag{4.3.9}$$

where $\{b_k\}$ and $\{\hat{f}_k\}$ are the *discrete* Fourier transforms of $\{u_n \simeq U(x_n)\}$ and $\{f_n = F(x_n)\}$, respectively. By insertion of these terms in the finite-difference equation (4.3.2), we get [instead of (4.3.6) in the exact case]

$$-\sum_{k=1}^{N} b_k\left[\frac{2\cos(k\pi h/l) - 2}{h^2}\right]\sin\left(\frac{k\pi x_n}{l}\right)$$

$$= \sum_{k=1}^{N} b_k\left[\frac{\sin(k\pi h/2l)}{k\pi h/2l}\right]^2\left(\frac{k\pi}{l}\right)^2 \sin\left(\frac{k\pi x_n}{l}\right)$$

$$= \sum_{k=1}^{N} \hat{f}_k \sin\left(\frac{k\pi x_n}{l}\right) \tag{4.3.10}$$

Because of the orthogonality relations (2.4.12), we may write term by term

$$b_k\left[\frac{\sin(k\pi h/2l)}{k\pi h/2l}\right]^2\left(\frac{k\pi}{l}\right)^2 = \hat{f}_k$$

or

$$b_k = \hat{f}_k\left(\frac{l}{k\pi}\right)^2\left[\frac{k\pi h/2l}{\sin(k\pi h/2l)}\right]^2 \tag{4.3.11}$$

The error

$$\{e_n\} = \{u_n - U(x_n)\} \tag{4.3.12}$$

may be expressed in discrete Fourier sine series form

$$e_n = \sum_{k=1}^{N} (b_k - B_k) \sin\left(\frac{k\pi x_n}{l}\right) - \sum_{k=N}^{\infty} B_k \sin\left(\frac{k\pi x_n}{l}\right) \tag{4.3.13}$$

The second term in the right-hand side reflects the presence of higher harmonics in the exact solution which are not present in the numerical solution. *Inasmuch as we shall be interested in the exact solution at the discrete points* x_n *only,* this solution is expressed exactly in those points by the frequency-limited representation [see Section 2.4, in particular equation (2.4.22)]. With $N' = N + 1$:

$$B_k^{N'} = B_k + \sum_{j=1}^{\infty} (B_{2jN'+k} - B_{2jN'-k}) \qquad (4.3.14)$$

In terms of the F_k, these are [using (4.3.7)]

$$B_k^{N'} = \hat{F}_k\left(\frac{l}{k\pi}\right)^2 + \sum_{j=1}^{\infty}\left[\hat{F}_{2jN'+k}\left(\frac{l}{2jN'+k}\right)^2 - \hat{F}_{2jN'-k}\left(\frac{l}{2jN'-k}\right)^2\right] \qquad (4.3.15)$$

Thence, an equivalent expression for (4.3.13) becomes

$$e_n = \sum_{k=1}^{N} (b_k - B_k^{N'}) \sin\left(\frac{k\pi x_n}{l}\right) \qquad (4.3.16)$$

Decomposition of the Error.   *The error e_n may be seen to consist of two parts:*

1. *Truncation error*, which is the error expressed by the replacement of

$$\left(\frac{k\pi}{l}\right)^2 \quad \text{by} \quad \left(\frac{k\pi}{l}\right)^2\left[\frac{\sin(k\pi h/2l)}{k\pi h/2l}\right]^2 \qquad (4.3.17)$$

   in (4.3.10)–(4.3.11), and which results in incorrect values of the ratios $b_k/\hat{f}_k$ in the numerical solution.

2. *Aliasing of the forcing function*, which is the error resulting from the fact that components of higher frequency [than $N\pi/l$] in $F(x)$ are folded or *aliased* into lower-frequency components by the sampling

$$F(x) \longrightarrow \{f_n = F(x_n)\}$$

   which results in

$$\hat{F}_k \longrightarrow \hat{f}_k = \hat{F}_k + \sum_{j=1}^{\infty} (\hat{F}_{2jN'+k} - \hat{F}_{2jN'-k}) \qquad (4.3.18)$$

Whereas the error due to truncation can (at least to a known degree) be reduced by improving the approximation (4.3.2), the error due to aliasing cannot be eliminated once $F(x)$ has been sampled into $\{f_n\}$, since *there is no way in which (4.3.18) can be solved from left to right [given $\{\hat{f}_k\}$, find $\{\hat{F}_k\}$].*

Reduction of the Aliasing Error.   The aliasing error is reduced when N is increased. It goes to zero as $N \rightarrow \infty$. For any practical application,

N should be chosen large enough that the aliasing error be tolerable. An alternative method which is sometimes recommended consists in using for $\{f_n\}$ in (4.3.2) the expression

$$f_n = \sum_{k=1}^{N} \hat{F}_k \sin\left(\frac{k\pi x_n}{l}\right)$$ (4.3.19)

[instead of $F(x_n)$] from which the higher-frequency terms

$$\hat{F}_{k'}; \qquad k' > N$$ (4.3.20)

have been eliminated altogether. There is no more aliasing, but simply elimination of higher frequencies. Of course, this requires that one be able to compute the $\{\hat{F}_k; k = 1, 2, \ldots, N\}$ by some analytic or numerical Fourier transform.

$\mathcal{L}_2$ **Norm of the Truncation Error.** The *truncation error* may be analyzed separately. Namely, we write (the subscript T stands for *truncation*):

$$e_{n,T} = \sum_{k=1}^{N} (b_k - B_k) \sin\left(\frac{k\pi x_n}{l}\right)$$ (4.3.21)

where $\{e_{n,T}\}$ is the *global error* due to truncation, and its discrete Fourier transform by

$$\mathcal{F}(\{e_{n,T}\}) = \hat{F}_k \left\{\left[\frac{k\pi h/2l}{\sin(k\pi h/2l)}\right]^2 - 1\right\}\left(\frac{l}{k\pi}\right)^2$$ (4.3.22)

We may now apply Parseval's theorem to obtain the discrete $\mathcal{L}_2$ norm of the error

$$\|\mathbf{e}_T\|_2^2 = \Delta x \sum_{n=1}^{N} e_{n,T}^2 = \frac{l}{2} \sum_{k=1}^{N} \hat{F}_k^2 \left\{\left[\frac{k\pi h/2l}{\sin(k\pi h/2l)}\right]^2 - 1\right\}^2\left(\frac{l}{k\pi}\right)^4$$ (4.3.23)

If we define the function

$$\epsilon_{k,T} = \left[\frac{k\pi h/2l}{\sin(k\pi h/2l)}\right]^2 - 1$$ (4.3.24)

then we may rewrite (4.3.22) and (4.3.23) as

$$\mathcal{F}(\{e_{n,T}\}) = \sum_{k=1}^{N} \epsilon_{k,T} \mathcal{F}(\{U_n\})$$ (4.3.25)

and

$$\| \mathbf{e}_T \|_2^2 = \frac{l}{2} \sum_{k=1}^{N} | \epsilon_{k,T} |^2 B_k^2 = \frac{l}{2} \sum_{k=1}^{N} | \epsilon_{k,T} |^2 \hat{F}_k^2 \left(\frac{l}{k\pi} \right)^4 \qquad (4.3.26)$$

The function $\epsilon_{k,T}$ is shown graphically in Figure 4.3.1. An important property displayed in (4.3.26) is that the $\mathcal{L}_2$ norm of the error is a sum of independent contributions at the different frequencies. This sum is not only a function of the difference approximation (4.3.2), but also of the set $\{\hat{F}_k\}$ [i.e., of the *energy distribution* of the forcing function $F(x)$]. Considering the ratio

$$\frac{\| \mathbf{e}_T \|_2^2}{\| \{ U_n \} \|_2^2} \qquad (4.3.27)$$

which expresses the global error due to truncation in relative form, this ratio is minimum when all the energy of F (and U) is in the lowest frequency, that is, when

$$\left. \begin{array}{ll} \hat{F}_k \neq 0 & \text{when } k = 1 \\ = 0 & \text{when } k \neq 1 \end{array} \right\} \longrightarrow \left\{ \begin{array}{ll} B_k \neq 0 & \text{when } k = 1 \\ = 0 & \text{when } k \neq 1 \end{array} \right.$$

and grows as more of the energy gets shifted toward the highest frequency $(N\pi/l)$.

As we shall see in Section 4.4, *trigonometric interpolation* completely eliminates the truncation error $\mathbf{e}_T$ (at the cost of increased complexity in the calculation), but of course does not eliminate the sampling/aliasing error.

Exercise

Find the equivalent of (4.3.23)–(4.3.24) for the Störmer–Numerov finite difference of (4.3.1) and show that the energy of the error due to truncation $\| \mathbf{e}_T \|_2^2$ is always less than for the approximation (4.3.2).

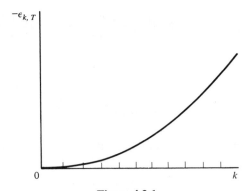

Figure 4.3.1

4.4 A FOURIER METHOD FOR TWO-POINT BOUNDARY-VALUE PROBLEMS

By contrast with the finite-difference method described in Section 4.1, we may seek an approximation to our two-point boundary-value problem

$$-\frac{d^2U}{dx^2} = F(x)$$

$$U(0) = U(l) = 0$$

(4.4.1)

using *trigonometric interpolation* instead of polynomial interpolation. To be specific, let $h = l/N$ (as with the finite-difference method) and

$$\{x_n = nh; \quad n = 0, 1, 2, \ldots, N + 1\}$$

be the set of grid points in which the solution of (4.4.1) is to be approximated by

$$u_n \simeq U(x_n)$$

To obtain this approximation, we assume that between those points u is interpolated by a Fourier series:

$$u(x) = \sum_{k=1}^{N} b_k \sin\left(\frac{k\pi x}{l}\right)$$

(4.4.2)

The $\{b_k\}$ are uniquely defined from the u_n (see Section 2.4):

$$b_k = \frac{2}{N+1} \sum_{n=1}^{N} u_n \sin\left(\frac{k\pi x_n}{l}\right); \quad k = 1, 2, \ldots, N$$

(4.4.3)

The second derivative of (4.4.2) is, *analytically*,

$$\frac{d^2u}{dx^2} = -\sum_k b_k \left(\frac{k\pi}{l}\right)^2 \sin\left(\frac{k\pi x}{l}\right)$$

(4.4.4)

Expressing the condition that (4.4.1) be satisfied exactly at the collocation points results in the N equations

$$\sum_k b_k \left(\frac{k\pi}{l}\right)^2 \sin\left(\frac{k\pi x_n}{l}\right) = f_n = F(x_n); \quad n = 1, 2, \ldots, N$$

(4.4.5)

which can be solved for the b_k:

$$b_k = \left(\frac{l}{k\pi}\right)^2 \hat{f}_k; \qquad k = 1, 2, \ldots, N \qquad (4.4.6)$$

where the $\hat{f}_k$ are the sine Fourier transform coefficients of $\{f_n\}$:

$$\hat{f}_k = \frac{2}{N+1} \sum_{n=1}^{N} f_n \sin\left(\frac{k\pi x_n}{l}\right) \qquad (4.4.7)$$

The nodal values $\{u_n\} \simeq \{U(x_n)\}$ may be computed from (4.4.2) as

$$u_n = \sum_{k=1}^{N} b_k \sin\left(\frac{k\pi x_n}{l}\right) \qquad (4.4.8)$$

On the other hand, $u(x)$ may be computed in any point in $(0, l)$ by the use of (4.4.2), not necessarily in one of the points where $F(x)$ was expressed.

4.4.1 ACCURACY

If accuracy is expressed by the $\mathfrak{L}_2$ norm of the error (as was done in Section 4.3 for the standard finite-difference method), *then it is found that this $\mathfrak{L}_2$ norm is identically zero.* Indeed, each numerical sinusoidal component, of the form

$$u_{k,n} = b_k \sin\left(\frac{k\pi x_n}{l}\right)$$

is exact, since its second derivative in (4.4.4) is exact. *The only error is that due to aliasing of the forcing function,* that is, in the folding of higher frequencies of $F(x)$ into the $(N-1)$ lower ones, as expressed by (4.3.18). If it is the case that $F(x)$ does not contain such higher frequencies, then the error is indeed equal to zero (i.e., the approximation yields the exact solution).

4.4.2 COMPUTING EFFORT
(OPERATIONS COUNT)

The computing effort of this method is not as high as one might suspect at first. Indeed, the calculation that consists in computing the $\{b_k\}$ [equation (4.4.3)] and the $\{u_n\}$ [equation (4.3.8)] should be implemented using the fast Fourier transform algorithm (Section 2.5).

4.5 INITIAL-VALUE METHODS

> "Begin at the beginning" the king said gravely "and go on till you come to the end: then stop."
>
> LEWIS CARROLL
> *Alice's Adventures in Wonderland* 12

Initial-value methods for the solution of two-point boundary-value problems consist (as the name suggests) in integrating the differential equation as one *or several* initial-value problems by *stepping* in the x direction.

Consider, for example, the equation

$$-\frac{d}{dx}\left(P(x)\frac{d}{dx}U\right) + Q(x)U = F(x) \tag{4.5.1}$$

with boundary conditions

$$U(0) = U(l) = 0 \tag{4.5.2}$$

4.5.1 SHOOTING METHOD

The most obvious initial-value method is the "shooting" method. Note first that if we want to integrate (4.5.1) numerically from $x = 0$ to $x = l$, then we are one initial condition short in $x = 0$ [the value of $(dU/dx)_0$]. We may:

1. Guess a value for this unknown parameter.

2. Integrate (4.5.1) from $x = 0$ to $x = l$ (e.g., by transforming it into the system of two ordinary differential equations,

$$\frac{dU}{dx} = \frac{Y}{P}$$

$$\frac{dY}{dx} = -F + QU \tag{4.5.3}$$

and then solving this system by a numerical integration method for ordinary differential equations of the initial value type).

3. Use the value of $[u(l)_{\text{computed}}]$ to modify $(dU/dx)_0$ iteratively until $[u(l)_{\text{computed}}]$ equals (nearly) the imposed boundary condition (zero).

Although obviously simple, this method is not always applicable: if both $P(x)$ and $Q(x)$ are positive, then (4.5.3) considered as an initial-value problem is an *ill-conditioned* problem (see Section 1.1). That is, *small* changes in the boundary value $(dU/dx)_0$ result in *large* changes in the solution.

This has the following consequence from the numerical error propagation viewpoint: When $P(x) > 0$ and $Q(x) > 0$, then small computing (truncation or roundoff) errors introduced in the calculation in $x_1 > 0$ produce an error in the numerical solution (= the propagated error) which grows exponentially with $x - x_1(x > x_1)$ and which may destroy any reasonable numerical accuracy.

Variants of the shooting method have been considered among others by Fox (1960), Morrison et al. (1962), and Keller (1968). Alternatives to shooting methods which are numerically *well conditioned* (i.e., small changes in the boundary values *and small errors in the calculation* do not grow in subsequent steps of the numerical calculation) have been developed. Some of these are described next.

4.5.2 THE OPERATOR DECOMPOSITION METHOD

Consider as an example the case

$$P(x) = Q(x) = 1 \tag{4.5.4}$$

in equation (4.5.1); that is,

$$LU \equiv -\frac{d^2U}{dx^2} + U = F(x) \tag{4.5.5}$$

The linear operator L may be *decomposed* in a product of two operators as follows (Vichnevetsky, 1968, 1971):

$$L = \left(\frac{d}{dx} + 1\right)\left(-\frac{d}{dx} + 1\right)$$
$$\equiv L_F L_B \tag{4.5.6}$$

Define a new variable $Y(x)$ and solve the initial-value equation:

$$L_F Y \equiv \frac{dY}{dx} + Y = F(x) \tag{4.5.7}$$

from $x = 0$ to $x = l$, followed by the solution of

$$L_B V = Y \quad \text{with } V(l) = 0 \tag{4.5.8}$$

We find that

$$LV = F \tag{4.5.9}$$

Thus, $V(x)$ satisfies equation (4.5.5). *An important property is this*: if, as indicated, (4.5.7) is integrated numerically in the *forward* direction ($x = 0$ to $x = l$) and (4.5.8) is integrated in the backward direction ($x = l$ to $x = 0$), then both integrations are *well conditioned* (the effect of small numerical errors *is not* amplified in the calculation; see, e.g., Vichnevetsky 1971). The boundary condition $U(0) = 0$ may be satisfied by a superposition of solutions as follows. Let $W(x)$ be the solution of

$$L_F Y \equiv \frac{dY'}{dx} + Y' = 0; \qquad Y'(0) = 1 \qquad (4.5.10)$$

$$L_B W \equiv -\frac{dW}{dx} + W = Y'; \qquad W(l) = 0 \qquad (4.5.11)$$

That is, $W(x)$ is a solution of the homogeneous equation

$$LW \equiv -\frac{d^2 W}{dx^2} + W = 0 \qquad (4.5.12)$$

obtained numerically by the well-conditioned forward and backward integrations of (4.5.10)–(4.5.11). Then, the sum

$$U(x) = V(x) + \alpha W(x) \qquad (4.5.13)$$

where

$$\alpha = -\frac{V(0)}{W(0)}$$

satisfies equations (4.5.5) and both boundary conditions (4.5.2) (i.e., is the solution).

4.5.3 RICCATI EQUATION

We now return to the more general problem where $P(x)$ and $Q(x)$ are not constant. A decomposition of the form (4.5.6) may still be obtained: for example, let L be decomposed as

$$L = \left[\frac{d}{dx} - \lambda_F(x)\right]\left[-P(x)\frac{d}{dx} + \lambda_B(x)\right] \qquad (4.5.14)$$

we find, by expanding this expression, that

$$L = -\frac{d}{dx}P(x)\frac{d}{dx} + (\lambda_F P + \lambda_B)\frac{d}{dx} - \left(\lambda_F \lambda_B + \frac{d\lambda_B}{dx}\right) \qquad (4.5.15)$$

For (4.5.15) to be identical to (4.5.1) (i.e., for $L_F L_B$ to be equal to L), we must have

$$\lambda_F P + \lambda_B = 0 \tag{4.5.16}$$

$$\frac{d\lambda_B}{dx} + \lambda_F \lambda_B = -Q(x) \tag{4.5.17}$$

or, eliminating λ_F,

$$\frac{d\lambda_B}{dx} - \frac{\lambda_B^2}{P(x)} = -Q(x) \tag{4.5.18}$$

This is a Riccati-type differential equation, which may be shown to be well conditioned from the computing error-propagation point of view when integrated numerically backward (i.e., in the direction $x = l$ to $x = 0$).

Thus, finding $V(x)$ is now obtained by solving successively:

1.   Equation (4.5.18) for $\lambda_B(x)$.
2.   Equation (4.5.16) for $\lambda_F(x) = -\lambda_B/P$.
3. $L_F Y = F$ for $Y(x)$.
4. $L_B V = Y$ for $V(x)$.

A homogeneous equation solution $W(x)$ is obtained by a similar sequence [the equivalent of equations (4.5.10)–(4.5.11)]. As in the constant P, constant Q case, $U(x)$ is then obtained by applying equation (4.5.13).

Other methods of decomposition leading to Riccati equations are described by Vichnevetsky (1970, 1971) and Andre (1970).

Invariant imbedding is another formalism leading to a similar decomposition of two-point boundary-values problems (see e.g., Bellman, Kagiwada, and Kalaba, 1967, Angel and Bellman, 1972, or Meyer, 1973, for details).

5

Finite-Difference Approximation
of Elliptic Equations

5.1 APPROXIMATION AND CONVERGENCE

5.1.1 INTRODUCTION

In Chapter 4 we have used two-point boundary-value problems as a convenient vehicle for the description of some of the basic concepts that apply to the finite-difference approximation of elliptic equations. In this chapter we describe and analyze the finite-difference approximation of elliptic equations in several dimensions. In addition to the standard equations of Laplace,

$$\nabla^2 U \equiv \frac{\partial^2 U}{\partial x^2} + \frac{\partial^2 U}{\partial y^2} = 0 \qquad (5.1.1)$$

and Poisson:

$$-\nabla^2 U = F$$

these methods of approximation also apply to the biharmonic equation:

$$\nabla^4 U \equiv \frac{\partial^4 U}{\partial x^4} + 2\frac{\partial^4 U}{\partial x^2 \partial y^2} + \frac{\partial^4 U}{\partial y^4} = F$$

which may be used, for instance, to describe the deformation of a thin, loaded elastic plate. The two questions of importance in the numerical solution of elliptic equations are (Figure 5.1.1):

1. The replacement of the equation by a discrete approximation (or "discretization" of the equation).

2. The setting up of a computer algorithm for the solution of the discrete algebraic equations generated by the discretization.

In this chapter we review techniques that are used in the *discretization* of elliptic equations by the method of finite-difference approximation. Numerical methods that are used to *solve* the algebraic equations resulting from the discretization are reviewed in Chapter 6.

Finite-difference methods are historically old, in fact almost inseparable from the development of partial differential equations in the eighteenth and nineteenth centuries. Another important method for the discretization of elliptic partial differential equations which was developed much more recently, and has displaced the use of finite-difference methods in many *applications, is the finite-element method.* This method is described in Chapter 8 and following.

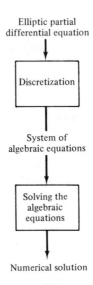

Elliptic partial differential equation

Discretization

System of algebraic equations

Solving the algebraic equations

Numerical solution

Figure 5.1.1

5.1.2 FINITE-DIFFERENCE APPROXIMATION

A simple approximation to Laplace's equation, written in Cartesian coordinates as in (5.1.1), is obtained by using simple three-point difference approximations to the derivatives; that is,

$$(\nabla^2 U)_{m,n} \simeq \frac{u_{m-1,n} - 2u_{m,n} + u_{m+1,n}}{\Delta x^2} + \frac{u_{m,n-1} - 2u_{m,n} + u_{m,n+1}}{\Delta y^2}$$

$$= \left(\frac{\delta_x^2}{\Delta x^2} + \frac{\delta_y^2}{\Delta y^2}\right) u_{m,n} = 0 \tag{5.1.2}$$

Here the

$$u_{m,n} \simeq U(m\,\Delta x, n\,\Delta y) = U_{m,n} \tag{5.1.3}$$

are representing values of a numerical solution meant to approximate the local value of $U(x, y)$ in the mesh points $x_m = m\,\Delta x;\ y_n = n\,\Delta y$ (Figure 5.1.2). When $\Delta x = \Delta y = h$, a simple approximation to Laplace's equation is thus given by

$$4u_P - (u_N + u_S + u_E + u_W) = 0 \tag{5.1.4}$$

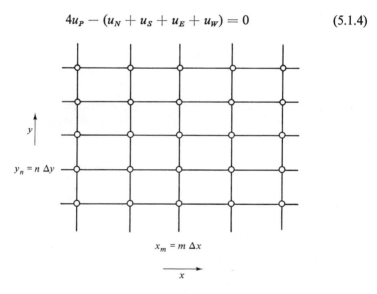

Figure 5.1.2   Rectangular grid for the finite-difference approximation of Laplace's equation.

where the subscripts N, S, E, and W denote the points located north, south, east, and west of P. The corresponding elementary "computing molecule" is that of Figure 5.1.3. We may rewrite (5.1.4) as

$$u_P = \tfrac{1}{4}(u_N + u_S + u_E + u_W) \tag{5.1.5}$$

and note that this relation is a finite-difference analog to the mean-value theorem.

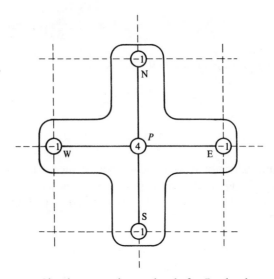

Figure 5.1.3 Simple computing molecule for Laplace's equation.

MEAN-VALUE THEOREM   The mean value of any function that satisfies Laplace's equation in two dimensions over a circle is equal to the value at the center[1]:

$$U_{\text{center}} = \int_0^{2\pi} U(R \cos \theta, R \sin \theta) \frac{d\theta}{2\pi} \qquad (5.1.6)$$

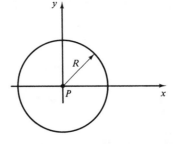

5.1.3 TRUNCATION ERROR

Before proceeding further with a description of finite-difference approximation to elliptic equations, it is useful to consider the problem of errors (most of the analysis given here is similar to that given in Section 4.3 in the

---

[1] A function that satisfies Laplace's equation is sometimes called a harmonic function, and Laplace's equation is referred to as the "harmonic equation."

context of two-point boundary-value problems). It will be convenient again to use the notation

$$L_h \mathbf{u} = 0 \qquad (5.1.7)$$

to express the difference approximation of the elliptic equation

$$LU = 0 \qquad (5.1.8)$$

in symbolic notation. Here h stands for a representative mesh dimension. In the example (5.1.2), we may, for instance, let $\Delta x = h$ and $\Delta y = gh$ such that both Δx and $\Delta y \longrightarrow 0$ with $\Delta y / \Delta x = g$ when $h \longrightarrow 0$. With this convention, the truncation error of the finite-difference approximation (5.1.7) to the elliptic equation (5.1.8) is by definition the value of $L_h U$, where U is an exact solution of the equation. Thus,

$$T_h \equiv L_h U \qquad (5.1.9)$$

where U is a solution of (5.1.8) ("sampled" at the mesh points). Alternatively, the truncation error may be defined as a property of the approximation of the operator L afforded by the operator L_h (rather than an approximation of the equation $LU = 0$ by the equation $L_h \mathbf{u} = 0$). This definition is

$$T_h = L_h U - LU \qquad (5.1.10)$$

where here U is any sufficiently smooth function. It is easily shown that when applied to the problem at hand, both definitions yield the same result. When we solve Poisson's equation

$$LU = F \qquad (5.1.11)$$

instead of Laplace's equation, then, as illustrated by equation (4.2.1) for a two-point boundary-value problem, the definition of the truncation error is easily extended to include the additional term.

The classical way to express the truncation error is by the use of Taylor series. Consider as an example the difference approximation (5.1.2) to Laplace's equation where

$$L_h \equiv \frac{\delta_x^2}{h^2} + \frac{\delta_y^2}{g^2 h^2}; \qquad \Delta x = h; \Delta y = gh \qquad (5.1.12)$$

To find the truncation error, we use Taylor's theorem to express

$$U_{m+r,\,n+s} = U_{m,n} + \left(\frac{\partial U}{\partial x}\right)_{m,n} r\,\Delta x + \left(\frac{\partial U}{\partial y}\right)_{m,n} s\,\Delta y$$

$$+ \frac{1}{2!}\left[\left(\frac{\partial^2 U}{\partial x^2}\right)_{m,n} r^2\,\Delta x^2 + 2\left(\frac{\partial^2 U}{\partial x\,\partial y}\right)_{m,n} rs\,\Delta x\,\Delta y + \left(\frac{\partial^2 U}{\partial y^2}\right)_{m,n} s^2\,\Delta y^2\right]$$

$$+ \frac{1}{3!}\left[\left(\frac{\partial^3 U}{\partial x^3}\right)_{m,n} r^3\,\Delta x^3 + 3\left(\frac{\partial^3 U}{\partial x^2\,\partial y}\right)_{m,n} r^2 s\,\Delta x^2\,\Delta y + \ldots\right] \quad (5.1.13)$$

$$+ \text{ higher-order terms}$$

Upon applying the definition (5.1.9), we find

$$T_h = L_h\{U_{m,n}\} = \left(\frac{\partial^2}{\partial x^2} + \frac{\partial^2}{\partial y^2}\right) U_{m,n}$$

$$+ \frac{1}{12}\left(\frac{\partial^4}{\partial x^4}\,\Delta x^2 + \frac{\partial^4}{\partial y^4}\,\Delta y^2\right) U_{m,n}$$

$$+ \text{ higher-order terms} \quad (5.1.14)$$

Since U is a solution of Laplace's equation, the first term on the right-hand side vanishes, and

$$T_h = \left(\frac{\partial^4 U}{\partial x^4} + g^2\frac{\partial^4 U}{\partial y^4}\right)_{m,n}\frac{h^2}{12}$$

$$+ \text{ higher-order terms} \quad (5.1.15)$$

Implicit in this notation is the fact that the truncation error is an array of values for all m and n:

$$T_h = \{(T_h)_{m,n}\}$$

The *order of accuracy* of the approximation is defined as the power of h in the principal part of the expression of the truncation error obtained by using the remainder of Taylor series in the manner just illustrated; that is, the approximation of the equation $\nabla^2 U = 0$ by (5.1.2) is of *second-order* accuracy. In general, we may write

$$T_h \le Kh^p + \text{ higher-order terms} \quad (5.1.16)$$

$$= O(h^p)$$

where K is a constant independent of h, containing higher derivatives of U, and p is the order of accuracy of the approximation of $LU = 0$ by $L_h\mathbf{u} = 0$. Formally, writing an inequality such as (5.1.16) demands that a norm be chosen. For example, if one uses the sup-norm, then the inequality must hold in every point (m, n) of the domain.

5.1.4 GLOBAL ERRORS

The truncation errors in the discrete approximation of an elliptic equation are discrepancies between local derivatives and their finite-difference approximation. What counts in actual calculations are the *global errors*, or discrepancies between the approximate and exact solution:

$$\mathbf{e} = \mathbf{u} - U \tag{5.1.17}$$

Implicit in this notation is that $\mathbf{e}$ is a vector of values with as many components as the numerical solution $\mathbf{u}$, and that, as before, U is to be interpreted as the vector of sampled values of $U(x, y)$ taken in each solution point (x_m, y_n). We obtain by subtraction of (5.1.9) from (5.1.7):

$$L_h(\mathbf{u} - U) = L_h\mathbf{e} = -T_h \tag{5.1.18}$$

which is the equation whereby the vector of global errors is related to the local truncation errors.

5.1.5 CONVERGENCE

When

$$\lim_{h \to 0} \|\mathbf{e}\| = 0 \tag{5.1.19}$$

(*according to some chosen norm*), then the difference approximation (5.1.7) to the elliptic problem is called *convergent*. As we may infer from (5.1.18), proof of convergence can be derived from an analysis of the norm of the operator L_h^{-1}, since

$$\mathbf{e} = -L_h^{-1}T_h \tag{5.1.20}$$

Proofs of convergence tend to follow the standard scenario:

1. Proof that the operator L_h^{-1} remains a bounded operator as $h \to 0$ (called the *stability* of L_h).
2. Proof that $\lim_{h \to 0} \|T_h\| = 0$ (called the *consistency* of the approximation of L by L_h).

Steps 1 and 2 imply that $\lim_{h \to 0} \|\mathbf{e}\| = 0$ (i.e., *convergence*). This trilogy,

---

> stability + consistency = convergence

---

is one of the keystones of theoretical numerical analysis. Global convergence as $h \rightarrow 0$ for difference approximations of elliptic equations was first proved for Laplace's equation on a square mesh by Phillips and Wiener (1923). Further refinements of the theory are found in the famous Courant, Friedrichs, and Lewy paper of 1928. It should be noted in passing that the aim of Phillips and Wiener and of Courant, Friedrichs, and Lewy was to establish existence theorems for solutions of Laplace's equation $\nabla^2 U = 0$ from *algebraic* existence theorems for $\nabla_h^2 u = 0$, *not* to analyze properties of computational algorithms. Other analyses of the convergence of numerical solutions of elliptic equations may be found in Gerschgorin (1930), Collatz (1933), and Wasow (1955). Reviews of this literature are given in Forsythe and Wasow (1960) and in Collatz (1960). An overall survey is given in an interesting monograph on this subject by Birkhoff (1972b).

5.1.6 THE PLACE
OF CONVERGENCE THEORIES

Convergence plays an important role in the theory of numerical analysis. The concept of convergence of a given method of approximation occurs in seeking to predict whether, when all increment sizes approach zero, the numerical solution will, in the limit, tend to the exact solution. The argument is completed by the remark that "if a method is convergent, then the practitioner (i.e., he who wants to solve real problems with the use of computers) can always choose increment sizes which are small enough to ensure any given accuracy." Moreover, since rates of convergence are proportional to the order of accuracy of the approximation ($=$ the power of h in the truncation error), maximizing the order of accuracy is often taken to be the (only) criterion by which different methods are compared.

One finds, though, that practitioners sometimes pay little attention to convergence theories. Complete textbooks intended for those practitioners, and a voluminous literature in technical journals concerned with applications, often go as far as ignoring the subject altogether.

5.2 HIGHER-ORDER APPROXIMATIONS

Higher-order finite-difference approximations of partial derivatives in elliptic problems may be obtained in a number of ways. We give as an example the derivation of approximations to the Laplacian operator ∇^2 and the biharmonic operator ∇^4 in Cartesian coordinates:

$$\nabla^2 \equiv \frac{\partial^2}{\partial x^2} + \frac{\partial^2}{\partial y^2}$$

$$\nabla^4 \equiv \frac{\partial^4}{\partial x^4} + 2\frac{\partial^4}{\partial x^2 \, \partial y^2} + \frac{\partial^4}{\partial y^4}$$

on a square lattice:

$$\Delta x = \Delta y = h$$

We label a typical point with the subscript 0 and neighboring points with 1, 2, 3, ..., as in Figure 5.2.1.

A convenient way to obtain difference approximation is by using the algebra of operators described in Section 3.2. We recall the symbolic expression for Taylor's series:

$$U(x + kh) = e^{khD_z}U(x) \tag{5.2.1}$$

where

$$D_x = \frac{\partial}{\partial x}$$

(and similarly in the y direction). Then we express the symmetric sum (see Bickley, 1948) obtained by taking values of U on a concentric circle:

$$
\begin{aligned}
S_1 &= U_1 + U_2 + U_3 + U_4 \\
&= (e^{hD_z} + e^{hD_v} + e^{-hD_z} + e^{-hD_v})U_0 \\
&= [2\cosh(hD_x) + 2\cosh(hD_y)]U_0 \\
&= \Big[4 + h^2(D_x^2 + D_y^2) + 2\frac{h^4}{4!}(D_x^4 + D_y^4) \\
&\quad + 2\frac{h^6}{6!}(D_x^6 + D_y^6) + \dots \Big]U_0
\end{aligned}
\tag{5.2.2}
$$

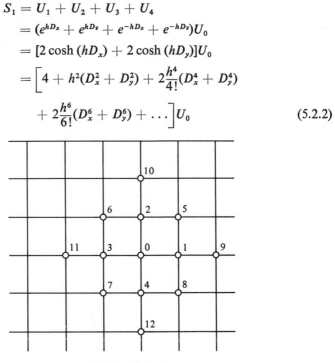

Figure 5.2.1

From this we derive the known formula

$$\frac{U_1 + U_2 + U_3 + U_4 - 4U_0}{h^2} = \nabla^2 U_0 + O(h^2) \tag{5.2.3}$$

In this expression U_n represents local values of a continuous function $U(x, y)$. We may thus express the approximation[2]

$$-\frac{u_1 + u_2 + u_3 + u_4 - 4u_0}{h^2} = f_0 \tag{5.2.4}$$

to Poisson's equation

$$-\nabla^2 U = F(x, y) \tag{5.2.5}$$

which, according to (5.2.3), has a truncation error of order $O(h^2)$. The next symmetric sum is

$$
\begin{aligned}
S_2 &= U_5 + U_6 + U_7 + U_8 \\
&= [e^{h(D_x + D_y)} + e^{h(-D_x + D_y)} + e^{h(-D_x - D_y)} + e^{(D_x - D_y)}]U_0 \\
&= 4[\cosh(hD_x)\cosh(hD_y)]U_0 \\
&= \Big[4 + \frac{4h^2}{2!}(D_x^2 + D_y^2) + \frac{4h^4}{4!}(D_x^4 + 6D_x^2 D_y^2 + D_y^4) \\
&\quad + \frac{4h^6}{6!}(D_x^6 + 15D_x^4 D_y^2 + 15D_x^2 D_y^4 + D_y^6) + \ldots\Big]U_0 \\
&= [4 + 2h^2\nabla^2 + \tfrac{1}{6}h^4(\nabla^4 + 4D_x^4 D_y^4) \\
&\quad + \tfrac{1}{180}h^6(\nabla^6 + 12\nabla^2 D_x^4 D_y^4) + \ldots]U_0 \\
&\quad + \text{higher-order terms}
\end{aligned} \tag{5.2.6}
$$

We may combine these expressions to obtain higher-order approximations. We find, for instance,

$$\frac{4S_1 + S_2 - 20U_0}{6h^2} = \nabla^2 U_0 + \tfrac{1}{12}h^2\nabla^4 U_0 + O(h^4) \tag{5.2.7}$$

If U is a solution of Poisson's equation (5.2.5), then

$$\nabla^4 U = \nabla^2 F$$

---

[2] As before, lower case u's represent values of a numerical approximation.

We may thus combine (5.2.4) and (5.2.7) as

$$-\frac{4S_1 + S_2 - 20U_0}{6h^2} = F_0 + \frac{1}{12}h^2\left(\frac{F_1 + F_2 + F_3 + F_4 - 4F_0}{h^2}\right) + O(h^4)$$

$$(5.2.8)$$

which gives us an approximation of Poisson's equation with $O(h^4)$ accuracy. The next sum is

$$
\begin{aligned}
S_3 &= U_9 + U_{10} + U_{11} + U_{12} \\
&= (e^{2hD_x} + e^{2hD_v} + e^{-2hD_x} + e^{-2hD_v})U_0 \\
&= [4 + 4h^2\nabla^2 + \tfrac{4}{3}h^4(\nabla^4 - 2D_x^4 D_y^4) \\
&\quad + \tfrac{8}{45}(\nabla^6 - 3\nabla^2 D_x^4 D_y^4) + \ldots]U_0 \\
&\quad + \text{higher-order terms}
\end{aligned}
\qquad (5.2.9)
$$

We may use $U_0, S_1, S_2,$ and S_3 to approximate $\nabla^4 U$ (see Figure 5.2.2):

$$\nabla^4 U_0 = \frac{20U_0 - 8S_1 + 2S_2 + S_3}{h^4} + O(h^2) \qquad (5.2.10)$$

and so on.

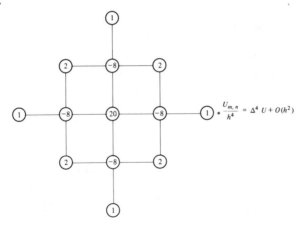

Figure 5.2.2 Computational molecule for $\nabla^4 u$.

5.2.1 APPROXIMATION ON A TRIANGULAR MESH

The preceding mathematics (i.e., the matching of terms in Taylor's series) lend themselves as well to approximations over other simple, *regular* meshes of nodal points. For example, the following stencil

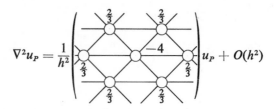

$$\nabla^2 u_P = \frac{1}{h^2} \left(\cdots \right) u_P + O(h^2)$$

describes a seven-point approximation to the Laplacian operator on a regular triangular mesh. Approximation of a higher order based on such a mesh may be found in tabulated form in Collatz (1960, pp. 545–546).

5.3   APPROXIMATION NEAR BOUNDARIES

5.3.1   CURVED BOUNDARIES

Curved boundaries must be dealt with separately, since standard finite-difference expressions are based on regular (usually rectangular) meshes. A general remark that applies here is that the approximation of elliptic equations near curved boundaries with finite differences is more of a collection of ad hoc procedures than a systematic approach. By contrast, the approximation near curved boundaries in the finite-element method (see, e.g., Chapter 10) is as systematic as the approximation in interior nodes. This point is often presented as one of the advantages that have been responsible for the successful takeover of the finite-element methods over finite differences in many fields of application. We illustrate some of these questions by considering the boundary-value problem of the first kind (U given on the boundary) for Laplace's equation. A first approach consists in deriving difference equations that make use of actual boundary points at mesh points. Consider, for instance, Figure 5.3.1, where U is specified on the boundary. We may easily derive the approximations [from (3.1.19)]

$$\left. \begin{array}{l} \left(\dfrac{\partial^2 U}{\partial x^2}\right)_P \simeq \dfrac{2}{h + \alpha h}\left(\dfrac{U_1 - u_P}{\alpha h} - \dfrac{u_P - u_3}{h}\right) \\[3mm] \left(\dfrac{\partial^2 U}{\partial y^2}\right)_P \simeq \dfrac{2}{h + \beta h}\left(\dfrac{U_2 - u_P}{\beta h} - \dfrac{u_P - u_4}{h}\right) \end{array} \right\} \tag{5.3.1}$$

whence in P,

$$(\nabla^2 U)_P \simeq \left(\frac{\partial^2 U}{\partial x^2}\right)_{P,\,\text{approx}} + \left(\frac{\partial^2 U}{\partial y^2}\right)_{P,\,\text{approx}} \tag{5.3.2}$$

where the right-hand side contains the boundary values U_1 and U_2, which are given.

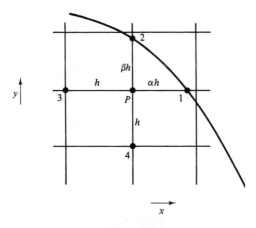

Figure 5.3.1

A second approach for curved boundaries consists in using interpolation to effectively displace the boundary. As a simple illustration, consider again the situation of Figure 5.3.1. We may express u_P as a linear interpolant between the points 2 and 4, thus:

$$u_P \simeq \frac{\beta h u_4 + h U_2}{h + \beta h}$$

where U_2 is a given boundary value.

The net result has been to replace the original boundary by a new boundary which passes through regular mesh points only.

5.3.2 BOUNDARY CONDITIONS OF THE SECOND AND THIRD KINDS

Dealing with conditions of the second and third kinds on a curved boundary may also be done by establishing ad hoc difference formulas for their approximation. Consider, for example, the case of Figure 5.3.2, where a boundary value of the third kind

$$\frac{\partial U}{\partial n} = a + bU \tag{5.3.3}$$

applies on the boundary ∂D. Let PAB be a perpendicular to the boundary that passes through the point A. We may write the simple approximations

$$u_P = \alpha u_D + \beta u_C \tag{5.3.4}$$

$$u_B = u_A + \frac{\delta}{\gamma}(u_A - u_P) \tag{5.3.5}$$

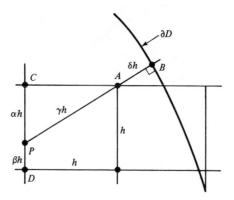

Figure 5.3.2

and

$$\left(\frac{\partial U}{\partial n}\right)_B \simeq \frac{u_B - u_P}{(\gamma + \delta)h} = a + bu_B \qquad (5.3.6)$$

These equations are to be used to determine u_A instead of the difference approximation that applies in the other (interior) mesh points.

More accurate formulas may of course be derived by bringing in more mesh points [see, e.g., Milne (1953, p. 149) for a formula using 11 points]. An alternative to using such higher-order formulas consists, of course, in using a finer mesh near a curved boundary.

6

Numerical Solution
of Algebraic Equations

6.1 DIRECT SOLUTION METHODS

6.1.1 SOLUTION OF ELLIPTIC
EQUATIONS

Elliptic linear boundary-value problems, when discretized by finite-difference or finite-element methods, result in systems of linear algebraic equations of the form

$$\mathbf{Au} = \mathbf{b}$$

where $\mathbf{u} = \{u_n\}$ is the numerical solution, in the form of a vector of N unknowns, $\mathbf{A}$ is an N by N matrix, and $\mathbf{b}$ is an N vector containing known source and boundary terms.

Finding the numerical solution $\mathbf{u}$ consists in solving the fundamental computational problem in linear algebra. There is a vast literature about such problems. Methods that are best suited to the capabilities of modern electronic computers have been developed over the past few decades. The items that may be considered as expensive in that respect and may thus serve as criteria to compare methods are:

1. *Computing effort*, measured by the number of computer operations needed to obtain a (sufficiently accurate) solution.

2. *Storage*, measured by the size of computer memory needed to store the elements of the system of equations, in particular to store the matrix **A** and other matrices occurring in the process of solving.

The two basic approaches for the solution of large systems of equations are direct or elimination methods and indirect or iterative methods. *Elimination (direct) methods* are typified by Gaussian elimination. This is a procedure wherein the matrix **A** is transformed to a triangular form which can then be solved directly for the unknowns. *Iterative (indirect) methods* consist in a sequence of successive corrections to an original estimate for the unknowns until the size of the necessary corrections becomes negligible.

Direct methods are discussed in the following pages. Iterative methods are described in Section 6.2.

Nonlinear Problems.   Many practical problems are of course *nonlinear*. The most common procedure for solving nonlinear elliptic problems is to formulate them as the solution of an iterative sequence of *linear* (or rather linearized) equations to which solution methods for linear problems still apply. This topic is covered in some detail in Section 6.4.

6.1.2 GAUSSIAN ELIMINATION

The best known and most widely used method for solving systems of linear algebraic equations is attributed to Gauss.[1] It is, basically, the elementary procedure in which the first equation is used to eliminate the first variable in the last $(N - 1)$ equations, the new second equation is used to eliminate the second variable from the last $(N - 2)$ equations, and so on. If $(N - 1)$ such eliminations can be performed, then the resulting linear system is triangular and is easily solved.

We proceed as follows: Let

$$
\mathbf{Au} \equiv
\begin{bmatrix}
a_{1,1} & a_{1,2} & \cdots & a_{1,N} \\
a_{2,1} & a_{2,2} & \cdots & a_{2,N} \\
\cdot & & & \\
\cdot & & & \\
\cdot & & & \\
a_{N,1} & & \cdots\cdots & a_{N,N}
\end{bmatrix}
\begin{bmatrix}
u_1 \\
u_2 \\
\cdot \\
\cdot \\
\cdot \\
u_N
\end{bmatrix}
=
\begin{bmatrix}
b_1 \\
b_2 \\
\cdot \\
\cdot \\
\cdot \\
b_N
\end{bmatrix}
\equiv \mathbf{b} \qquad (6.1.1)
$$

be the system of equations. We assume in the following that the matrix **A** is nonsingular. Then the system has a unique solution. We also assume that

---

[1] Carl Friedrich Gauss (1777–1855), a German mathematician, was called the "prince of mathematics" in his time.

the diagonal elements are nonzero:

$$a_{ii} \neq 0 \qquad (6.1.2)$$

We first eliminate u_1 from the last $(N - 1)$ equations by subtracting from the ith equation:

$$\left(\frac{a_{i,1}}{a_{1,1}} \text{ times the first equation}\right) \qquad i = 2, \ldots, N$$

Then the last $(N - 1)$ equations become

$$\left.\begin{array}{l} a_{2,2}^2 u_2 + a_{2,3}^2 u_3 + \ldots = b_2^2 \\ a_{3,2}^2 u_2 + a_{3,3}^2 u_3 + \ldots = b_3^2 \\ \qquad \cdot \\ \qquad \cdot \\ \qquad \cdot \\ a_{N,2}^2 u_2 + a_{N,3}^2 u_3 + \ldots = b_N^2 \end{array}\right\} \qquad (6.1.3)$$

Here, $a_{i,j}^2$ and b_i^2 are the value of $a_{i,j}$ and b_i after the first step of the elimination procedure [and, generally, we shall denote by $a_{i,j}^k$ and b_i^k the value of those coefficients after $(k - 1)$ steps of elimination]. The new coefficients are given by

$$a_{i,j}^2 = a_{i,j} - \left(\frac{a_{i,1}}{a_{1,1}}\right) a_{1,j}; \qquad b_i^2 = b_i - \left(\frac{a_{i,1}}{a_{1,1}}\right) b_1; \qquad i = 2, 3, \ldots, N$$

This is a system of $(N - 1)$ equations in the $(N - 1)$ unknowns

$$u_2, u_3, \ldots, u_N$$

We may repeat the process to eliminate u_2 from the last $(N - 2)$ of these equations. We then obtain a system of $(N - 2)$ equations in the $(N - 2)$ unknowns $u_3, u_4, \ldots, u_N$. This system is of the form

$$\left.\begin{array}{l} a_{3,3}^3 u_3 + a_{3,4}^3 u_4 + \ldots = b_3^3 \\ a_{4,3}^3 u_3 + a_{4,4}^3 u_4 + \ldots = b_4^3 \\ \qquad \cdot \\ \qquad \cdot \\ \qquad \cdot \\ a_{N,3}^3 u_3 + \ldots \qquad\quad = b_N^3 \end{array}\right\} \qquad (6.1.4)$$

After $(N - 1)$ steps we are left with one equation,

$$a_{N,N}^N u_N = b_N^N \qquad (6.1.5)$$

We now collect the first equation from each step to get

$$\left.\begin{array}{c} a_{1,1}u_1 + a_{1,2}u_2 + \ldots\ldots\ldots + a_{1,N}u_N = b_1 \\ a_{2,2}^2 u_2 + a_{2,3}^2 u_3 + \ldots + a_{2,N}^2 u_N = b_2^2 \\ \cdot\quad\cdot \\ \cdot\quad\quad\cdot \\ \cdot\quad\quad\quad\cdot \\ a_{N,N}^N u_N = b_N^N \end{array}\right\} \tag{6.1.6}$$

Thus, we have reduced the original system (6.1.1) to the triangular system (6.1.6). Solving this system is easy: we start from the bottom, and solve for u_N:

$$u_N = \frac{b_N^N}{a_{N,N}^N} \tag{6.1.7}$$

and then proceed upward, solving for one unknown per step:

$$u_{N-1} = \frac{b_{N-1}^{N-1} - a_{N-1,N}^{N-1} u_N}{a_{N-1,N-1}^{N-1}} \tag{6.1.8}$$

and so on.

The method of Gaussian elimination as just described allows us to compute the solution $\mathbf{u}$ to the linear system (6.1.1) for a given vector $\mathbf{b}$. For a new vector $\mathbf{b}$, the whole process is to be repeated entirely. If we need solve the system more than once with the same matrix $\mathbf{A}$ but different $\mathbf{b}$'s, it may be more advantageous to proceed somewhat differently.

One procedure consists in computing and storing the inverse matrix $\mathbf{A}^{-1}$, and then obtaining for each different $\mathbf{b}$ the solution by multiplication:

$$\mathbf{u} = \mathbf{A}^{-1}\mathbf{b} \tag{6.1.9}$$

A second procedure, called **LU** decomposition, will be shown to be more advantageous both in terms of storage (when $\mathbf{A}$ is sparse) and computing effort.

Both procedures (computing $\mathbf{A}^{-1}$ and effecting the **LU** decomposition of $\mathbf{A}$) are by-products of the basic Gaussian elimination method and will be described below.

6.1.3 COMPUTING $\mathbf{A}^{-1}$

The Gaussian elimination procedure can be applied to several vectors simultaneously. That is, if we put side by side two vectors $\mathbf{b}_1$ and $\mathbf{b}_2$ as a 2 by N matrix and write

$$\mathbf{A}[\mathbf{u}_1, \quad \mathbf{u}_2] = [\mathbf{b}_1, \quad \mathbf{b}_2] \tag{6.1.10}$$

then we may apply to this system the formal steps of Gaussian elimination to obtain the solution $[\mathbf{u}_1 \quad \mathbf{u}_2]$, where $\mathbf{u}_1$ and $\mathbf{u}_2$ are solutions of

$$\mathbf{A}\mathbf{u}_1 = \mathbf{b}_1; \qquad \mathbf{A}\mathbf{u}_2 = \mathbf{b}_2 \qquad (6.1.11)$$

respectively. If we now put side by side the N simple right-hand side vectors

$$\mathbf{e}_n \equiv \begin{bmatrix} 0 \\ 0 \\ \cdot \\ \cdot \\ \cdot \\ 1 \\ 0 \\ \cdot \\ \cdot \\ \cdot \\ 0 \end{bmatrix} \longleftarrow n\text{th line}$$

then we may observe that the resulting matrix

$$[\mathbf{e}_1, \quad \mathbf{e}_2, \quad \ldots, \quad \mathbf{e}_N] = \mathbf{I} \qquad (6.1.12)$$

is simply the identity matrix. Proceeding as above, we solve by Gaussian elimination:

$$\mathbf{A}[\mathbf{v}_1, \quad \mathbf{v}_2, \quad \ldots, \quad \mathbf{v}_N] = [\mathbf{e}_1, \quad \mathbf{e}_2, \quad \ldots, \quad \mathbf{e}_N] \qquad (6.1.13)$$

and observe that the solution is simply $\mathbf{A}^{-1}$:

$$[\mathbf{v}_1, \quad \mathbf{v}_2, \quad \ldots, \quad \mathbf{v}_N] = \mathbf{A}^{-1} \qquad (6.1.14)$$

Even if, as is the case with elliptic problems, $\mathbf{A}$ itself is sparse, the inverse matrix $\mathbf{A}^{-1}$ is a full matrix. For even moderately sized problems, the storage requirements for $\mathbf{A}^{-1}$ may become unbearable.

Rather than explicitly computing $\mathbf{A}^{-1}$, we shall now look at two classes of methods for solving system (6.1.1), where the storage requirement in intermediate calculations remains reasonably small (for sparse matrices). These are iterative methods (see Section 6.2); and direct methods, known as "decomposition" or "double-sweep" methods, which are variations of the standard Gauss elimination, in which one stores intermediate matrices that may retain the characteristic sparsity of the original matrix whenever $\mathbf{A}$ is not only sparse but is *banded*.

6.1.4 LU DECOMPOSITION

As previously stated, this method is a more economical substitute for computing A^{-1}, typically to be used when $Au = b$ is to be solved several times for different values of b.

It proceeds as follows. Suppose that we find a way to *decompose* or *factor* the matrix A into the product of a lower-triangular matrix L and an upper-triangular matrix U, such that

$$LU = A \tag{6.1.15}$$

where

$$L = \begin{bmatrix} l_{1,1} & & & & 0 \\ l_{2,1} & l_{2,2} & & & \\ \cdot & & \cdot & & \\ \cdot & & & \cdot & \\ \cdot & & & & \cdot \\ l_{N,1} & & \cdots & & l_{N,N} \end{bmatrix} \tag{6.1.16}$$

and

$$U = \begin{bmatrix} m_{1,1} & m_{1,2} & \cdots & & m_{1,N} \\ & m_{2,2} & \cdots & & m_{2,N} \\ & & \cdot & & \cdot \\ & & & \cdot & \cdot \\ 0 & & & \cdot & \cdot \\ & & & & m_{N,N} \end{bmatrix} \tag{6.1.17}$$

Then the solution of system (6.1.1) can be achieved in two steps:

Step 1. Define the intermediate vector $y = \{y_i\}$ as the solution of

$$Ly = b \tag{6.1.18}$$

Step 2. Having thus computed y, it may be observed that u is the solution of

$$Uu = y \tag{6.1.19}$$

By way of proof, it suffices to multiply (6.1.19) by L:

$$LUu = Au = Ly = b \tag{6.1.20}$$

which is identical to (6.1.1).

One important element in this procedure is that since L and U are both triangular, the solution of (6.1.18) consists in a simple explicit forward substitution. Solve for y_1:

$$y_1 = \frac{b_1}{l_{1,1}}$$

Then solve for y_2:

$$y_2 = \frac{b_2 - l_{2,1}y_1}{l_{2,2}} \tag{6.1.21}$$

and so on, down to y_n.

The solution of (6.1.19) consists in a simple (explicit) backward substitution. Solve for u_N:

$$u_N = \frac{y_N}{m_{N,N}}$$

Then solve for u_{N-1}:

$$u_{N-1} = \frac{y_{N-1} - m_{N-1,N}y_N}{m_{N-1,N-1}} \tag{6.1.22}$$

and so on, up to u_1.

Thus, having derived and stored $\mathbf{L}$ and $\mathbf{U}$, the actual solution of (6.1.1) consists in two simple sweeps.

A second important aspect of this procedure is that when the matrix $\mathbf{A}$ is *banded*, then both of the matrices $\mathbf{L}$ and $\mathbf{U}$ are also banded, with a bandwidth which does not exceed that of the original matrix $\mathbf{A}$. This has important consequences in storage-space requirements for large systems. Recall that, in contrast to this, $\mathbf{A}^{-1}$ is *full* matrix.

Also, as we shall see, the LU decomposition procedure requires slightly less work than that required to form $\mathbf{A}^{-1}$ and compute $\mathbf{A}^{-1}\mathbf{b}$.

Forming L and U. The matrices $\mathbf{L}$ and $\mathbf{U}$ may be generated in the regular process of Gaussian elimination. Let

$$\mathbf{A} = \mathbf{A}^{(1)} = \begin{bmatrix} a^1_{1,1} & a^1_{1,2} & \cdots & a^1_{1,N} \\ a^1_{2,1} & a^1_{2,2} & & \vdots \\ \vdots & \vdots & & \vdots \\ \vdots & \vdots & & a^1_{N,N} \end{bmatrix} \tag{6.1.23}$$

be the original matrix, and denote by

$$\mathbf{A}^{(2)} = \begin{bmatrix} a^1_{1,1} & a^1_{1,2} & \cdots & a^1_{1,N} \\ 0 & a^2_{2,1} & & a^2_{2,N} \\ 0 & a^2_{3,1} & & \vdots \\ \vdots & \vdots & & \vdots \\ \vdots & \vdots & & a^2_{N,N} \end{bmatrix} \tag{6.1.24}$$

the transformed matrix after the first step of elimination, and generally after $(k-1)$ steps:

$$
\mathbf{A}^{(k)} =
\begin{bmatrix}
a^1_{1,1} & a^1_{1,2} & & & \cdots & & & a^1_{1,N} \\
0 & a^2_{2,2} & & & & & & a^2_{2,N} \\
0 & 0 & a^3_{3,3} & & \cdots & & & a^3_{3,N} \\
& & & \ddots & & & & \\
& & & & a^k_{k,k} & \cdots & & a^k_{k,N} \\
& & & & a^k_{k-1,k} & \cdots & & a^k_{k+1,N} \\
& & 0 & & \vdots & & & \vdots \\
& & & & a^k_{N,k} & \cdots & & a^k_{N,N}
\end{bmatrix}
\tag{6.1.25}
$$

Then, it may be verified that the matrices $\mathbf{L}$ and $\mathbf{U}$ are related to the factors generated above as

$$
\mathbf{L} =
\begin{bmatrix}
1 & & & & & \\
\dfrac{a^1_{2,1}}{a^1_{1,1}} & 1 & & & 0 & \\
\dfrac{a^1_{3,1}}{a^1_{1,1}} & \dfrac{a^2_{3,2}}{a^2_{2,2}} & 1 & & & \\
& \vdots & & \ddots & & \\
\dfrac{a^1_{N,1}}{a^1_{1,1}} & & & & 1
\end{bmatrix}
\tag{6.1.26}
$$

$$
\mathbf{U} =
\begin{bmatrix}
a^1_{1,1} & a^1_{1,2} & \cdots & & a^1_{1,N} \\
& a^2_{2,2} & & & a^2_{2,N} \\
& & \ddots & \\
& 0 & & & \\
& & & & a^N_{N,N}
\end{bmatrix}
\tag{6.1.27}
$$

That $\mathbf{LU} = \mathbf{A}$ can be verified by multiplication. The off-diagonal elements of $\mathbf{L}$ can be stored in the lower zero elements of $\mathbf{A}^{(k)}$ as they are formed.

As with all direct methods of solution, even if $\mathbf{A}$ is a *general* sparse matrix, $\mathbf{L}$ and $\mathbf{U}$ will be *full* (triangular) matrices. But if $\mathbf{A}$ is *banded*, then the bandwidth of $\mathbf{L}$ and $\mathbf{U}$ will not exceed that of $\mathbf{A}$, as will be discussed below. Since banded matrices occur naturally in many instances of finite-difference and finite-element approximations to elliptic equations, the **LU** decomposition offers an appealing solution method for those cases and is used frequently in large computer codes.

Operation Counts.   Consider first the case of a general N by N matrix **A**. To form **L** requires

$$1 + 2 + \ldots (N - 1) = \frac{N(N - 1)}{2} \tag{6.1.28}$$

divisions.

U is formed in $(N + 1)$ steps. In step 1 the last $(N - 1)$ elements of the first row of **A** are multiplied by a distinct factor $(a_{i,1}/a_{1,1})$ for each row and added to the corresponding elements in each of the other $(N - 1)$ rows. This requires $(N - 1)^2$ multiplications and $(N - 1)^2$ additions. This step is then applied successively on the smaller matrix formed by omitting the first row and column from the previous matrix and hence (Jolley, 1961, p. 5)

$$\sum_{k=1}^{N-1} k^2 = \frac{(N - 1)(N)(2N - 1)}{6} \tag{6.1.29}$$

multiplications and a similar number of additions are required. To solve **Ly** $=$ **b** requires $[N(N - 1)]/2$ multiplications and $[N(N - 1)]/2$ additions and to solve **Uu** $=$ **y** requires $[(N + 1)N]/2$ multiplications and $[N(N - 1)]/2$ additions. Hence to solve **Au** $=$ **b** using **LU** decomposition requires $(N^3/3) + O(N^2)$ multiplications and a similar number of additions. For $N = 5000$ as in our earlier example, this is clearly not feasible because of practical limitations on both time and storage.

6.1.5 DECOMPOSITION OF SYMMETRIC MATRICES (THE CHOLESKY METHOD)

If **A** is a symmetric positive-definite matrix, then the diagonal elements of U turn out to be all positive. Let **D** be the matrix formed by those diagonal elements

$$\mathbf{D} = \begin{bmatrix} a_{1,1}^1 & & & 0 \\ & a_{2,2}^2 & & \\ & & \ddots & \\ 0 & & & a_{N,N}^N \end{bmatrix} \tag{6.1.30}$$

The matrix

$$\mathbf{L}' = \mathbf{L}\mathbf{D}^{1/2} \tag{6.1.31}$$

is obtained by multiplying the nth line of $\mathbf{L}$ by $(a_{n,n}^n)^{1/2}$. The matrix

$$\mathbf{U}' = \mathbf{D}^{-1/2}\mathbf{U} \qquad (6.1.32)$$

is obtained by dividing the nth column of $\mathbf{U}$ by $(a_{n,n}^n)^{1/2}$. We may then observe [from equations (6.1.26) and (6.1.27)] that as before we have a decomposition of $\mathbf{A}$,

$$\mathbf{A} = \mathbf{L}'\mathbf{U}' \qquad (6.1.33)$$

and more interestingly, that

$$(\mathbf{L}')^T = \mathbf{U} \qquad (6.1.34)$$

(i.e., $\mathbf{L}'$ and $\mathbf{U}'$ are the transpose of one another). Thus, only one of the two matrices need be stored, cutting the storage requirement by about half. This method, called the *square-root* or *Cholesky method*, is very popular because many elliptic problems lead to matrices $\mathbf{A}$ which are symmetric-positive-definite.

6.1.6 BANDED MATRICES

In many important applications $\mathbf{A}$ is either banded or can be put in banded form with appropriate row and column interchanges: that is,

$$a_{i,j} = 0 \qquad \text{for } |i - j| > B$$

Here B is the bandwidth and those elements lying outside the $2B + 1$ diagonal bands are 0 (Figure 6.1.1). In this case the matrices $\mathbf{L}$ and $\mathbf{U}$ formed by the decomposition are also banded of width B, (Figure 6.1.2) and the

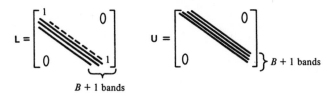

Figure 6.1.1

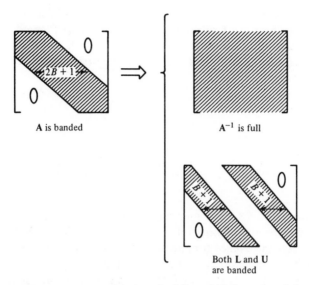

Figure 6.1.2 Denseness of A^{-1} and of L and U for a banded sparse matrix.

work of decomposing A into L and U is greatly reduced (i.e., it becomes linear in N if $B \ll N$). By contrast, the LU decomposition of a sparse, but not banded, matrix does not benefit from such an advantage (Figure 6.1.3).

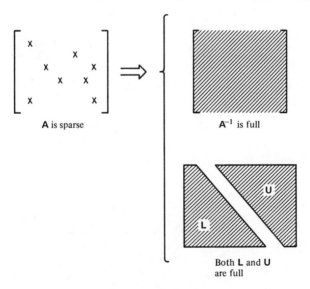

Figure 6.1.3 Denseness of A^{-1} and L and U for a general sparse matrix.

To form **L** requires $B(N - B - 1)$ divisions, which is easily derived by noting that there are 1's on the diagonal, which is of length N, leaving B bands to be computed, and each of these is of length N less the elements in the triangular area at each end (i.e., 2 for the first band, 4 for the second, etc.); see Figure 6.1.4. Hence,

$$NB - 2(1 + 2 + \ldots + B) = NB - 2\frac{B(B + 1)}{2}$$

elements must be computed.

To form **U** requires

$$B^2(N - B - 1) + \frac{B(B + 1)(2B - 1)}{6}$$

multiplications and a similar number of additions. This is derived by noting that for each step the elimination procedure is applied only to the B rows below the diagonal. Hence for the first $(N - B - 1)$ steps, only the B non-zero elements in the first row of the matrix (actually submatrix for the step) are multiplied by the appropriate factor for the row and added to the corresponding elements in the next B rows. For the last B steps the matrix is full and hence

$$B^2(B - 2)^2 + \ldots + 1 = \frac{B(B + 1)(2B - 1)}{6}$$

multiplications and additions are required (Figure 6.1.5).

To solve the lower-triangular system requires

$$\frac{B(B + 1)}{2} + B[N - (B + 1)] = B\left(N - \frac{B + 1}{2}\right)$$

additions and multiplications and similarly to solve the upper triangular system requires

$$(B + 1)\left(N - \frac{B - 1}{2}\right)$$

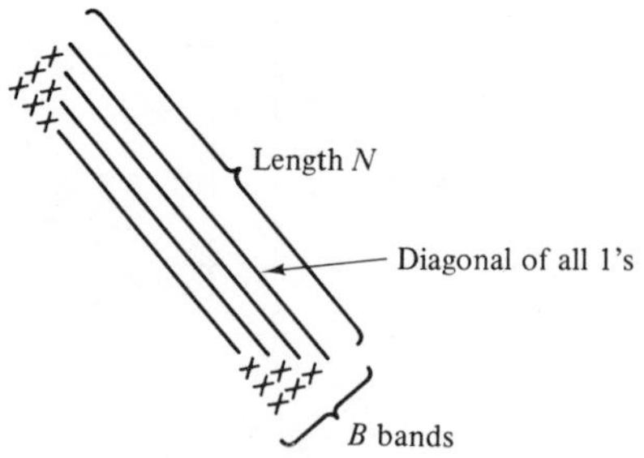

Length N

Diagonal of all 1's

B bands

Figure 6.1.4

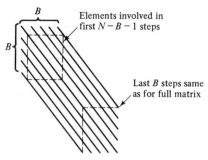

B

B

Elements involved in
first $N - B - 1$ steps

Last B steps same
as for full matrix

Figure 6.1.5

multiplications and

$$B\left(N - \frac{B+1}{2}\right)$$

additions. These can be derived in a manner similar to that used for the previous operational counts.

Hence, to solve the system $\mathbf{Au} = \mathbf{b}$ in the banded case requires $NB^2 + O(NB)$ multiplications and a similar number of additions, assuming that $B \ll N$.

6.1.7 IMPLEMENTATION

An algorithm for solving the linear system $\mathbf{Au} = \mathbf{b}$ when $\mathbf{A}$ is a banded matrix with bandwidth B is given next.

Form **LU**: For each column $j = 1, \ldots, N$

For each row $i = j + 1, \ldots, \min\{N, j + B\}$

$$a_{i,j} \longleftarrow \frac{a_{i,j}}{a_{j,j}}$$

For each column $k = j + 1, \ldots, \min\{N, j + B\}$

$$a_{i,k} \longleftarrow a_{i,k} - a_{i,j} * a_{j,k}$$

Solve $\mathbf{Ly} = \mathbf{b}$: $y_1 \longleftarrow b_1$

For each row $i = 2, \ldots, N$

$$y_i \longleftarrow b_i - \sum_{\substack{j=i-B \\ j \geq 1}}^{i=1} a_{i,j} * y_j$$

Solve $\mathbf{Uu} = \mathbf{y}$: $u_N \longleftarrow \dfrac{y_N}{a_{N,N}}$

For each row $i = N - 1, \ldots, 1$

$$u_i \longleftarrow \frac{1}{a_{i,i}}\left(y_i - \sum_{\substack{j=i-1 \\ j \leq N}}^{i+B} a_{i,j} * u_i\right)$$

Here we have assumed that the pivot elements $a_{j,j}^{(j)} \neq 0$ and that the elements $a_{i,j}$ are easily accessed.

Partial Pivoting. We note in passing that if row interchanges are performed so that the pivot element is the largest in the column, then the width of the band above the diagonal (B) will be at most doubled.

6.1.8 LU DECOMPOSITION FOR TRIDIAGONAL MATRICES

Band matrices are frequently stored as linear arrays with each array corresponding to one nonzero band.[2] The preceding algorithm becomes especially simple using this storage scheme in the special case where $\mathbf{A}$ is tridiagonal. In the following algorithm, $\mathbf{A}$ is assumed to be stored in the vectors $\mathbf{p}$, $\mathbf{q}$, and $\mathbf{r}$ corresponding to the lower, diagonal, and upper bands of $\mathbf{A}$.

Form $\mathbf{LU}$: For each column $j = 2, \ldots, N$

$$p_j \longleftarrow \frac{p_j}{v_{j-1}}$$

$$q_j \longleftarrow q_j - r_{j-1} * p_j$$

Solve $\mathbf{Ly} = \mathbf{b}$: $y_1 \longleftarrow b_1$

For each row $i = 2, \ldots, N$

$$y_i \longleftarrow b_i - p_i * y_{i-1}$$

Solve $\mathbf{Uu} = \mathbf{y}$: $u_N \longleftarrow \dfrac{y_N}{q_N}$

For each row $i = N - 1, \ldots, 1$

$$u_i \longleftarrow \frac{1}{q_i}(y_i - r_i * u_{i+1})$$

---

[2]See also Section 6.3.3.

A slightly different version of the **LU** decomposition of a tridiagonal system was given as an example in Section 4.1.

6.2 ITERATIVE SOLUTION METHODS

The methods described in Section 6.1 for the solution of systems of linear equations

$$\mathbf{Au} = \mathbf{b} \tag{6.2.1}$$

were *direct* methods, involving a fixed number of operations. Unlike these, *iterative methods* (to be described here) start from a first approximation, which is successively improved until a sufficiently accurate solution is obtained.

6.2.1 JACOBI[3] OR SIMULTANEOUS DISPLACEMENT METHOD

Let $a_{n,m}$ be the elements of **A** and b_n those of **b**. We assume that the diagonal elements are nonzero:

$$a_{n,n} \neq 0; \quad n = 1, 2, \ldots, N$$

(This is true for all regular systems, possibly at the cost of a reordering of the variables.) The nth line of (6.2.1) may be written as

$$a_{n,1}u_1 + a_{n,2}u_2 + \ldots + a_{n,N}u_N = b_n \tag{6.2.2}$$

Let

$$\mathbf{u}^k = \begin{bmatrix} u_1^k \\ u_2^k \\ \cdot \\ \cdot \\ \cdot \\ u_N^k \end{bmatrix} \tag{6.2.3}$$

be the vector of an approximate solution obtained at the end of the kth iteration cycle, starting from some initial estimate $\mathbf{u}^0$ in the manner described next.

---

[3]Carl Gustav Jacob Jacobi (1804–1851, German mathematician. Often labelled a "pure mathematician," he criticized Fourier for having been too "applied."

Having $\mathbf{u}^k$ (initially $\mathbf{u}^0$), we construct $\mathbf{u}^{k+1}$ (initially $\mathbf{u}^1$) as follows: Solve (6.2.2) for $u_1^{k+1}, u_2^{k+1}, \ldots, u_N^{k+1}$ as

Jacobi iteration:

$$\mathbf{u}_n^{k+1} = \frac{b_n - \sum_{i \neq n} a_{n,i} u_i^k}{a_{n,n}}; \qquad n = 1, 2, \ldots, N \qquad (6.2.4)$$

This procedure is repeated (for $k = 0, 1, 2, \ldots$) until it *converges*[4] to a stationary solution for which

$$\mathbf{u}^{k+1} - \mathbf{u}^k \simeq 0 \qquad (6.2.5)$$

according to some norm. It is of course easily verified that if for any k, $\mathbf{u}^{k+1} - \mathbf{u}^k = 0$, then $\mathbf{u}^{k+1} = \mathbf{u}^k = \mathbf{u}$; that is, $\mathbf{u}^k$ is *the* solution of (6.2.1).

The iterative process just described is called *Jacobi iteration* or the *simultaneous-displacements method*.

6.2.2 GAUSS–SEIDEL OR SUCCESSIVE-DISPLACEMENTS METHOD

If *new values* u_n^{k+1} are inserted in the remaining equations as they become available when (6.2.4) is solved line by line, then we have, instead of (6.2.4):

Gauss–Seidel iteration:

$$u_n^{k+1} = \frac{b_n - \sum_{i < n} a_{n,i} u_i^{k+1} - \sum_{i > n} a_{n,i} u_i^k}{a_{n,n}}; \qquad n = 1, 2, \ldots, N \qquad (6.2.6)$$

This iterative process is called *Gauss–Seidel iteration* or the *successive-displacements method*.

6.2.3 MATRIX NOTATION

We may decompose the matrix $\mathbf{A}$ into its diagonal, lower-off-diagonal and upper-off-diagonal parts:

$$\mathbf{A} = \mathbf{L} + \mathbf{D} + \mathbf{U} \qquad (6.2.7)$$

---

[4]If it converges, which is not always the case.

where

$$
L = \begin{bmatrix}
0 & & & & \\
a_{2,1} & 0 & & \text{\Large 0} & \\
a_{3,1} & a_{3,2} & 0 & & \\
& & & \ddots & \\
a_{N,1} & a_{N,2} & & a_{N,N-1} & 0
\end{bmatrix} ;
$$

$$
D = \begin{bmatrix}
a_{1,1} & & & \\
& a_{2,2} & & \text{\Large 0} \\
& & \ddots & \\
\text{\Large 0} & & & \ddots \\
& & & & a_{N,N}
\end{bmatrix} ; \qquad (6.2.8)
$$

$$
U = \begin{bmatrix}
0 & a_{1,2} & a_{1,3} & \cdots & a_{1,N} \\
& & a_{2,3} & \cdots & a_{2,N} \\
& & \ddots & & \\
\text{\Large 0} & & & \ddots & a_{N-1,N} \\
& & & & 0
\end{bmatrix}
$$

Then (6.2.4) may be expressed as:

Jacobi iteration:

$$
\mathbf{u}^{k+1} = \mathbf{D}^{-1}[\mathbf{b} - (\mathbf{L} + \mathbf{U})\mathbf{u}^k] \qquad (6.2.4a)
$$

Similarly, (6.2.6) may be written as:

Gauss–Seidel iteration:

$$
\mathbf{u}^{k+1} = (\mathbf{D} + \mathbf{L})^{-1}(\mathbf{b} - \mathbf{U}\mathbf{u}^k) \qquad (6.2.6a)
$$

6.2.4 GRADIENT METHOD (AND ITS RELATION WITH JACOBI'S METHOD)

Let us assume that $\mathbf{A}$ is a positive-definite matrix.[5] Consider then the real-valued quadratic function:

$$
Q(\mathbf{u}) = \mathbf{u}^T(\tfrac{1}{2}\mathbf{A}\mathbf{u} - \mathbf{b}) \qquad (6.2.9)
$$

---

[5] A matrix $\mathbf{A}$ is called positive-definite when the associated quadratic form $(\mathbf{u}^T\mathbf{A}\mathbf{u})$ is positive for all $|\mathbf{u}| \neq 0$.

One sees that Q attains its minimum when

$$\nabla Q \equiv \left\{\frac{\partial Q}{\partial u_n}\right\} = \mathbf{A}\mathbf{u} - \mathbf{b} = 0 \qquad (6.2.10)$$

(∇Q is a vector, the "gradient" of Q.) Thus, finding the solution of (6.2.1) is identical to finding the vector $\mathbf{u}$ which mimimizes $Q(\mathbf{u})$. Finding such a minimum iteratively may be carried out by the "gradient method," which has found applications in a variety of problems in applied science.

Let $\mathbf{u}^0$ be an initial guess of $\mathbf{u}$, and consider the following iteration for computing the sequence $\{\mathbf{u}^k\}$. Given a current approximate solution $\mathbf{u}^k$, the gradient of $Q(\mathbf{u})$ (a vector pointing in the direction of *steepest ascent* of Q) is

$$\nabla Q^k \equiv \left\{\frac{\partial Q}{\partial u_n}\right\}^k = \mathbf{A}\mathbf{u}^k - \mathbf{b} = \mathbf{r}^k \qquad (6.2.11)$$

where $\mathbf{r}^k$ is the residual vector of the system $\mathbf{A}\mathbf{u} - \mathbf{b}$ at $\mathbf{u} = \mathbf{u}^k$. Since Q increases in the direction of ∇Q^k, it is reasonable to choose

$$\mathbf{u}^{k+1} = \mathbf{u}^k - \alpha^k \, \nabla Q^k$$
$$= \mathbf{u}^k - \alpha^k \mathbf{r}^k \qquad (6.2.12)$$

where α^k is a positive number yet to be chosen. The displacement in (6.2.12) is in the direction of *steepest descent* of the quadratic function Q.

Suppose that $\mathbf{A}$ has been normalized by predividing each equation by $(a_{n,n})$; that is, all diagonal elements are $1's$. Then (6.2.12) with the choice

$$\alpha^k = \text{constant} = 1$$

becomes

$$\mathbf{u}^{k+1} = -(\mathbf{L} + \mathbf{U})\mathbf{u}^k + \mathbf{b} \qquad (6.2.13)$$

which is identical to Jacobi iteration.

Other values, including "optimal" choices for α, are discussed in Forsythe (1953), Stiefel (1955), Forsythe and Wasow (1960), and Forsythe and Moler (1967).

We may also make the following observation: when the matrix $\mathbf{A}$ is not sign-definite, then the quadratic form $Q(\mathbf{u})$ is stationary but does not attain an extremum when $\mathbf{A}\mathbf{u} = \mathbf{b}$. Since Jacobi's method (6.2.13) is a gradient or "steepest-slope" method, it will thus in general not converge when $\mathbf{A}$ is not sign-definite.

6.2.5 ACCELERATION
(OVERRELAXATION METHODS)

Starting from the basic Jacobi and Gauss–Seidel iterations, one may generate families of iterative procedures by inserting an additional parameter in the calculation with the intent of accelerating the rate of convergence[6]; the corresponding methods are called *overrelaxation* methods.

JACOBI METHOD WITH ACCELERATION

Rewrite the expression (6.2.4a) of Jacobi's iteration as

$$\mathbf{u}^{k+1} = \mathbf{u}^k + \{\mathbf{D}^{-1}[\mathbf{b} - (\mathbf{L} + \mathbf{u})\mathbf{u}^k] - \mathbf{u}^k\}$$
$$= \mathbf{u}^k + \mathbf{r}^k \tag{6.2.14}$$

where $\mathbf{r}^k$, the term in braces, $\{ \cdot \}$, is seen to be the correction in $\mathbf{u}$ in the $(k + 1)$st iteration cycle.

To generate a family of iterative procedures, we multiply this correction by the scalar $\bar{\omega}$, called the acceleration factor:

$$u^{k+1} = \mathbf{u}^k + \bar{\omega}\mathbf{r}^k$$
$$= \mathbf{u}^k + \bar{\omega}\{\mathbf{D}^{-1}[\mathbf{b} - (\mathbf{L} + \mathbf{U})\mathbf{u}^k] - \mathbf{u}^k\} \tag{6.2.15}$$

which may also be written as

$$\mathbf{u}^{k+1} = (1 - \bar{\omega})\mathbf{u}^k + \bar{\omega}\mathbf{D}^{-1}[\mathbf{b} - (\mathbf{L} + \mathbf{U})\mathbf{u}^k] \tag{6.2.15a}$$

When $\bar{\omega} = 1$, this expression is the Jacobi iteration or simultaneous-displacements method. When $\bar{\omega} > 1$, (6.2.15) is called overrelaxation, and when $\bar{\omega} < 1$, it is called underrelaxation. On a line-by-line basis, (6.2.15a) may be written as

$$u_n^{k+1} = (1 - \bar{\omega})u_n^k + \bar{\omega}\frac{b_n - \sum_{i \neq n} a_{n,i}u_i^k}{a_{n,n}} \tag{6.2.16}$$

We also note that (6.2.15) is equivalent to the following two steps:

$$\left. \begin{array}{l} 1. \quad \mathbf{u}_*^{k+1} = \mathbf{D}^{-1}[\mathbf{b} - (\mathbf{L} + \mathbf{U})\mathbf{u}^k] \\ 2. \quad \mathbf{u}^{k+1} = (1 - \bar{\omega})\mathbf{u}^k + \bar{\omega}\mathbf{u}_*^{k+1} \end{array} \right\} \tag{6.2.17}$$

---

[6]Note that the word "convergence" is used here with a quite different meaning from that defined in Sections 4.2 and 5.1.

where u_*^{k+1} is the normal Jacobi result. Similarly, on a line-by-line basis, we may write (6.2.17) as

$$
\left.
\begin{aligned}
&1. \quad u_{*,n}^{k+1} = \frac{b_n - \sum\limits_{i \neq n} a_{n,i} u_i^k}{a_{n,n}} \\
&2. \quad u_n^{k+1} = (1 - \bar{\omega}) u_n^k + \bar{\omega} u_{*,n}^{k+1}
\end{aligned}
\right\}
\qquad (6.2.17a)
$$

6.2.6   GAUSS–SEIDEL WITH ACCELERATION OR SUCCESSIVE OVERRELAXATION (SOR) METHOD

Of more interest is the Gauss–Seidel method with acceleration (the method is then called the *successive overrelaxation* or *SOR method*). Instead of (6.2.6), acceleration is effected after each line as

$$
\begin{aligned}
u_n^{k+1} &= u_n^k + \bar{\omega} \frac{b_n - \sum\limits_{i<n} a_{n,i} u_i^{k+1} - \sum\limits_{i>n} a_{n,i} u_i^k}{a_{n,n} - u_n^k} \\
&= (1 - \bar{\omega}) u_n^k + \bar{\omega} \frac{b_n - \sum\limits_{i<n} a_{n,i} u_i^{k+1} - \sum\limits_{i>n} a_{n,i} u_i^k}{a_{n,n}}
\end{aligned}
\qquad (6.2.18)
$$

which may be expressed in matrix form as

$$
\mathbf{u}^{k+1} = (1 - \bar{\omega}) \mathbf{u}^k + \bar{\omega} \mathbf{D}^{-1} (\mathbf{b} - \mathbf{L} \mathbf{u}^{k+1} - \mathbf{U} \mathbf{u}^k)
\qquad (6.2.19)
$$

or

$$
\mathbf{u}^{k+1} = (\mathbf{I} + \bar{\omega} \mathbf{D}^{-1} \mathbf{L})^{-1} \{[(1 - \bar{\omega}) \mathbf{I} + \bar{\omega} \mathbf{D}^{-1} \mathbf{U}] + \bar{\omega} \mathbf{D}^{-1} \mathbf{b}\}
\qquad (6.2.20)
$$

6.2.7   HISTORICAL

Iterative methods of the kind described here were first devised for pencil-and-paper computing, originally under such names as "method of successive corrections" and the like. Its first published accounts are those of Gauss (1823), who described a basic method; Seidel (1874), who proved its convergence for linear systems with positive definite matrices; Jacobi (1845), who described the method of simultaneous displacements as a "method of least squares for degenerate linear equations"; and Cross (1932), who devised iterative methods for engineering calculations.

The name "relaxation" was given by Southwell in the 1930s. The justification for this name is best described by an example. Let

$$
-\frac{u_{n-1} - 2u_n + u_{n+1}}{h^2} = f_n
\qquad (6.2.21)
$$

be the discretized form of the equation for the position of a loaded, taut string. With an approximation $\mathbf{u}^k$ to the solution of (6.2.21), we may write for each n:

$$r_n^k = -\frac{u_{n-1}^k - 2u_n^k + u_{n+1}^k}{h^2} - f_n \tag{6.2.22}$$

These residuals have the same dimension as the f_n; that is, they are localized spurious forces or loads which result from the inaccuracy of $\mathbf{u}^k$. When we proceed to improve the $\{u_n^k\}$ iteratively by changing one u_n^k at a time as

$$u_n^{k+1} = \frac{u_{n-1}^k + u_{n+1}^k + h^2 f_n}{2} \tag{6.2.23}$$

then we have in effect removed the spurious force r_n^k or "relaxed" the solution. When all r_n^k are zero, then the corresponding $\mathbf{u}^k$ is the intended solution of (6.2.21).

Other than Southwell and Cross, the names of Richardson and Liebmann are attached to the development of iterative or relaxation methods in the first decades of the twentieth century.

6.2.8   CONVERGENCE RATES

We shall now discuss *convergence* rates for those iterative methods. Here, convergence is the property that the error

$$\boldsymbol{\epsilon}^k = \mathbf{u}^k - \mathbf{u} \tag{6.2.24}$$

(where $\mathbf{u}$ is the exact solution of $\mathbf{Au} = \mathbf{b}$) tends to zero as $k \longrightarrow \infty$. For linear processes, this property is independent of the initial vector $\mathbf{u}^0$.

Analysis of convergence is an important concern, because there is no a priori indication that any of these methods should converge at all. Moreover, we shall see that the rate of convergence for methods with acceleration depends (as expected) on the acceleration factor $\bar{\omega}$, and that one can accordingly choose it optimally.

We may express both Jacobi and Gauss–Seidel iterations in the common form

$$\mathbf{u}^{k+1} = \mathbf{B}\mathbf{u}^k + \mathbf{c} \tag{6.2.25}$$

where $\mathbf{B}$ is for the basic methods:

Jacobi:

$$\mathbf{B_J} = -\mathbf{D}^{-1}(\mathbf{L} + \mathbf{U}) \tag{6.2.26}$$

Gauss–Seidel:

$$\mathbf{B_{GS}} = -(\mathbf{D} + \mathbf{L})^{-1}\mathbf{U} \tag{6.2.27}$$

and for the Gauss–Seidel method with over relaxation (SOR):

$$\mathbf{B}_{\bar{\omega}} = (\mathbf{I} - \bar{\omega}\mathbf{D}^{-1}\mathbf{L})^{-1}[(1 + \bar{\omega})\mathbf{I} + \bar{\omega}\mathbf{D}^{-1}\mathbf{U}] \qquad (6.2.28)$$

B is called the iteration matrix for the corresponding procedure.

A relation between the error in successive approximations can be derived by subtracting from (6.2.25) the equation

$$\mathbf{u} = \mathbf{Bu} + \mathbf{c} \qquad (6.2.29)$$

which is satisfied by **u**. We thus obtain

$$\begin{aligned}
\boldsymbol{\epsilon}^k &= \mathbf{B}\boldsymbol{\epsilon}^{k-1} \\
&= \mathbf{B}^2\boldsymbol{\epsilon}^{k-2} \\
&\qquad \cdot \\
&\qquad \cdot \\
&\qquad \cdot \\
&= \mathbf{B}^k\boldsymbol{\epsilon}^0
\end{aligned} \qquad (6.2.30)$$

Now, let **B** have eigenvalues $\lambda_1, \lambda_2, \ldots, \lambda_N$ and assume that the corresponding eigenvectors $\mathbf{v}^0, \mathbf{v}^1, \ldots, \mathbf{v}^N$ are linearly independent. Then we can expand the initial error as

$$\boldsymbol{\epsilon}^0 = \gamma_1\mathbf{v}^1 + \gamma_2\mathbf{v}^2 + \ldots + \gamma_N\mathbf{v}^N \qquad (6.2.31)$$

and thus

$$\boldsymbol{\epsilon}^k = \gamma_1\lambda_1^k\mathbf{v}_1 + \gamma_2\lambda_2^k\mathbf{v}^2 + \ldots + \gamma_N\lambda_N^k\mathbf{v}^k \qquad (6.2.32)$$

From this it follows that the iteration will converge from an arbitrary initial $\mathbf{u}^0$ if and only if the eigenvalues of **B** satisfy

$$|\lambda_n| < 1; \qquad n = 1, 2, \ldots, N \qquad (6.2.33)$$

It can be shown (Varga, 1962) that this result holds without the assumption of independence on the eigenvectors. We thus have the following theorem:

THEOREM A necessary and sufficient condition for the iterative method $\mathbf{u}^{k+1} = \mathbf{Bu}^k + \mathbf{c}$ to converge for an arbitrary starting approximation $\mathbf{u}^0$ is that

$$\rho(\mathbf{B}) = \max_n |\lambda_n| < 1 \qquad (6.2.34)$$

In classical linear algebra, $\rho(\mathbf{B})$ is called the *spectral* radius of the matrix $\mathbf{B}$.

One defines quantitatively the rate of convergence of (6.2.25) as follows. For k large, the expression (6.2.32) will be dominated by the eigenvalue of largest modulus. To reduce the magnitude of the corresponding term in (6.2.32),

$$\gamma_i \lambda_i^k \tag{6.2.35}$$

by a factor of 10, k must be the smallest integer for which

$$|\lambda_i^k| = [\rho(\mathbf{B})]^k \leq 10^{-1} \tag{6.2.36}$$

that is,

$$k \geq \frac{1}{-\log_{10} \rho(\mathbf{B})} \tag{6.2.37}$$

The positive quantity

$$R = -\log_{10} \rho(\mathbf{B}) \tag{6.2.38}$$

is called the *asymptotic rate of convergence* (Young, 1954a) of the iterative procedure (6.2.25). Its inverse is equal to the number of iterations needed (asymptotically) to reduce the residual error by a factor of 10.

6.2.9 COMPARISON OF RATES OF CONVERGENCE

We shall now establish the following important property:

THEOREM   For the iteration matrix in the SOR method, we have

$$\rho(\mathbf{B}_{\bar\omega}) \geq |\bar\omega - 1| \tag{6.2.39}$$

So the method can only converge for $0 < \bar\omega < 2$.

Proof.   We use the standard result from matrix theory that the product of the eigenvalues of a matrix is equal to its determinant:

$$\lambda_1 \lambda_2 \ldots \lambda_N = \det \mathbf{B}_{\bar\omega}$$
$$= \det (\mathbf{I} - \bar\omega \mathbf{D}^{-1} \mathbf{L}) \det [(1 + \bar\omega)\mathbf{I} + \bar\omega \mathbf{D}^{-1} \mathbf{U}] \tag{6.2.40}$$

Both matrices appearing in the above are triangular (the inverse of a triangular matrix is also a triangular matrix); and the determinant of a triangular matrix is equal to the product of its diagonal elements. Thus,

$$\lambda_1\lambda_2 \ldots \lambda_N = (1 - \bar{\omega})^N \qquad (6.2.41)$$

whence

$$\max_i |\lambda_i| \geq |1 - \omega| \qquad (6.2.42)$$

which proves the theorem.   □

Further results and optimal values for $\bar{\omega}$ can only be derived by being more specific about the nature of the matrix $\mathbf{A}$ of the problem (6.2.1). One simple result follows.

Definition 1.   A matrix $\mathbf{A} = \{a_{n,m}\}$ is called *diagonally dominant* if

$$\sum_{m \neq n} |a_{n,m}| \leq |a_{n,n}|; \qquad n = 1, 2, \ldots, N \qquad (6.2.43)$$

with strict inequality for at least one n.

Definition 2.   If a matrix $\mathbf{A}$ can be reduced by row, column permutations to the form

$$\mathbf{B} = \begin{bmatrix} \mathbf{A}_1 & 0 \\ \mathbf{A}_2 & \mathbf{A}_3 \end{bmatrix} \qquad (6.2.44)$$

then $\mathbf{A}$ is called *reducible*.

If $\mathbf{A}$ is reducible with $\mathbf{A}_1$ of order $M < N$, then this means that in the linear system with matrix (6.2.44),

$$\mathbf{B}\mathbf{v} = Q$$

the first M equations can be partitioned out and solved independently of the other equations.

A matrix $\mathbf{A}$ that is not reducible is called *irreducible*.

THEOREM   (Collatz, 1950) If $\mathbf{A}$ is diagonally dominant and is not reducible, then the Jacobi method of iteration converges.

It will be sufficient to prove, by (6.2.33), that all eigenvalues of the iteration matrix

$$\mathbf{B}_J = -\mathbf{D}^{-1}(\mathbf{L} + \mathbf{U}) \qquad (6.2.45)$$

satisfy $|\lambda_n| < 1$. The characteristic equation of $\mathbf{B}_J$ may be written as

$$\det(\mathbf{L} + \lambda\mathbf{D} + \mathbf{U}) = 0 \qquad (6.2.46)$$

With any $|\lambda| \geq 1$, the matrix $(\mathbf{L} + \lambda\mathbf{D} + \mathbf{U})$ is clearly diagonally dominant and irreducible. A classical theorem of linear algebra states that this implies that $\det(\mathbf{L} + \lambda\mathbf{D} + \mathbf{U}) \neq 0$. Thus, for λ to be a root of (6.2.46), we must have $|\lambda| < 1$. ☐

Another significant property is stated in the following:

THEOREM (Reich, 1949) If $\mathbf{A}$ is a symmetric, nonsingular matrix and all $a_{n,n} > 0$, then the Gauss–Seidel method of iteration converges for all initial vectors $\mathbf{u}^0$ if and only if $\mathbf{A}$ is positive-definite.

Proofs may be found in Reich (1949) and in Forsythe and Wasow (1960, p. 237).

6.2.10 THE YOUNG–FRANKEL THEORY OF SOR

An analysis of the conditions under which SOR methods are profitable over the basic Jacobi (simultaneous displacements) and Gauss–Seidel (successive displacement) methods was the subject of Young's dissertation of 1950 (Young, 1954a). Much of the same theory had been given by Frankel (1950) for the special case of Laplaces's difference equation over a rectangle. A complete account of this theory has been given recently in Young (1971). An earlier but enlightening discussion may be found in Forsythe and Wasow (1960, pp. 242 et seq.).

In brief, Young's answer to the question of when overrelaxation pays is that it pays very well for a class of matrices which possess what he calls property (A), a property associated with many boundary-value problems for elliptic difference operators.

Describing this theory is somewhat beyond the scope of this book. We shall simply state without proof the following important results due to Young:

Definition 1. A square matrix **A** of order N is said to be *block-tridiagonal* if it is of the form

$$
\mathbf{A} =
\begin{bmatrix}
\mathbf{D}_1 & \mathbf{U}_1 & & & \mathbf{0} \\
\mathbf{L}_2 & \mathbf{D}_2 & \mathbf{U}_2 & & \\
& \mathbf{L}_3 & \mathbf{D}_3 & \cdot & \\
& & \cdot & \cdot & \cdot \\
\mathbf{0} & & & \cdot & \cdot
\end{bmatrix}
\qquad (6.2.47)
$$

where the $\mathbf{D}_i$, $\mathbf{L}_i$, and $\mathbf{U}_i$ are square matrices of order $m \geq 2$.

Definition 2. If **A** is block-tridiagonal and if, moreover, each $\mathbf{D}_i$ is a *diagonal* matrix, then **A** will be called *diagonally-block-tridiagonal*.

Definition 3. A square matrix **A** of order N is said to have *property* (A) if it can be transformed to diagonally-block-tridiagonal form by permutation of rows and columns.

If **A** has property (A), there ordinarily exist more than one diagonally-block-tridiagonal form into which it can be transformed.

THEOREM (Young)   Let **A** be of diagonally-block-tridiagonal form. Then,

$$
\rho(\mathbf{B}_{\text{GS}}) = \rho^2(\mathbf{B}_{\text{J}}) \qquad (6.2.48)
$$

This theorem relates the spectral radius of the Jacobi (simultaneous displacements) and Gauss–Seidel (successive displacements) methods. In particular, if **A** is symmetric-positive-definite, then by Reich's theorem and by (6.2.48), *both* Jacobi and Gauss–Seidel methods are convergent.

THEOREM   Let **A** by symmetric, positive definite, and of diagonally-block-tridiagonal form. Then the optimal relaxation factor for Gauss–Seidel SOR iteration is given by

$$
\bar{\omega}_{\text{opt}} = \frac{2}{1 + [1 - \rho(\mathbf{B}_{\text{GS}})]^{1/2}} \qquad (6.2.49)
$$

and the corresponding optimal spectral radius is

$$
\rho(\mathbf{B}_{\bar{\omega}})_{\text{opt}} = \bar{\omega}_{\text{opt}} - 1 \qquad (6.2.50)
$$

These properties remain true for matrices that are not diagonally-block-tridiagonal but have property (A), provided that the order in which the iteration is carried out line by line is *consistent* with the tridiagonal representation (6.2.47) of the matrix.

6.2.11   OTHER ITERATIVE METHODS

Other iterative methods, more specialized than those discussed in this section, have been developed and used successfully for elliptic equations. Among those is the alternating directions implicit (ADI) class of methods (Peaceman and Rachford, 1955) the "splitting" techniques, developed mostly in the USSR: see, for example, Janenko (1967), and, more recently, multi-level adaptive methods, in which meshes of different fineness are used alternately to increase rates of convergence (see, e.g., Brandt, 1972, 1976, and Gaur and Brandt, 1977).

6.3   SPARSE MATRICES

The formulation of finite-difference and finite-element approximations to elliptic problems results in systems of linear equations which are *large* as a rule (i.e., the number of nodal unknowns is a large number). As an example, a discrete approximation of Poisson's equation in a three-dimensional domain covered by a grid of 20 by 20 by 20 points results in the formulation of 8000 (linear) equations in 8000 unknowns! The problem's matrix is formally 8000 by 8000 and would require 64×10^6 memory locations if stored as a regular, full square matrix. Among other inconveniences, this makes it impossible to bring the matrix at once into the working memory of *any* of existing large computers.

But the matrix associated with the equations resulting from the finite-difference or finite-element approximation of elliptic equations are typically *sparse:* by this it is meant that many of the elements of the matrix are zeros. If, for example, we approximate Poisson's equation in three dimensions with finite differences using the simple seven-point formula

$$\nabla^2 u_{m,n,k} = \frac{1}{h^2} \left(\begin{array}{c} +1 \\ +1 \\ +1 \quad -6 \quad +1 \\ +1 \\ +1 \end{array} \right) u_{m,n,k} = f_{m,n,k}$$

then only about $7 \times 8000 \simeq 56{,}000$ of the matrix's elements are nonzero. Whereas storing in a computer's memory the full 8000 by 8000 matrix was out of the question, storing 56,000 words is within the reach of most medium-size machines.

This brings to the fore a fundamental problem in the computer solution of (moderately large) elliptic equations: that of the need for methods that economize computer memory in storing large sparse matrices.

6.3.1 PACKED FORM OF STORAGE OF SPARSE MATRICES

The only practical way to store large sparse matrices in a computer is in *packed form*, in which only (or almost only) the nonzero elements are stored. We shall at first look at *banded matrices*, for which economical forms of storage suggest themselves naturally. Then we shall look at *general sparse matrices*, for which special methods of storage must be devised.

6.3.2 BANDED MATRICES

A desirable form for sparse matrices, both for storage and solution, is as a *limited bandwidth* or *banded* matrix: If all elements $a_{i,j}$ that are away from the main diagonal by more than B positions are zero, that is, if

$$a_{ij} = 0 \qquad \text{if } |i - j| > B$$

then the matrix $\mathbf{A} \equiv \{a_{i,j}\}$ is called *banded*, with bandwidth $(2B + 1)$; see Figure 6.3.1.

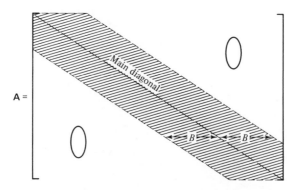

Figure 6.3.1 Banded matrix of bandwidth $(2B + 1)$.

6.3.3 TRIDIAGONAL MATRICES

Tridiagonal matrices are simple examples of banded matrices, with $B = 1$. They occur, for instance, in the discrete (finite-difference or finite-element) approximation of two-point boundary-value problems and in the implicit approximation of parabolic and hyperbolic equations. For example, in Section 1.2 we have approximated the second-order equation (1.2.13) by the system

$$\mathbf{Au} = \mathbf{b} \tag{6.3.1}$$

where $\mathbf{A}$ is the tridiagonal matrix

$$\mathbf{A} \equiv \frac{1}{h^2} \begin{bmatrix} -2 & 1 & & & & & \\ 1 & -2 & 1 & & & \mathbf{0} & \\ & 1 & -2 & 1 & & & \\ & & & \ddots & \ddots & \ddots & \\ & & & & \ddots & \ddots & 1 \\ \mathbf{0} & & & & & 1 & -2 \end{bmatrix} \tag{6.3.2}$$

A convenient method of computer storage for such matrices is (in FORTRAN notations)

$$\boxed{\text{DIMENSION} \quad A(3, N)}$$

which corresponds to

$$\mathbf{A} \equiv \begin{bmatrix} A(2, 1) & A(3, 1) & & & \\ A(1, 2) & A(2, 2) & A(3, 2) & & \\ & A(1, 3) & A(2, 3) & A(3, 3) & \\ & & A(1, 4) & A(2, 4) & \cdot & \cdot \\ & & & \cdot & \cdot & \cdot & \cdot \\ & & & & & & \text{etc.} \end{bmatrix} \tag{6.3.3}$$

If, moreover, we know that the matrix is symmetric, then we may delete the third diagonal from storage:

$$\text{DIMENSION} \quad A(2, N)$$

which, save for the omission of one diagonal, is to be interpreted as above.

6.3.4 BLOCK-TRIDIAGONAL MATRICES

Block-tridiagonal matrices are matrices of the form (6.3.2), but where the elements are themselves m by m square matrices (with m generally a small number). An example of the occurrence of a block-tridiagonal matrix is described in the following.

EXAMPLE

Consider the simple elliptic problem

$$-\nabla^2 U = F(x, y) \qquad (6.3.4)$$

in a two-dimensional domain. Whereas two-point boundary-value problems (which are one-dimensional elliptic problems) do lead to systems of linear equations with symmetric, *tridiagonal* matrices, it is not evident that a similar property shall apply to the two-dimensional case. In fact, similar properties of symmetry may apply if the ordering of nodes is done correctly.

To illustrate this, consider equation (6.3.4) to be solved over a rectangular domain. We assume for simplicity that $\Delta x = \Delta y = h$ and that (6.3.4) is approximated by the simple five-point formula

$$-\frac{u_{m+1,n} + u_{m-1,n} + u_{m,n+1} + u_{m,n-1} - 4u_{m,n}}{h^2} = f_{m,n};$$

$$m = 1, 2, \ldots, M; \quad n = 1, 2, \ldots, N \qquad (6.3.5)$$

This system of equations may be written in matrix form as

$$A\{u_{m,n}\} = \{f_{m,n}\} \qquad (6.3.6)$$

where, in spite of the double index, $\{u_{m,n}\}$ and $\{f_{m,n}\}$ are one-dimensional vectors.

When the ordering is chosen "one x line at a time," in the manner illustrated in Figure 6.3.2, then the structure of **A** is *block-tridiagonal*, with blocks of size $N \times N$, respectively. Indeed, the reader may easily verify that, given such a one-line-at-a-time ordering of the nodes, the following

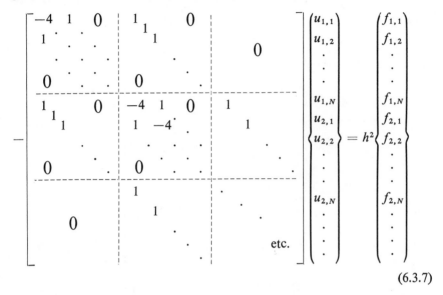

$$(6.3.7)$$

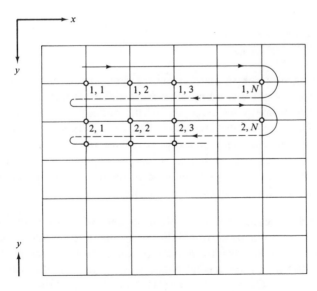

Figure 6.3.2   Line-by-line ordering of points in a rectangular domain.

is equivalent to (6.3.5). The matrix $\mathbf{A}$ may be written as

$$
\mathbf{A} \equiv - \begin{vmatrix} \mathbf{B} & \mathbf{I} & & \mathbf{0} \\ \mathbf{I} & \mathbf{B} & \mathbf{I} & \\ & . & . & . \\ \mathbf{0} & & . & . & . \end{vmatrix} \tag{6.3.8}
$$

where $\mathbf{B}$ and $\mathbf{I}$ are the $N \times N$ matrices

$$
\mathbf{B} \equiv \begin{bmatrix} -4 & 1 & & & \mathbf{0} \\ 1 & -4 & & 1 & \\ & 1 & . & & . \\ & & . & . & & . \\ & & & . & . & & . \\ \mathbf{0} & & & & . \end{bmatrix} ;
$$

$$
\mathbf{I} = \begin{bmatrix} 1 & & & & \mathbf{0} \\ & . & & & \\ & & . & & \\ & & & . & \\ \mathbf{0} & & & & 1 \end{bmatrix} \tag{6.3.9}
$$

6.3.5 GENERAL PACKING SCHEMES FOR SPARSE MATRICES

The storage of sparse matrices where symmetry of the kind found in banded matrices is not present requires more elaborate techniques. Whereas in the storage schemes considered so far, the indices of an element are implicit in the location of the element in the array where it is stored, such information is not available for truly sparse matrices. It is then the case that one must store, with each element, some explicit *address information* from which the element's indices may be recovered. An example of such a method is the "profile storage scheme," which we describe next.

Profile Storage Scheme. It is often the case with finite-difference and finite-element approximations of elliptic problems that the matrix $\mathbf{A}$ is symmetric, with limited bandwidth, *but the bandwidth differs greatly from row to row.* An efficient storage scheme that applies here is the following *profile storage scheme* (see George, 1971, and Dahlquist and Bjorck, 1974).

The lower triangle of the symmetric matrix is stored row by row in a one-dimensional array. In each row, all elements *from the first nonzero entry to the diagonal* are stored (including the zeros contained between these two elements).

The storage consists of two one-dimensional arrays, one of them being the vector of *values* $\{v_i\}$, the other a vector of *pointers* $\{p_j\}$. *The pointer* p_j *(an integer) is the column location of the first nonzero element in the j*th *row.*

The scheme is best illustrated by an example. Let

$$\mathbf{A} = \begin{bmatrix} 21 & 22 & 0 & 0 & 0 & 0 \\ 22 & 23 & 24 & 25 & 0 & 0 \\ 0 & 24 & 6 & 0 & 0 & 0 \\ 0 & 25 & 0 & 7 & 0 & 0 \\ 0 & 0 & 0 & 0 & 28 & 17 \\ 0 & 0 & 0 & 0 & 17 & 39 \end{bmatrix} \tag{6.3.10}$$

Then the two arrays are:

Values: $\quad \mathbf{V} = \{v_i\} = (21, 22, 23, 24, 6, 25, 0, 7, 28, 17, 39)$

$\qquad\qquad\qquad\quad \uparrow \ \uparrow \qquad\quad \uparrow \qquad \uparrow \qquad\quad \uparrow \ \uparrow$

Pointers: $\quad \mathbf{P} = \{p_j\} = (\ 1, \ 1, \qquad 2, \qquad 2, \qquad\quad 5, \ 5 \ )$

An interesting property of the profile storage scheme is that when **LU** decomposition (without pivoting) is applied (or more precisely the Cholesky [square-root] version of **LU** decomposition), then the reduced matrix **L** may be stored exactly in **V**, and the zeros that were originally unused are now filled in this process. The **LU** decomposition may thus be effected in place, without requiring additional storage. The only "overhead" storage needed is that for the N (integer) pointers $\{p_j\}$.

Linked Lists. A common form of storage for sparse matrices is that using "linked lists"—several versions of such techniques are described in the specialized literature on sparse matrices—for example, Rose and Willoughby (1972), Tewarson (1973), and Baker (1976).

6.4 NONLINEAR EQUATIONS AND QUASILINEARIZATION

Nonlinear elliptic problems are generally solved by transforming them into an *iterative* sequence of linear problems which can each be solved by the standard solution methods that are available for linear elliptic systems.

The transformation of a nonlinear problem into a sequence of linear problems is sometimes referred to as "quasilinearization." Some of the basic methods that fall in this category will be described in this section.

EXAMPLE

A simple example will be useful to clarify basic ideas before analyzing solution methods. Suppose that we wish to solve the simple *nonlinear* two-point boundary-value problem (remember that all cases treated in Chapter 4 were *linear*)

$$-P(U)\frac{d^2U}{dx^2} = F(x)$$

$$U(0) = U(l) = 0$$

$$(6.4.1)$$

The nonlinearity here stems from the presence of U in P. If we follow the finite-difference procedure of Section 4.1 to approximate this equation, we obtain

$$-P(u_n)\left(\frac{u_{n-1} - 2u_n + u_{n+1}}{h^2}\right) = F(x_n);$$

$$n = 1, 2, \ldots, N; \quad u_0 = u_{N+1} = 0 \qquad (6.4.2)$$

where

$$\{u_n\} \equiv \mathbf{u} \simeq \{U(x_n)\} \qquad (6.4.3)$$

is the desired numerical solution. But (6.4.2) is now a system of *nonlinear* algebraic equations, to which the solution techniques of Sections 6.1 and 6.2 *do not* apply. However, if we accept to solve (6.4.2) iteratively, we may still be able to use the techniques of Sections 6.1 and 6.2, which were devised for linear problems. Let

$$\mathbf{u}^k \equiv \begin{Bmatrix} u_1^k \\ u_2^k \\ \cdot \\ \cdot \\ \cdot \\ u_N^k \end{Bmatrix}; \quad k = 0, 1, 2, \ldots \qquad (6.4.4)$$

be successive approximations to $\mathbf{u}$, starting from some first guess $\mathbf{u}^0$. There is more than one way to build an iterative scheme. For instance, the successive iterates $\mathbf{u}^k$ may be defined as the solution of

$$-P(u_n^k)\left(\frac{u_{n-1}^{k+1} - 2u_n^{k+1} + u_{n+1}^{k+1}}{h^2}\right) = F(x_n); \quad n = 0, 1, 2, \ldots, N \qquad (6.4.5)$$

which is a system of *linear* equations in $\mathbf{u}^{k+1} = \{u_n^{k+1}\}$ [since the u_n^k appearing as arguments of $P(\cdot)$ are known].

There is, of course, the question of how fast, or even whether, the iteration converges. Analyzing this is beyond our scope in this example. We simply wish to point out that if P were not dependent on U (i.e., $P = $ constant), then the problem would be linear, and (6.4.5) would produce the *exact* numerical solution $\mathbf{u}$ at the end of the first iteration. What this suggests is that the speed of convergence of the iteration is somehow related to how little P depends on U (or, in other terms, how *weakly* nonlinear the problem is).

6.4.1 BASIC ITERATIVE METHODS

Some of the basic iterative methods will be described here. What all have in common is that each step of the iteration is a set of *linear* equations in the next iterate, $\mathbf{u}^{k+1} = \{u_n^{k+1}\}$. One form in which the nonlinear equations may appear is

$$A(\mathbf{u})\mathbf{u} = \mathbf{b}(\mathbf{u}) \qquad (6.4.6)$$

Here $\mathbf{u}$ is the N vector of unknowns $\{u_n\}$, $\mathbf{b}$ is an N vector, and $\mathbf{A}$ is an N by N matrix whose elements are algebraic functions of the unknowns. Equation (6.4.2) in the preceding example falls in this category. A simple iterative procedure that applies to this case is

$$A(\mathbf{u}^k)\mathbf{u}^{k+1} = \mathbf{b}(\mathbf{u}^k); \qquad k = 0, 1, 2, \ldots \qquad (6.4.7)$$

starting from some initial guess $\mathbf{u}^0$. An example is given by (6.4.5). If $\mathbf{A}$ and $\mathbf{b}$ were constant (i.e., if the problem were linear), then, as previously remarked, the solution $\mathbf{u}$ would be obtained in one iteration step. It is often the case in the discrete approximation of elliptic equations that $\mathbf{A}$ and $\mathbf{b}$ are only weakly dependent on $\mathbf{u}$. The procedure (6.4.7) may then be expected to converge rapidly.

Another form in which the nonlinear equations may appear is

$$\mathbf{u} = G(\mathbf{u}) \qquad (6.4.8)$$

Here $\mathbf{G}$ is an N vector of functions of $\mathbf{u}$.

The simplest procedure for finding a solution of this system of equations is known as *fixed-point iteration*, also called sometimes *functional iteration* and *Picard iteration*. It proceeds as follows. From some initial guess u^0 at the solution, the sequence of iterates $\{u^k; k = 1, 2, \ldots\}$ is defined by the relation

$$u^{k+1} = G(u^k) \tag{6.4.9}$$

Although simple, this scheme is occasionally quite practical. Convergence of this procedure is governed by the *contraction mapping theorem*.

THEOREM   If $G(u)$ satisfies the condition (called the Lipschitz condition)

$$\| G(u) - G(v) \| \leq \lambda \,|u - v|; \qquad |\lambda| < 1 \tag{6.4.10}$$

according to some norm $\| \cdot \|$, then the sequence $\{u^k\}$ converges to the solution.

The formal proof may be found in many textbooks in numerical analysis (e.g., Isaacson and Keller, 1966). A simple interpretation of condition (6.4.10) may be derived geometrically for a single equation (Figure 6.4.1).

Yet another form of the equations may be:

$$F(u) = 0 \tag{6.4.11}$$

If $B(u)$ is any Nth-order matrix, then we may transform (6.4.11) into the form (6.4.8) by

$$G(u) = u - B(u)F(u) \tag{6.4.12}$$

(Verification of this is trivial.) A particularly effective procedure of this form is known as *Newton's* method or the *Newton–Raphson* method. It consists in taking $B = J^{-1}$, where $J(u)$ is the Jacobian matrix

$$J = \frac{\partial F}{\partial u} = \left\{ \frac{\partial F_m}{\partial u_n} \right\} \tag{6.4.13}$$

The iteration is then

$$u^{k+1} = u^k - J^{-1}(u^k)F(u^k) \tag{6.4.14}$$

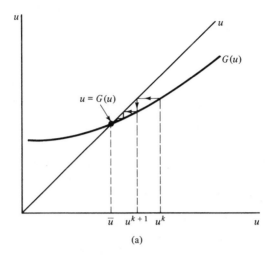

(a)

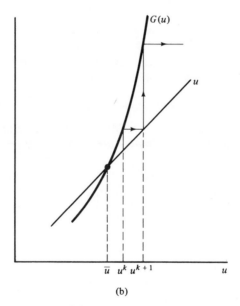

(b)

Figure 6.4.1 (a) Graphical construction of fixed-point iteration u^{k+1} $= G(u^k)$, convergent case. The mean slope $\lambda = \dfrac{|G(u) - G(v)|}{|u - v|}$ is everywhere < 1; (b) divergent case, the mean slope $\lambda = \dfrac{|G(u) - G(v)|}{|u - v|}$ is > 1 near the solution u.

Of course, the computation is not carried out in this form, but rather by solving the system of linear equations

$$\mathbf{J}(\mathbf{u}^{k+1} - \mathbf{u}^k) = -\mathbf{F}(\mathbf{u}^k) \qquad (6.4.15)$$

at each step of the iteration.

The justification for Newton's method is found with Taylor's theorem. The linear Taylor expansion of $\mathbf{F}$ in a point $\mathbf{u}^k$ that lies in the neighborhood of a solution $\bar{\mathbf{u}}$ may be expressed as

$$\mathbf{F}(\bar{\mathbf{u}}) = \mathbf{F}(\mathbf{u}^k) + \mathbf{J}(\mathbf{u}^k)(\bar{\mathbf{u}} - \mathbf{u}^k) + \text{higher-order terms} \qquad (6.4.16)$$

But since $\bar{\mathbf{u}}$ is a root, $\mathbf{F}(\bar{\mathbf{u}}) = 0$, whence

$$0 \simeq \mathbf{F}(\mathbf{u}^k) + \mathbf{J}(\mathbf{u}^k)(\bar{\mathbf{u}} - \mathbf{u}^k) \qquad (6.4.17)$$

This approximation may be solved for the unknown $\bar{\mathbf{u}}$, giving precisely the formula (6.4.14). It is noted here also that if $\mathbf{F}$ is *linear* in the neighborhood of $\bar{\mathbf{u}}$, then the iteration yields the solution $\bar{\mathbf{u}}$ in exactly one step.

Note.   Inserting the exact value of the Jacobian matrix $\mathbf{J}(\mathbf{u})$ in (6.4.15) is not an absolute requisite for the procedure to converge. In those cases where evaluation of this matrix requires much time-consuming algebra, one may achieve economy by such modifications as:

1. Replacing $\mathbf{J}$ by a simpler approximation, or
2. Recomputing (numerically) the elements of the matrix $\mathbf{J}(\mathbf{u})$ only once in a while, using the same matrix for several iterations.

7

Special Tools

7.1 SPECIAL POISSON SOLVERS

We describe in this section special methods for the efficient solution of Poisson's equation in simple geometries. These methods were initially developed (in the mid-1960s) by workers in applied physics who were concerned with plasma dynamics. More refinements and applications to other fields came in subsequent years.

Before entering in the detail of these methods, we describe briefly some of their context.

7.1.1 SOME PROBLEMS IN THE SIMULATION OF CHARGED PARTICLE DYNAMICS

There are important applications of computation in physics, or in "computational physics" as it is getting to be called, in which Poisson's equation needs being solved repeatedly.

An example is that of simulation of the dynamics of systems of charged particles. Problems of this kind are often related to electrical plasmas or other charged particle clouds associated with the development of nuclear-fusion energy production devices.

Typically, one is concerned with following the evolution of a large number of particles subjected to several types of forces, including that due to the particle's self-generated electrostatic field. This field is the gradient of the electrical potential,

$$E = -\nabla V \qquad (7.1.1)$$

The electrical potential is, in turn, the solution of Poisson's equation

$$\nabla^2 V = -4\pi\rho \qquad (7.1.2)$$

where ρ is the sum of the electrical charges carried by the particles. In practice, calculations are carried out over a discrete mesh in which the solution of (7.1.2) is approximated numerically.

Particles themselves are, for reasons of economy, represented numerically by superparticles (of the order of 10^{10} particles each) which are assumed to occupy a small but finite region in space.

A typical calculation may proceed as follows:

1. *Charge distribution:* At the beginning of each time step the position of each particle is examined, and the charge of each particle is distributed according to distance to the nearest mesh points (or cell centers).

2. *Potential:* The charge distribution found in step 1 is used as the source term or right-hand side of *Poisson's equation, the numerical solution of which gives the electrostatic potential in the region.*

3. *Acceleration:* The gradient of the potential distribution found in step 2 is computed by numerical differentiation to give the electrostatic field. This field (together with others, such as the magnetic field) is allowed to accelerate each particle for a short time interval Δt (i.e., integrating over one Δt the system of *ordinary* differential equations for the particles positions and velocities).

The cycle is then repeated starting with 1.

Useful simulations related to important problems in particle dynamics may involve on the order of 10^4 moving superparticles, followed as many as 10^4 steps in time. The needed resolution of the electrostatic potential may demand on the order of 10^3 to 10^4 spatial mesh points.

Solving Poisson's equation several thousand times within a single simulation represents an overwhelming task, which has fostered the search for efficient Poisson solvers. Luckily, the boundary conditions are often relatively simple, and the regions of interest are reasonably regular (i.e., rectangular, cylindrical, etc.), or can be so approximated.

Some of the "special" techniques for such efficient solutions of Poisson's equation are described next. Our presentation is limited to two-dimensional problems. Extension to three spatial dimensions is relatively straightforward.

7.1.2 A DOUBLE FOURIER SERIES METHOD

Consider Poisson's equation

$$\frac{\partial^2 U}{\partial x^2} + \frac{\partial^2 U}{\partial y^2} = F(x, y) \qquad (7.1.3)$$

over the rectangular domain D:

$$x \in [0, x_{\max}]$$
$$y \in [0, y_{\max}]$$

associated to the simple Dirichlet boundary conditions:

$$U = 0 \qquad \text{on } \partial D$$

As a general approach to the *analytic* solution of this problem, we may seek an expression of U in the form of a double Fourier series:

$$U(x, y) = \sum_{p,q} \hat{U}_{p,q} \sin\left(\frac{p\pi x}{x_{\max}}\right) \sin\left(\frac{q\pi y}{y_{\max}}\right) \qquad (7.1.4)$$

The corresponding trigonometric basis functions

$$\varphi_{p,q}(x, y) = \sin\left(\frac{p\pi x}{x_{\max}}\right) \sin\left(\frac{q\pi y}{y_{\max}}\right) \qquad (7.1.5)$$

possess the property of orthogonality:

$$\langle \varphi_{p,q}, \varphi_{p',q'} \rangle = \int_D \varphi_{p,q} \varphi_{p',q'} \, dx \, dy$$

$$= \begin{cases} \dfrac{S}{4} & \text{if } p = p' \text{ and } q = q' \\ 0 & \text{otherwise} \end{cases} \qquad (7.1.6)$$

where $S = x_{\max} y_{\max}$ is the area of D. Inserting (7.1.4) into (7.1.3) results in the identity

$$-\sum_{p,q} \varphi_{p,q} \hat{U}_{p,q} \left[\left(\frac{p\pi}{x_{\max}}\right)^2 + \left(\frac{q\pi}{y_{\max}}\right)^2 \right] = \sum_{p,q} \varphi_{p,q} \hat{F}_{p,q} \qquad (7.1.7)$$

where $\hat{F}_{p,q}$ are the coefficients of the double Fourier transform of F:

$$F(x, y) = \sum_{p,q} \varphi_{p,q} \hat{F}_{p,q} \qquad (7.1.8)$$

149

Solving (7.1.7) term by term, we obtain

$$\hat{U}_{p,q} = \frac{-\hat{F}_{p,q}}{-[(p\pi/x_{max})^2 + (q\pi/y_{max})^2]} \tag{7.1.9}$$

The $\hat{F}_{p,q}$ may be computed by the classical Fourier transform

$$\hat{F}_{p,q} = \frac{4}{x_{max}y_{max}}\langle \varphi_{p,q}, F(x,y)\rangle$$

$$= \frac{4}{x_{max}y_{max}}\int_D \sin\left(\frac{p\pi x}{x_{max}}\right)\sin\left(\frac{q\pi y}{y_{max}}\right)dx\,dy \tag{7.1.10}$$

The solution $U(x, y)$ may then be obtained point by point by the implementation of (7.1.4). This, we are reminded, is an analytic treatment of the problem. This analytic treatment serves as a basis for the following method of numerical approximation.

Numerical Approximation. Numerical implementation of the foregoing requires approximations of different kinds:

1. The number of trigonometric modes (p, q) must be limited to a finite number.

2. The (x, y) representation of both $F(x, y)$ and $U(x, y)$ must be restricted to local values at a finite number (say $M \times N$) of interior grid points, with, for instance:

$$\left.\begin{array}{l} f_{m,n} = F(m\,\Delta x, n\,\Delta y) = F(x_m, y_n) \\ u_{m,n} \simeq U(m\,\Delta x, n\,\Delta y) = U(x_m, y_n) \end{array}\right\}$$

$$\Delta x = \frac{x_{max}}{M+1}; \qquad \Delta y\,\frac{y_{max}}{N+1} \tag{7.1.11}$$

The approximate problem is then complete if we take M and N to be the maximum values of p and q, respectively.

The exact integral (7.1.10) is replaced by the *discrete Fourier transform*

$$\hat{F}_{p,q} \simeq \hat{f}_{p,q} = \frac{4\,\Delta x\,\Delta y}{x_{max}y_{max}}\sum_{m=1}^{M}\sin\left(\frac{p\pi m\,\Delta x}{x_{max}}\right)\sum_{n=1}^{N}\sin\left(\frac{q\pi n\,\Delta y}{y_{max}}\right)f_{m,n};$$

$$p = 1, 2, \ldots, M; \quad q = 1, 2, \ldots, N \tag{7.1.12}$$

Obtaining a numerical solution thus requires the implementation of two two-dimensional Fourier transforms. For instance, the calculations described by (7.1.12) imply Fourier transforms in x followed by Fourier transforms in y. Each should make use of the "fast" Fourier method (Section 2.5). And the same holds for the discrete equivalent of (7.1.4).

A specific sequence is thus:

1. Compute the coefficients $\hat{f}_{p,q}$ by a fast Fourier transform implementation of (7.1.12).

2. Compute the coefficients $\hat{u}_{p,q}$ by the application of (7.1.9).

3. Compute the local values of the solution u by a fast Fourier transform implementation of (7.1.4).

Note that this method is an extension to two dimensions of the Fourier method described in Section 4.4 for two-point boundary-value problems.

7.1.3 HOCKNEY'S METHOD[1]

This method uses typical Fourier transformation in one direction (say x) and finite differences in the other (say y). Solution of the finite-difference equations is implemented by means of a special technique, called *odd–even reduction*. The analytic relations that underlie Hockney's methods are as follows. For the problem (7.1.3), let the solution be expressed in the partial Fourier expansion

$$U(x, y) = \sum_p \hat{U}_p(y) \sin \left(\frac{p\pi x}{x_{\max}}\right) \tag{7.1.13}$$

By insertion in (7.1.3) and term-by-term identification, it is found that

$$\frac{d^2 \hat{U}_p}{dy^2} - \left(\frac{p\pi}{x_{\max}}\right)^2 \hat{U}_p = \hat{F}_p(y) \tag{7.1.14}$$

where

$$\hat{F}_p(y) = \frac{2}{x_{\max}} \int_0^{x_{\max}} F(x, y) \sin \left(\frac{p\pi x}{x_{\max}}\right) dx \tag{7.1.15}$$

A finite-difference approximation of (7.1.14) is

$$\hat{u}_{p, n-1} - (2 + R)\hat{u}_{p, n} + \hat{u}_{p, n+1} = \Delta y^2 \hat{f}_{p, n} \tag{7.1.16}$$

with

$$R = \left(\frac{p\pi \, \Delta y}{x_{\max}}\right)^2 \tag{7.1.17}$$

---

[1]See Hockney (1965).

and where

$$\hat{f}_{p,n} \simeq \hat{F}_p(y_n) \tag{7.1.18}$$

is a usual *discrete* Fourier transform approximation of $\hat{F}$.

ODD-EVEN REDUCTION

Assume that N is chosen an even number. Premultiplying the nth equation (7.1.16) by $(2 + R)$ and adding to the $(n + 1)$st and $(n - 1)$st equation results in

$$\hat{u}_{p,n-2} + \lambda\hat{u}_{p,n} + \hat{u}_{p,n+2} = \Delta y^2 \, \hat{f}^*_{p,n} \tag{7.1.19}$$

with

$$\lambda = 2 - (2 + R)^2 \tag{7.1.20}$$

and

$$\hat{f}^*_{p,n} = \hat{f}_{p,n-1} + (2 + R)\hat{f}_{p,n} + \hat{f}_{p,n+1}$$

This is a self-contained tridiagonal system of $N/2$ difference equations (instead of the original N equations). When (7.1.19) is solved (giving the solution along the even lines), the solution along the odd lines may be obtained explicitly thereafter, solving the basic equation (7.1.16) (for n even)

$$\hat{u}_{p,n} = \frac{\hat{u}_{p,n-1} + \hat{u}_{p,n+1} - \Delta y^2 \hat{f}_{p,n}}{2 + R} \tag{7.1.21}$$

RECURSIVE CYCLIC REDUCTION

If N is chosen a power of 2 times 3 (e.g., $N = 48 = 2^4 \times 3$), then the process of odd-even reduction can be repeated until what remains is a single equation. This procedure, called *recursive cyclic reduction*, is attributed to Golub by Hockney (1965) in the first published description of his method.[2]

The general concept of cyclic reduction exemplified by the above is susceptible to several variations,

7.1.4 A DOUBLE-SWEEP METHOD[3]

As with Hockney's method, we seek an approximation to the solution of Poisson's equation in a rectangle in the form of a Fourier series in x:

$$u(x, y) = \sum_p \hat{u}_p(y) \sin\left(\frac{p\pi x}{x_{\max}}\right) \tag{7.1.13}$$

---

[2]Original work along these lines has also been done by Buneman (1969).

[3]See Vichnevetsky (1975a) for additional details.

where the functions $\hat{u}_p(y)$ satisfy (7.1.14)–(7.1.15). *We now assume that the source is concentrated at the nodes:*

$$F(x, y) \simeq f(x, y) = \sum_{m, n} f_{m, n} \, \delta(x - x_m) \, \delta(y - y_n) \qquad (7.1.22)$$

Unlike in Hockney's method, the points y_n need not be equidistant. We only assume a division of $[0, y_{\max}]$ such that

$$y_0 = 0 < y_1 < y_2 < \ldots < y_{N-1} < y_N = y_{\max}$$

The assumption (7.1.22) may be interpreted physically by using the analogy of an electrical field, where the charge that creates the field consists of "point charges" located at the nodes. (This is a fairly common procedure in computational physics.) There are, moreover, many computing procedures in the realm of charged particle dynamics calculations where this nodal concentration of the forcing function occurs naturally in the formulation of the problem.

Finally, we should point out that such an assumption, whose description depends on physical intuition more than mathematics, is not dissimilar from the kind of mathematical approximation that is contained, for instance, in (7.1.12), and we find in a form similar to (7.1.14),

$$\frac{d^2 \hat{u}_p}{dy^2} - \left(\frac{p\pi}{x_{\max}}\right)^2 \hat{u}_p = \sum_n \hat{f}_{p, n} \, \delta(y - y_n) \qquad (7.1.23)$$

where the $\hat{f}_{p, n}$ are the discrete Fourier transforms:

$$\hat{f}_{p, n} = \frac{2}{x_{\max}} \sum_m f_{m, n} \sin\left(\frac{p\pi x_m}{x_{\max}}\right) \qquad (7.1.24)$$

To solve (7.1.23) we seek a solution that is *analytic* in each interval (y_{n-1}, y_n):

$$\hat{u}_p(y) = a_{p, n} \sinh(ky) + b_{p, n} \cosh(ky); \quad y \in (y_{n-1}, y_n); \quad k = \frac{p\pi}{x_{\max}}$$

$$(7.1.25)$$

It is easily verified that this expression satisfies the equation

$$\frac{d^2 \hat{u}_p}{dy^2} - k^2 \hat{u}_p = 0$$

which is the form taken by (7.1.23) in (y_{n-1}, y_n) (i.e., between mesh points).

Recurrence relations for $a_{k, n}$ and $b_{k, n}$ are easily derived. We have:

Continuity of $\hat{u}_p$ across node points:

$$\hat{u}_p(y_{n-}) = \hat{u}_p(y_{n+})$$

or

$$a_{p,n} \sinh(ky_n) + b_{p,n} \cosh(ky_n) = a_{p,n+1} \sinh(ky_n) + b_{p,n+1} \cosh(ky_n)$$
$$(7.1.26)$$

Integration of (7.1.23) *across node points:*

$$\left(\frac{d\hat{u}_p}{dy}\right)_{n_+} - \left(\frac{d\hat{u}_p}{dy}\right)_{n_-} = \hat{f}_{p,n}$$

or

$$k[(a_{p,n+1} - a_{p,n}) \cosh(ky_n) + (b_{p,n+1} - b_{p,n}) \sinh(ky_n)] = \hat{f}_{p,n} \quad (7.1.27)$$

Adding $k \sinh(ky_n)$ times (7.1.26) to $\cosh(ky_n)$ times (7.1.27) gives

$$k[\sinh^2(ky_n) - \cosh^2(ky_n)](a_{p,n} - a_{p,n+1}) = \hat{f}_{p,n} \cosh(ky)$$

or

$$a_{p,n+1} - a_{p,n} = \frac{\hat{f}_{p,n} \cosh(ky_n)}{k} \quad (7.1.28)$$

And, by a similar process:

$$b_{p,n+1} - b_{p,n} = \frac{\hat{f}_{p,n} \sinh(ky_n)}{k} \quad (7.1.29)$$

Relations (7.1.28) and (7.1.29) may now be used to derive the $\{a_{p,n}\}$ and $\{b_{p,n}\}$ as follows. Starting in $y = 0$ the boundary condition $\hat{u}_p(0) = 0$ becomes

$$b_{p,0} = 0 \quad (7.1.30)$$

We then proceed, using (7.1.29), to compute successively $b_{p,1}, b_{p,2}, \ldots$ until the last value $b_{p,N}$. The boundary condition $\hat{u}_p(y_{\max}) = 0$ is then used to compute $a_{p,N}$:

$$a_{p,N} = -b_{p,N} \frac{\cosh(ky_{\max})}{\sinh(ky_{\max})} \quad (7.1.31)$$

Relation (7.1.28) is then used backward, recursively, to derive $a_{p,N-1}, a_{p,N-2}, \ldots$ down to $a_{p,0}$.

Having the $\{a_{p,n}\}$ and $\{b_{p,n}\}$, equations (7.1.13) and (7.1.25) may be used to compute the numerical solution u in any point in (x, y) (not just at the mesh points).

7.1.5 A DOUBLE-SWEEP METHOD
IN CIRCULAR COORDINATES

A similar "efficient Poisson's equation solving" method can be implemented in a circle. We seek a solution to

$$\nabla^2 U = \left(\frac{\partial^2}{\partial r^2} + \frac{1}{r} \frac{\partial}{\partial r} + \frac{1}{r^2} \frac{\partial^2}{\partial \alpha^2} \right) U(r, \alpha) = F(r, \alpha) \qquad (7.1.32)$$

over the circular domain

$$D \equiv r \leq R$$

with Dirichlet boundary conditions

$$U(R, \alpha) = 0$$

The domain D is divided by a symmetric grid consisting of $2M$ equally
spaced radii and N concentric circles at arbitrary distances from the center
(Figure 7.1.1):

$$0 < r_1 < r_2 < \ldots < r_n < \ldots < r_N = R \qquad (7.1.33)$$

As in the rectangular coordinates case, the computing procedure that we
shall describe uses the assumption that the right-hand side of (7.1.32) has

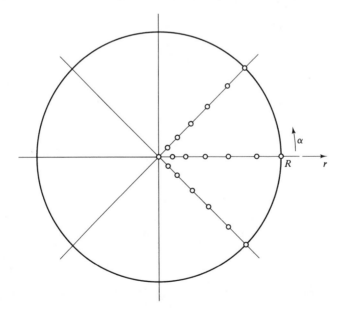

Figure 7.1.1

been reduced to weighted Dirac delta functions located at the nodes of the circular grid. Thus, (7.1.32) may be approximated by

$$\nabla^2 u = \sum_{m,n} f_{m,n}\, \delta(r - r_m)\, \delta(\alpha - \alpha_n) \equiv f \qquad (7.1.34)$$

where

$$f_{m,n} \simeq \int_{D_{m,n}} F(r, \alpha)\, dD \qquad (7.1.35)$$

and $D_{m,n}$ is the element of area D associated with the node (r_m, α_n).

We seek an expansion of the solution in terms of shape functions which are sinusoidal in α: that is,

$$u(r, \alpha) = \sum_{k=0}^{M} A_k(r) \cos(k\alpha) + B_k(r) \sin(k\alpha) \qquad (7.1.36)$$

Substituting in (7.1.34), we find the differential equations

$$\sum_{k} \left[\left(\frac{d^2 A_k}{dr^2} + \frac{1}{r} \frac{dA_k}{dr} - \frac{k^2}{r^2} A_k \right) \cos(k\alpha) \right.$$
$$\left. + \left(\frac{d^2 B_k}{dr^2} + \frac{1}{r} \frac{dB_k}{dr} - \frac{k^2}{r^2} B_k \right) \sin(k\alpha) \right] = f \qquad (7.1.37)$$

Multiplying by $\cos(k\alpha)$, $\sin(k\alpha)$, and integrating over $\alpha \in (0, 2\pi)$ results in the familiar decomposition for $k > 0$:

$$\frac{d^2 A_k}{dr^2} + \frac{1}{r} \frac{dA_k}{dr} - \frac{k^2}{r^2} A_k = \frac{1}{\pi} \sum_{m,n} f_{m,n} \cos(k\alpha_m)\, \delta(r - r_n)$$
$$= \sum_{n} C_{k,n}\, \delta(r - r_n) \qquad (7.1.38)$$

$$\frac{d^2 B_k}{dr^2} + \frac{1}{r} \frac{dB_k}{dr} - \frac{k^2}{r^2} B_k = \frac{1}{\pi} \sum f_{m,n} \sin(k\alpha_m)\, \delta(r - r_n)$$
$$= \sum_{n} S_{k,n}\, \delta(r - r_n) \qquad (7.1.39)$$

with

$$C_{k,n} = \frac{1}{\pi} \sum_{m} f_{m,n} \cos(k\alpha_m)$$
$$S_{k,n} = \frac{1}{\pi} \sum_{m} f_{m,n} \sin(k\alpha_m) \qquad (7.1.40)$$

and for $k = 0$:

$$\frac{d^2 A_0}{dr^2} + \frac{1}{r} \frac{dA_0}{dr} = \frac{1}{2\pi} \sum_{m,n} f_{m,n}\, \delta(r - r_n)$$
$$= \sum_{n} C_{0,n}\, \delta(r - r_n) \qquad (7.1.41)$$

with

$$C_{0,n} = \frac{1}{2\pi} \sum_m f_{m,n} \tag{7.1.42}$$

These equations can be integrated analytically between nodes. The general solution is, for $k > 0$,

$$A_k(r) = a_k^+ r^k + a_k^- r^{-k}$$
$$B_k(r) = b_k^+ r^k + b_k^- r^{-k} \tag{7.1.43}$$

and for $k = 0$,

$$A_0(r) = a_0^+ + a_0^- \ln(r) \tag{7.1.44}$$

The relations that hold across nodes are obtained by integrating (7.1.38) twice between $r_n - \epsilon$ and $r_n + \epsilon$, yielding (for all k)

$$\frac{dA_k}{dr}(r_n + \epsilon) - \frac{dA_k}{dr}(r_n - \epsilon) = C_{k,n} \tag{7.1.45}$$

$$A_k(r_n + \epsilon) - A_k(r_n - \epsilon) = 0 \tag{7.1.46}$$

and similarly for the $B_k(r)$.

The $A_k(r)$ and $B_k(r)$ may thus be expressed in a piecewise analytic form as

$$A_k(r) = a_{k,n}^+ r^k + a_{k,n}^- r^{-k} \left.\right\} \quad k \neq 0 \left.\right\} \tag{7.1.47}$$
$$B_k(r) = b_{k,n}^+ r^k + b_{k,n}^- r^{-k} \tag{7.1.48}$$
$$A_0(r) = a_{0,n}^+ + a_{0,n}^- \ln r \tag{7.1.49}$$

The continuity relations (7.1.46) lead to the relations (for $k > 0$)

$$a_{k,n-1}^+ r_n^k + a_{k,n-1}^- r_n^{-k} = a_{k,n}^+ r_n^k + a_{k,n}^- r_n^{-k} \tag{7.1.50}$$

and the derivative jump conditions (7.1.45) become

$$\frac{k}{r_n}[(a_{k,n-1}^+ - a_{k,n}^+)r_n^k - (a_{k,n-1}^- - a_{k,n}^-)r_n^{-k}] = C_{k,n} \tag{7.1.51}$$

Combining these two expressions results in two equivalent relations:

$$2k r_n^{k-1}(a_{k,n-1}^+ - a_{k,n}^+) = C_{k,n} \tag{7.1.52}$$
$$2k r^{-k-1}(a_{k,n-1}^- - a_{k,n}^-) = -C_{k,n} \tag{7.1.53}$$

which are more convenient for computation.

The known property that $U(0) \neq \infty$ implies that $a_{k,0}^- = 0$. This suggests the following computing sequence.

1. Starting with $a_{k,0}^- = 0$, we compute

$$a_{k,n}^- = a_{k,n-1}^- + \frac{C_{k,n-1} r_n^{k+1}}{2k} \qquad \text{for } n = 1, 2, \ldots, N$$

2. Using the boundary condition $A_k(R) = 0$, we find, from (7.1.47),

$$a_{k,N}^+ = -a_{k,N}^- R^{-2k}$$

3. Starting with this value of $A_{k,N}^+$, we compute

$$a_{k,n}^+ = a_{k,n+1}^+ - \frac{C_{k,n+1}}{2k r_{n+1}^{k-1}}; \qquad n = N - 1, N - 2, \ldots, 0$$

Similar sequences apply to the $a_{0,n}$ and $b_{k,n}$ coefficients.

Having obtained all the coefficients $a_{k,n}$ and $b_{k,n}$, values of the solution are then simply obtained by computing (7.1.36) pointwise. An example of a numerical solution so obtained is shown in Figure 7.1.2.

7.1.6 EXTENSIONS

There is a considerable literature describing extensions of those ideas to other geometries and other types of boundary conditions. See Swartztrauber (1974) for an example of the former and Mayo (1979) for an example of the latter.

7.2 NETWORKS AND ANALOGS

Networks of resistors and other configurations of electrical components establish implicit relationships between electrical potentials and currents. For example, if we denote by u_n the voltages at the nodes of the elementary electrical network of Figure 7.2.1, then Kirchhoff's first law (akin to Euler's

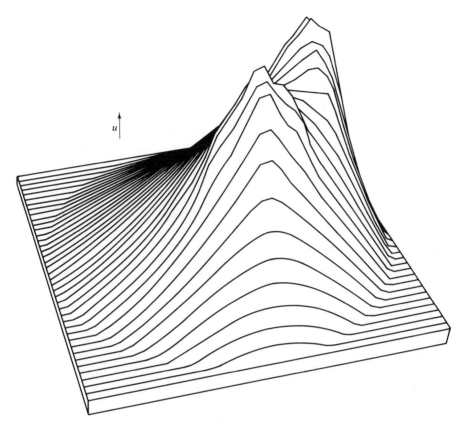

Figure 7.1.2 Solution of Poisson's equation in a cylinder.

equation of continuity in fluid mechanics) states that

$$u_1 + u_2 + u_3 + u_4 - 4u_5 = 0 \qquad (7.2.1)$$

The similarity of equation (7.2.1) with the finite-difference approximation to a solution of Laplace's equation

$$\frac{\partial^2 U}{\partial x^2} + \frac{\partial^2 U}{\partial y^2} \equiv \nabla^2 U = 0 \qquad (7.2.2)$$

is evident. If an electrical source is added as in Figure 7.2.2, then the relation becomes

$$u_1 + u_2 + u_3 + u_4 - 4u_5 = Rs \qquad (7.2.3)$$

which is a discrete approximation to Poisson's equation (with $R \sim \Delta x^2 = \Delta y^2$)

$$\frac{\partial^2 U}{\partial x^2} + \frac{\partial^2 U}{\partial y^2} = s \qquad (7.2.4)$$

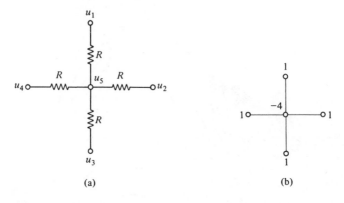

Figure 7.2.1 (a) Elementary *cell* of a network of electrical resistors; (b) computing *molecule* for Laplace's equation.

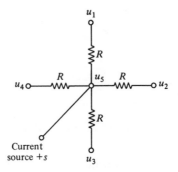

Figure 7.2.2

One may assemble such elementary networks of resistors into larger networks covering an entire domain D on which Laplace's equation is to be solved much in the manner in which one would divide the same domain into elementary cells to construct an approximation by finite differences.

7.2.1 USE OF DISTRIBUTED CONDUCTING MEDIA

The potential distribution in a uniform sheet of a solid conductive material is governed by Laplace's equation in two dimensions. The same is true of a liquid conductive medium contained in a shallow tank. In three dimensions, the potential distribution in a body of solid conducting material or in a conductive liquid is also governed by Laplace's equation. Thus, particular solutions of Laplace's equation (also called Laplacian field problems) may be solved in the following three steps:

1. A conductive solid or liquid of the same geometrical shape as the field under study is devised.

2. The boundary conditions of the original field are simulated in the electrical analog by means of suitable voltage and/or current sources.

3. By means of suitable sensing equipment, the voltage distribution within the conductive medium is measured and recorded. The voltage obtained in this manner can be interpreted directly as the potential distribution in the original system under study.

The measurement of voltages in a two-dimensional continuous analog presents no difficulty. A chief difficulty in the use of three-dimensional continuous solid analogs is that it is difficult to measure the potential distribution at interior points. The use of such analogs has therefore been mostly restricted to problems in which only the potential distribution at the exterior surfaces of the physical system is required.

7.2.2 HISTORICAL DEVELOPMENTS

Physical analogs for the solution of Laplace and Poisson's equations have been in use for a long time. One finds reference to the use of electrolytic tanks for this purpose in Langmuir, Adams, and Meikle (1913).

The first systematic discription of lumped electrical networks to find approximate solutions to elliptic equations seems to be that of Beuken (1936). His description is not limited to elliptic equations, though; he remarked that the network shown in Figure 7.2.3 is governed by the relation

$$\frac{u_1 + u_2 + u_3 + u_4 - 5u_5}{R} = C\frac{du_5}{dt} \qquad (7.2.5)$$

which is an elementary analog of the diffusion equation

$$\sigma\left(\frac{\partial^2 U}{\partial x^2} + \frac{\partial^2 U}{\partial y^2}\right) = \frac{\partial U}{\partial t}; \qquad \sigma = \frac{\Delta x^2 = \Delta y^2}{RC} \qquad (7.2.6)$$

Assembly of such elementary modules allows for the construction of physical analogs of Fourier's equation.

Considerable work using such devices has taken place in the following few decades. Among the representative publications of that period are those of Beuken (1936) and Liebmann (1954, 1955) in western Europe, of Karplus (1954, 1956, 1958) and Paschkis and Ryder (1968) in the United States, of

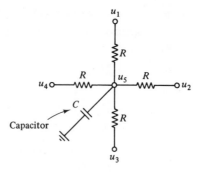

Figure 7.2.3 Elementary cell for an electrical analog of Fourier's equation.

Kantorovich and Krylov (1964) of Guthenmaker (1949), and of Volinskii and Bukhman (1965) in the USSR.

An interesting publication of that time is that of MacNeal (1953). What he wanted to establish was a general method for the derivation of resistor values in the construction of an asymmetric network analog to Laplace's equation. The method of approach consists in attaching a sub-domain to each node of the network, and the resulting mathematics have a flavor that is reminiscent of the finite-element method, which of course did not appear significantly in numerical analysis until several years later.

Before the advent of large-scale digital computers, electrical analogs for the solution of elliptic problems have held a reasonably important place in applied research. To wit, an extended list of publications by Paschkis, Karplus, and others is concerned with applications to heat transfer in solids, water flow in the underground, and the like.

Much of this has changed in the past 15 years or so. The power and flexibility of electronic digital computers has resulted in a gradual replacement of analog networks for the solution of elliptic equations by purely digital methods.

Specialized computers using these techniques are still used occasionally, though; see, for example, Caussade and Renard (1977) or Czerwinska (1979).

III

The
Finite-Element
Method

8

Rayleigh-Ritz and Galerkin Methods

8.1 THE CALCULUS OF VARIATIONS

For since the fabric of the universe is most perfect and the work of a most wise Creator, nothing at all takes place in the universe in which some rule of maximum or minimum does not appear.

LEONHARD EULER

8.1.1 DEFINITIONS AND PRINCIPLES

The calculus of variations holds an important place in theoretical expositions of the finite-element method.

A short definition of what the name "calculus of variations" means is "the study of variations (= derivatives) of functionals." Functionals are defined first.

Functionals. A standard definition of a *functional* is "an operator that carries functions into numbers." The typical expression of a functional is

$$J(U) = \int_0^l \Phi(U, U', x)\, dx \qquad (8.1.1)$$

where $U(x)$ plays the role of the *independent variable* (it is a function of x in $x \in [0, l]$) and J is the dependent variable (a scalar).

EXAMPLE

Consider a path joining two points A and B whose horizontal distance from one another is l (Figure 8.1.1). It is assumed that the path goes monotonically from left to right, as illustrated. The length of the path is, from simple geometrical considerations:

$$\text{length} = \int_0^l [1 + (U')^2]^{1/2}\, dx \qquad (8.1.2)$$

which is a simple functional of the form (8.1.1).

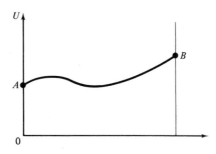

Figure 8.1.1

Variation of a Functional.   The notion of the variation of a functional is the extension to functional analysis (or analysis of functionals) of the notion of derivative in classical function theory.

Given a function $U(x)$ in $[0, l]$, to which corresponds the functional

$$J(U) \equiv \int_0^l \Phi(U, U', x)\, dx \qquad (8.1.3)$$

we propose to seek the difference

$$\delta J \equiv J(U + \delta U) - J(U) \qquad (8.1.4)$$

corresponding to a "small" variation $\delta U(x)$ in the argument U (Figure 8.1.2). [Note that it is assumed that the support $x \in (0, l)$ is kept constant.]

δJ is called the *variation* of J. We may write formally, using Taylor's theorem:

$$J(U + \delta U) = \int_0^l \Phi(U + \delta U, U' + \delta U', x)\, dx$$

$$= \int_0^l \left(\Phi(U, U', x) + \frac{\partial \Phi}{\partial U} \delta U + \frac{\partial \Phi}{\partial U} \delta U' \right) dx + O(\delta U^2) \qquad (8.1.5)$$

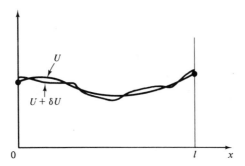

Figure 8.1.2

[the $O(\delta U^2)$ term shall be omitted, since we are interested in δJ when $\delta U \to 0$]. Note that in (Figure 8.1.3)

$$\delta U' = \frac{d}{dx}(\delta U) \qquad (8.1.6)$$

Therefore,

$$J(U + \delta U) = \int_0^l \left[\Phi(U, U', x) + \frac{\partial \Phi}{\partial U}\delta U + \frac{\partial \Phi}{\partial U'}\left(\frac{d}{dx}\delta U\right) \right] dx \quad (8.1.7)$$

The last term of this expression may be integrated by parts:

$$\int_0^l \frac{\partial \Phi}{\partial U'}\left(\frac{d}{dx}\delta U\right) dx = \left[\frac{\partial \Phi}{\partial U'}\delta U\right]_0^l - \int_0^l \frac{d}{dx}\left(\frac{\partial \Phi}{\partial U'}\right)\delta U\, dx \qquad (8.1.8)$$

whence, subtracting (8.1.3) from (8.1.7),

$$\delta J = J(U + \delta U) - J(U) = \int_0^l \left[\frac{\partial \Phi}{\partial U} - \frac{d}{dx}\left(\frac{\partial \Phi}{\partial U'}\right)\right]\delta U(x)\, dx$$

$$+ \left[\frac{\partial \Phi}{\partial U'}\delta U\right]_0^l = \delta J \qquad (8.1.9)$$

If we limit ourselves to cases where $U(0)$ and $U(l)$ are fixed, then $\delta U(0) = \delta U(l) = 0$ and

$$\delta J = \int_0^l \left[\frac{\partial \Phi}{\partial U} - \frac{d}{dx}\left(\frac{\partial \Phi}{\partial U'}\right)\right]\delta U(x)\, dx \qquad (8.1.10)$$

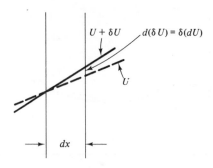

Figure 8.1.3

Extremal Curves. When $U(x)$ is such that $J(U)$ reaches a maximum or minimum, then $U(x)$ is called an *extremal function* or *extremal curve* of the functional $J(U)$. A necessary condition for $U(x)$ to be extremal is that $\delta J = 0$ for *any* small, permissible displacement $\delta U(x)$. Returning to (8.1.10), we see that this condition will require that[1]:

$$\frac{\partial \Phi}{\partial U} - \frac{d}{dx}\left(\frac{\partial \Phi}{\partial U'}\right) = 0 \qquad (8.1.11)$$

everywhere in $x \in [0, l]$. If the extremities are free [i.e., $\delta U(0)$ and $\delta U(l) \neq 0$], then it is also necessary that end conditions be satisfied:

$$\left(\frac{\partial \Phi}{\partial U'}\right)_0 = \left(\frac{\partial \Phi}{\partial U'}\right)_l = 0 \qquad (8.1.12)$$

The ordinary differential equation (8.1.11) is known as the *Euler–Lagrange equation* for the given extremum problem.

EXAMPLE 1

What is the minimum distance path between two fixed points (Figure 8.1.4)?
We have seen previously that

$$\text{distance} = \int_0^l [(1 + (U')^2]^{1/2}\, dx$$

---

[1]Note that the condition $\delta J = 0$ does not define only an extremal curve: there are cases for which J is merely *stationary*. This is why the condition is necessary and not sufficient.

168

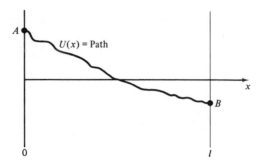

Figure 8.1.4

Thus, the minimum distance problem may be formulated as a calculus of variation problem, where

$$\Phi = [1 + (U')^2]^{1/2} \tag{8.1.13}$$

The corresponding Euler–Lagrange equation is

$$\frac{d}{dx}\left(\frac{\partial\Phi}{\partial U'}\right) = U''[1 + (U')^2]^{-3/2} = 0 \tag{8.1.14}$$

One solution is

$$U'' = 0$$

that is,

$$U' = \text{constant}$$

or

$$U(x) = a + bx = \text{straight line} \tag{8.1.15}$$

We have thus determined that a straight line is a stationary distance path between two fixed points A and B in the plane.[2]

The other part of (8.1.14) is

$$[1 + (U')^2]^{-3/2} = 0$$

---

[2]It is, of course, not only stationary but is the *least* distance path. We must take this for granted since we did not prove it.

which does not have a *finite* solution and corresponds to the maximum of distance.

If boundaries are not fixed, then we have to satisfy the additional end conditions

$$\frac{\partial \Phi}{\partial U'}\bigg|_{x=0} = \frac{\partial \Phi}{\partial U'}\bigg|_{x=l} = 0 \tag{8.1.16}$$

or

$$U'(0) = 0; \qquad U'(l) = 0 \tag{8.1.17}$$

(i.e., the solution is then a *horizontal* straight line). It is easily verified that such a line is indeed that of shortest distance between two verticals.

EXAMPLE 2

Consider the problem of finding the displacement $U(x)$ of a loaded string (Figure 8.1.5). The assumptions are:

1. *Fixed extremities:*

$$U(0) = U(l) = 0$$

2. *Small displacements:*

$$\tau = \text{tension in the string} = \text{constant}$$

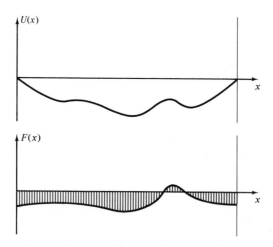

Figure 8.1.5

We shall derive the differential equation describing the displacement by two methods—*geometrical derivation* and *derivation from the calculus of variations*—and show that both lead to the same result.

Geometrical Derivation. Consider an element of string between x and $(x + dx)$ (Figure 8.1.6). The equilibrium law states that the sum of vertical forces acting on that element is equal to zero. This translates into:

$$-\tau \frac{dU}{dx}\bigg|_{x} \qquad \text{left force}$$

$$+\tau \frac{dU}{dx}\bigg|_{x+dx} \qquad \text{right force}$$

$$+F(x)\, dx \qquad \text{load on the element}$$

$$= 0$$

Using Taylor's theorem, we may express this as

$$-\tau \left(\frac{dU}{dx}\right)_{x} + \tau \left[\left(\frac{dU}{dx}\right)_{x} + \left(\frac{d^2U}{dx^2}\right)_{x} dx\right] + F\, dx = 0$$

or

$$\tau \frac{d^2U}{dx^2} + F(x) = 0 \qquad\qquad (8.1.18)$$

which is the equation of the position of the string.

Variational Derivation. One may derive the equation of the loaded string by invoking more general principles. The principle that applies here is called the principle of virtual works. It states that for any small displacement

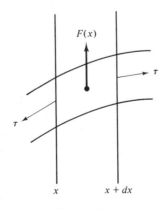

$F(x)$

τ

τ

x $x + dx$

Figure 8.1.6

$\delta U(x)$ from the position of equilibrium, the internal work and external work must be equal. Another way to state this principle is, since work is the change of energy, to say that the "variation" of total energy of the system (note that the energy is a functional) must be equal to zero. The shape of equilibrium of the string corresponds to a *minimum of the total energy.*

External Work and Energy.   The virtual work of external forces w_E is

$$w_E = \int_0^l \delta U F\, dx \tag{8.1.19}$$

This work is expanded at the expense of the external energy. Thus, the total external energy may be obtained by integration (assigning to it arbitrarily the value 0 when $U = 0$):

$$external\ energy = Q = -\int_0^U w_E\, d(\delta U) = -\int_0^l UF\, dx \tag{8.1.20}$$

Internal Work and Energy.   A small element dx of the string contains internal energy. This energy is that needed to bring this element of string from an unloaded position (Figure 8.1.7a) to a loaded position (Figure 8.1.7b). The energy needed to go from Figure 8.1.7a to b can be obtained by integration. The vertical force in B is

$$\tau \frac{dU}{dx} \tag{8.1.21}$$

Thus, integrating (force times vertical displacement in B from $B = B'$ to $B = B''$), we obtain

$$\begin{aligned} internal\ energy = \mathcal{P} &= \int_0^l \int_0^{dU/dx} \tau \frac{dU}{dx} d(\text{vertical displacement}) \\ &= \int_0^l \int_0^{dU/dx} \tau \left(\frac{dU}{dx}\right) d\left(\frac{dU}{dx}\right) dx \\ &= \int_0^l \frac{\tau}{2}\left(\frac{dU}{dx}\right)^2 dx \end{aligned} \tag{8.1.22}$$

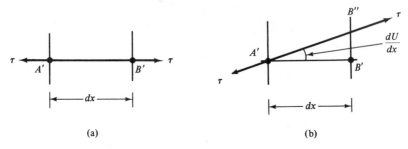

(a)   (b)

Figure 8.1.7

Thus,

$$total\ energy = \mathcal{P} + \mathcal{Q}$$

$$= \int_0^l \left[\frac{\tau}{2} \left(\frac{dU}{dx} \right)^2 - UF \right] dx \qquad (8.1.23)$$

The *principle of minimum energy* states that the actual position of the string minimizes this quantity:

$$\delta \int_0^l \left[\frac{\tau}{2} \left(\frac{dU}{dx} \right)^2 - UF \right] dx = 0 \qquad (8.1.24)$$

The corresponding Euler–Lagrange equation,

$$-\left(\tau \frac{d^2U}{dx^2} + F \right) = 0 \qquad (8.1.25)$$

which must be satisfied by the position of the string, is seen to be identical to (8.1.18) derived from purely geometrical considerations.

HISTORICAL NOTE

The invention of the calculus of variations and its introduction in physics was one of the crown pieces of eighteenth-century mathematics. In particular, the replacement of geometrical constructions by analytic calculations to derive the equations of mechanics (as illustrated by the preceding and other examples in this section) was systematized by Lagrange (1788), who says about this:

> The methods which I present here do not require either constructions or reasonings of geometrical or mechanical nature, but only algebraic operations proceeding after a regular and uniform plan. Those who love the calculus, will see with pleasure Mechanics made a branch of it and will be grateful to me for having thus extended its domain.

8.1.2 FUNCTIONALS WITH HIGHER-ORDER DERIVATIVES

Consider the functional

$$J = \int_0^l \Phi(U, U', \dots, U^{(r)}, x)\, dx \qquad (8.1.26)$$

where the function Φ contains higher derivatives of U with respect to x. The variation of J resulting from a (small) variation in U is

$$\delta J = \int_0^l \left(\frac{\partial \Phi}{\partial U} \delta U + \frac{\partial \Phi}{\partial U'} \delta U' + \dots + \frac{\partial \Phi}{\partial U^{(r)}} \partial U^{(r)} \right) dx \qquad (8.1.27)$$

Since

$$
\left.
\begin{aligned}
\delta U' &= (\delta U)' \\
\delta U'' &= (\delta U)'' \\
&\quad \cdot \\
&\quad \cdot \\
&\quad \cdot
\end{aligned}
\right\}
\tag{8.1.28}
$$

it results that we may also write

$$
\delta J = \int_0^l \left[\frac{\partial \Phi}{\partial U} \, \partial U + \frac{\partial \Phi}{\partial U'}(\delta U)' + \ldots + \frac{\partial \Phi}{\partial U^{(r)}}(\delta U)^{(r)} \right] dx
\tag{8.1.29}
$$

Integration by parts gives us

$$
\int_0^l \frac{\partial \Phi}{\partial U'}(\delta U)' \, dx = \left[\frac{\partial \Phi}{\partial U'} \, \delta U \right]_0^l - \int_0^l \left(\frac{\partial \Phi}{\partial U'} \right)' \delta U \, dx
$$

$$
\begin{aligned}
\int_0^l \frac{\partial \Phi}{\partial U''}(\delta U)'' \, dx &= \left[\frac{\partial \Phi}{\partial U''} \, \delta U' \right]_0^l - \int_0^l \left(\frac{\partial \Phi}{\partial U''} \right)' (\delta U)' \, dx \\
&= \left[\frac{\partial \Phi}{\partial U''} \, \delta U' \right]_0^l - \left[\left(\frac{\partial \Phi}{\partial U''} \right)' \delta U \right]_0^l \\
&\quad + \int_0^l \left(\frac{\partial \Phi}{\partial U''} \right)'' \delta U \, dx
\end{aligned}
\tag{8.1.30}
$$

$$
\begin{aligned}
&\quad \cdot \\
&\quad \cdot \\
&\quad \cdot
\end{aligned}
$$

Substituting back in (8.1.27), we find

$$
\begin{aligned}
\delta J = \int_0^l &\left[\left(\frac{\partial \Phi}{\partial U} \right) - \left(\frac{\partial \Phi}{\partial U'} \right)' + \left(\frac{\partial \Phi}{\partial U''} \right)'' + \ldots \right. \\
&\left. + (-1)^r \left(\frac{\partial \Phi}{\partial U^{(r)}} \right)^{(r)} \right] \delta U \, \delta x \\
&+ \left[\left(\frac{\partial \Phi}{\partial U'} - \left(\frac{\partial \Phi}{\partial U''} \right)' + \ldots \right) \delta U \right. \\
&\left. + \left(\frac{\partial \Phi}{\partial U''} - \ldots \right) \delta U' + \ldots \right]_0^l
\end{aligned}
\tag{8.1.31}
$$

which is, again, in the form of an integral term in $\delta U(x)$, and boundary contributions in $\delta U, \delta U', \ldots$.

EXAMPLE 1

Consider a loaded beam with fixed extremities (Figure 8.1.8). The boundary conditions are:

Fixed end positions:

$$U(0) = 0; \qquad U(l) = 0 \qquad\qquad (8.1.32)$$

Zero torque applied at the ends:

$$U''(0) = 0; \qquad U''(l) = 0 \qquad\qquad (8.1.33)$$

The principle of virtual work

$$\delta(\mathcal{P} + \mathcal{Q}) = 0 \qquad\qquad (8.1.34)$$

where $\mathcal{P}$ is the internal energy and $\mathcal{Q}$ the external energy, will lead to the differential equation to be satisfied by $U(x)$. The external energy created by work of the load is identical to that of a loaded string [see (8.1.20)], thus:

$$\mathcal{Q} = \int_0^l -F(x)U(x)\,dx \qquad\qquad (8.1.35)$$

The internal energy stored in the beam is that created by work due to bending (= work of internal stress). This is expressed as

$$\mathcal{P} = \int_0^l \tfrac{1}{2}\mathbb{E}\mathbb{I}(U'')^2\,dx \qquad\qquad (8.1.36)$$

where $\mathbb{E}$ is Young's elasticity modulus of the material and $\mathbb{I}$ the moment of

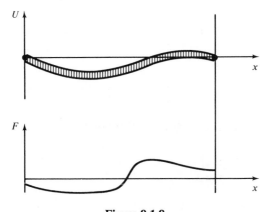

Figure 8.1.8

inertia of the cross section of the beam. Thus, the functional of this problem is

$$J = \mathcal{P} + \mathcal{Q}$$

$$= \int_0^l [\tfrac{1}{2}EI(U'')^2 - FU]\, dx \qquad (8.1.37)$$

The corresponding Euler–Lagrange equation is

$$\frac{\partial \Phi}{\partial U} - \left(\frac{\partial \Phi}{\partial U'}\right)' + \left(\frac{\partial \Phi}{\partial U''}\right)'' = 0$$

$$\Phi = \tfrac{1}{2}EI(U'')^2 - FU$$

or

$$\frac{d^2}{dx^2}\left(EI\frac{d^2 U}{dx^2}\right) = F \qquad (8.1.38)$$

It is the differential equation of the beam.

It is of fourth order and thus requires four boundary conditions. The first boundary conditions (8.1.32) are geometrical and straightforward. The two additional conditions may be shown to imply a minimum of energy. Indeed, the variation of (8.1.37) is [from (8.1.31)], taking into account the fact that $\delta U_0 = \delta U_l = 0$,

$$\delta J = \int_0^l \left[\frac{d^2}{dx^2}\left(EI\frac{d^2 U}{dx^2}\right) - F\right]\delta U\, dx + \left[\frac{\partial \Phi}{\partial U''}\delta U'\right]_0^l \qquad (8.1.39)$$

This variation will be zero for all $\delta U'$ if, in addition to satisfying (8.1.38), the solution also satisfies

$$U_0'' = U_l'' = 0 \qquad (8.1.40)$$

EXAMPLE 2

Pursuing the mathematics of Example 1, consider now a cantilever, loaded beam (Figure 8.1.9). The only imposed boundary conditions are geometrical, in $x = 0$:

$$U(0) = \frac{dU}{dx}(0) = 0 \qquad (8.1.41)$$

The variation of the functional (8.1.37) becomes in this case

$$\delta J = \int_0^L \left[\frac{d^2}{dx^2}\left(EI\frac{d^2 U}{dx^2}\right) - F\right]\delta U\, dx$$

$$- \left[\left(\frac{\partial \Phi}{\partial U''}\right)'\delta U + \left(\frac{\partial \Phi}{\partial U''}\right)\delta U'\right]_0^l \qquad (8.1.42)$$

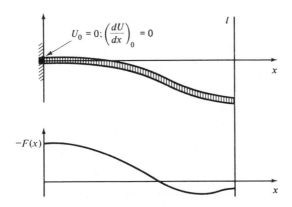

Figure 8.1.9 Cantilever beam.

Since δU_l and $\delta U_l'$ are both arbitrary, the boundary conditions that will make $\delta J = 0$ are thus

$$\left(\frac{\partial \Phi}{\partial U''}\right)_l = 0 \quad \text{and} \quad \left(\frac{\partial \Phi}{\partial U''}\right)_l' = 0 \qquad (8.1.43)$$

which translates into

$$\left(\frac{d^2 U}{dx^2}\right)_l = 0 \qquad (8.1.44\text{a})$$

and

$$\left[\frac{d}{dx}\left(EI\,\frac{d^2 U}{dx^2}\right)\right]_l = 0 \qquad (8.1.44\text{b})$$

The physical interpretation of these conditions is that the beam's *bending moment* and *shear stress* are both zero at the extremity $x = l$. They are called *natural boundary conditions* for the differential equation (8.1.38).

8.1.3 FUNCTIONALS IN SEVERAL-DIMENSIONAL SPACES

The extension of functionals to several-dimensional spaces is straightforward: for example,

$$J = \int_D \Phi\left(U, \frac{\partial U}{\partial x}, \frac{\partial U}{\partial y}, x, y\right) dD \qquad (8.1.45)$$

is a functional of the function $U(x, y)$. The domain D of U in (x, y) is generally a closed domain of the kind illustrated in Figure 8.1.10, and $dD = dx\,dy$

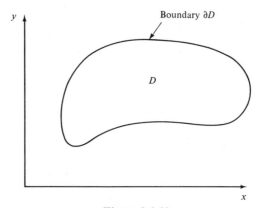

Figure 8.1.10

is the element of area. J may also be expressed as

$$J = \int_D \Phi(U, \nabla U, x, y)\, dD \qquad (8.1.46)$$

The variation of J is

$$\delta J = \int_D \left[\left(\frac{\partial \Phi}{\partial U}\right) \delta U + \left(\frac{\partial \Phi}{\partial \nabla U}\right) \delta_\Delta U \right] dD$$

$$= \int_D \left[\left(\frac{\partial \Phi}{\partial U}\right) \delta U + \left(\frac{\partial \Phi}{\partial \nabla U}\right) \nabla \delta U \right] dD \qquad (8.1.47)$$

After integration by parts, this becomes

$$\delta J = \int_D \left[\frac{\partial \Phi}{\partial U} - \nabla\left(\frac{\partial \Phi}{\partial \nabla U}\right) \right] \delta U\, dD + \int_{\partial D} \frac{\partial}{\partial n}\left(\frac{\partial \Phi}{\partial \nabla U}\right) \delta U\, ds \qquad (8.1.48)$$

The second integral is a contour integral on the boundary δD of the domain D. The term $\dfrac{\partial}{\partial n}(\cdot)$ is the derivative of $(\cdot)$ in the direction of the external normal to ∂D. This term may also be written as

$$\frac{\partial}{\partial n}\left(\frac{\partial \Phi}{\partial \nabla U}\right) = \left[\nabla\left(\frac{\partial \Phi}{\partial \nabla U}\right)\right]^T \bar{1}_n \qquad (8.1.49)$$

where $\bar{1}_n$ is the unit vector normal to ∂D.

To make the variation $\delta J = 0$ for all variations $\delta U(x, y)$, the function $U(x, y)$ must satisfy the corresponding Euler–Lagrange equation [from (8.1.48)]:

$$\frac{\partial \Phi}{\partial U} - \nabla\left(\frac{\partial \Phi}{\partial \nabla U}\right) = 0 \qquad (8.1.50)$$

If $\delta U = 0$ on the boundary, then satisfaction of (8.1.50) is sufficient for δJ to be zero for all δU (J is then stationary).

Examples of other boundary conditions are given in Chapter 10.

8.2 THE RAYLEIGH–RITZ METHOD

In the first half of the eighteenth century, mathematicians were concerned with finding analytic solutions to the vibrating-string equation

$$\frac{\partial^2 U}{\partial t^2} = c^2 \frac{\partial^2 U}{\partial x^2} \tag{8.2.1}$$

which had been discovered by Jean le Rond d'Alembert (1717–1783). Brook Taylor (1685–1731) had suggested that with the particular initial condition

$$U(x, 0) = a \sin \left(\frac{\pi x}{l} \right) \tag{8.2.2}$$

one could obtain an analytic solution that was sinusoidal in time as well as in space.

Leonhard Euler (1707–1783), Lagrange (1736–1813), and Daniel Bernouilli (1700–1782) considered *superpositions of sinusoidal solutions*, such as

$$U(x, t) = \sum_k a_k(t) \sin \left(\frac{k\pi x}{l} \right) \tag{8.2.3}$$

as solutions of (8.2.1). What is characteristic of this expression is that it is in the form of a *function series* or *function expansion*. Function series were to become an extremely important tool in the search for analytic solutions of partial differential equations. The generalization of (8.2.3) is

$$U = \sum_k \varphi_k(x) a_k \tag{8.2.4}$$

where the set of functions $\{\varphi_k(x)\}$ forms the *basis* of a subspace in which the solution of a differential equation is to be expressed.

The fact that trigonometric series [such as (8.2.3)] are known today as *Fourier series* results from the fact that Fourier used them extensively (a half century later) to express solutions to the heat equation (which he had just discovered) and that along the way he raised important questions about convergence.

The specific use of function expansions to obtain *approximate* solutions of partial differential equations started by the end of the nineteenth century and is illustrated by the work of *Rayleigh*,[3] extended by *Ritz* (1908). Their work was mostly concerned with finding solutions of steady-state or stationary problems described by *elliptic* partial-differential equations: in many such cases, the solution U minimizes some functional $J(U)$, and this fact was used as the starting point of a new class of methods, which we describe below.

Another important basic method in which approximations in the form of sums of functions are sought is that generally known as the *Galerkin* or *Ritz–Galerkin* method. It has more general applicability than the Rayleigh–Ritz method. Its essence was first described by Galerkin in a paper published in 1915, with emphasis on applications taken from structural engineering, such as plate and beam bending and vibrations problems.

The *finite-element method* entered the scene in the 1940s and 1950s. It is a hybrid between function expansion methods [representations of the form (8.2.4) are used] and finite-difference methods [the a_k in (8.2.4) are, as with finite differences, taken to be local or point values of the approximate solution].

One of the first theoretical descriptions of the finite-element method, based on the calculus of variations of the Rayleigh–Ritz method was given by Courant in 1943. The introduction to his paper contains interesting historical comments:

> Since Gauss and W. Thompson, the equivalence between boundary value problems of partial differential equations on the one hand and problems of the calculus of variations on the other hand has been a central point in analysis. At first, the theoretical interest in existence proofs dominated and only much later were practical applications envisaged by two physicists, Lord Rayleigh and Walther Ritz; they independently conceived the idea of utilizing this equivalence for numerical calculation of the solutions, by substituting for the variational problems simpler approximating extremum problems in which but a finite number of parameters need be determined. Rayleigh, in his classical work—*Theory of Sound*—and in other publications, was the first to use such a procedure. But only the spectacular success of Walther Ritz[4] and its tragic circumstances caught the general interest. In two publications of 1908 and 1909, Ritz, conscious of his imminent death from consumption, gave a masterly account of the theory, and at the same time applied his method to the calculation of the nodal lines of vibrating plates, a problem of classical physics that previously had not been satisfactorily treated.
>
> Thus methods emerged which could not fail to attract engineers and physicists; after all, the minimum principles of mechanics are more suggestive than the differential equations. Great successes in applica-

---

[3]J. W. Strutt, 3d Baron Rayleigh (1842–1919), British physicist. His famous treatise *Theory of Sound* was published in 1877. He belonged to the great Cambridge school of mathematical physicists which included also Sir William Thomson and, Clerk Maxwell.

[4]W. Ritz was a Swiss mathematician. He was 31 years old when he died in 1909.

tions were soon followed by further progress in the understanding of the theoretical background, and such progress in turn must result in advantages for the applications.

Somewhat independently, the finite-element method was developed by structural engineers in the 1950s. Their approach (one of the first descriptions was given by Argyris in 1954) borrows from the mathematics of equilibrium mechanics, originally with little or no reference to the calculus of variations (see the end of Section 8.6 for more historical comments on this).

With the advent of large electronic computers in the 1960s, the finite-element method has become a major tool in the computer solution of practical problems formulated not only by elliptic, but also (to a lesser extent) by parabolic and hyperbolic equations.

8.2.1   THE RAYLEIGH–RITZ CLASS
OF METHODS (VARIATIONAL METHODS)

As we have seen in Section 8.1, certain problems in ordinary and partial differential equations may be formulated in variational form. The specific part of the calculus of variations that is invoked is illustrated as follows.

Let J be a functional of the function $U(x)$ of the form

$$J(U) = \int_0^l \Phi\left(U, \frac{dU}{dx}, x\right) dx \qquad (8.2.5)$$

where $U(x)$ belongs to a class of twice-differentiable functions in $x \in [0, l]$, restricted to fixed end values $U(0)$ and $U(l)$. One may seek among these functions those for which $J(U)$ is stationary (in most cases an extremum). The definition of stationarity is that the first variation of J,

$$\delta J = \int_0^l \Phi\left(U + \delta U, \frac{d(U + \delta U)}{dx}, x\right) dx - \int_0^l \Phi\left(U, \frac{dU}{dx}, x\right) dx \qquad (8.2.6)$$

is equal to zero [i.e., δJ is $O(\delta U^r)$; $r \geq 2$].

We have seen in Section 8.1 that for J to be stationary, $U(x)$ must satisfy the associated *Euler–Lagrange* equation in $x \in (0, l)$:

$$\frac{d}{dx}\left(\frac{\partial \Phi}{\partial(dU/dx)}\right) - \frac{\partial \Phi}{\partial U} = 0 \qquad (8.2.7)$$

which is an *ordinary differential equation.*

Thus, an *exact* solution of equation (8.2.7) is obtained by finding a function $U(x)$ for which $J(U)$ is stationary (an extremum). And plausibly, an approximate solution may by analogy be found by:

1. Selecting an arbitrary form for the approximation which contains N free parameters

$$\left. \begin{array}{l} U(x) \simeq u(x, a_1, a_2, \ldots, a_N) \\ \left. \begin{array}{l} u(0) = U(0) \\ u(l) = U(l) \end{array} \right\} \text{ for all } \{a_k\} \end{array} \right\} \qquad (8.2.8)$$

2. Inserting it in (8.2.5) and seeking a solution to the N necessary conditions for an extremum in parameter space:

$$\frac{\partial J(u)}{\partial a_1} = \frac{\partial J(u)}{\partial a_2} = \cdots = \frac{\partial J(u)}{\partial a_N} = 0 \qquad (8.2.9)$$

The same holds for several-dimensional problems; for example, in two dimensions (8.2.5) becomes

$$J = \int_D \Phi\left(U, \frac{\partial U}{\partial x}, \frac{\partial U}{\partial y}, x, y\right) dD$$

$$= \int_D \Phi(U, \nabla U, x, y)\, dD \qquad (8.2.10)$$

and the corresponding Euler–Lagrange equation

$$\frac{\partial}{\partial x}\left(\frac{\partial \Phi}{\partial(\partial U/\partial x)}\right) + \frac{\partial}{\partial y}\left(\frac{\partial \Phi}{\partial(\partial U/\partial y)}\right) - \frac{\partial \Phi}{\partial U} \equiv \nabla\left(\frac{\partial \Phi}{\partial(\nabla U)}\right) - \frac{\partial \Phi}{\partial U} = 0 \qquad (8.2.11)$$

is a *partial differential equation* that must be satisfied by $U(x, y)$ for J to be stationary (extremal). Again, finding the values of $a_1, a_2, \ldots, a_N$ that make

$$J(u(x, y, a_1, a_2, \ldots, a_N))$$

stationary (with the form of u specified, and satisfying the boundary conditions for all values of $\{a_k\}$) is equivalent to finding an approximation u of an exact solution U of the partial differential equation (8.2.11).

This connection between differential equations and extremum properties of functionals is what has been taken advantage of by Rayleigh in the development of a general philosophy of approximation. This was further applied by Ritz into what is generally referred to as the *Rayleigh–Ritz method*.

The following example will illustrate these concepts.

EXAMPLE

Consider the equation describing the equilibrium position of a loaded string:

$$\tau \frac{d^2U}{dx^2} + F(x) = 0 \qquad (8.2.12)$$

$$U(0) = U(l) = 0 \qquad (8.2.13)$$

for which a variational principle was derived in Section 8.1 (see Figure 8.2.1). Equation (8.2.12) is the Euler–Lagrange equation of the functional

$$J = \int_0^l \left[-\frac{\tau}{2} \left(\frac{dU}{dx}\right)^2 + FU \right] dx \qquad (8.2.14)$$

This leads to the following.

Approximation Procedure.   If we want to find an approximate solution of the differential equation (8.2.12) we may do so by invoking its variational property and proceed as follows:

1.   Express $u(x)$, the intended approximation of $U(x)$, as a function that contains a finite number (say N) of undetermined parameters.

2.   Then, an approximation of the true solution will be obtained by minimizing $J(u)$ with respect to the undetermined parameters $a_1, a_2, \ldots, a_N$. The necessary conditions for such a minimum are expressed by (8.2.9).

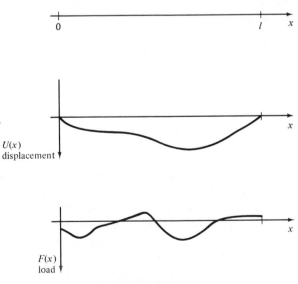

Figure 8.2.1

Among the functions $u(x, a_1, a_2, \ldots, a_N)$ that one may choose, a linear combination or expansion of the form

$$u(x) = \sum_{k=1}^{N} \varphi_k(x) a_k = \{\varphi_k\}^T \{a_k\} \tag{8.2.15}$$

where the φ_k are prescribed functions of x that satisfy the boundary conditions (8.2.13)

$$\varphi_k(0) = \varphi_k(l) = 0; \quad k = 1, 2, \ldots, N$$

is among the easiest to handle. We shall restrict ourselves to this choice, which is by far the most common.

We insert (8.2.15) into the expression of the functional (8.2.14) to obtain

$$J(u) = \int_0^l \left[-\frac{\tau}{2} \left(\sum_{k=1}^{N} \frac{d\varphi_k}{dx} a_k \right)^2 + F \sum_{k=1}^{N} \varphi_k a_k \right] dx \tag{8.2.16}$$

The N derivatives are equated to zero:

$$0 = \frac{\partial J(u)}{\partial a_k} = \int_0^l \left(-\tau \frac{d\varphi_k}{dx} \sum_{k'=1}^{N} \frac{d\varphi_{k'}}{dx} a_k + F\varphi_k \right) dx;$$

$$k = 1, 2, \ldots, N \tag{8.2.17}$$

We may observe that the two terms in this equation are [from equations (8.1.20) and (8.1.22)]

$$\int_0^l \tau \frac{d\varphi_k}{dx} \sum_{k'=1}^{N} \frac{d\varphi_{k'}}{dx} a_k \, dx = \frac{\partial \mathcal{P}(u)}{\partial a_k}$$

and

$$-\int_0^l F\varphi_k \, dx = \frac{\partial \mathcal{Q}(u)}{\partial a_k}$$

where $\mathcal{P}(u)$ and $\mathcal{Q}(u)$ are the internal and external energy of the string when it is assuming the shape (8.2.15). Thus, the interpretation of (8.2.17) is that the numerical solution should satisfy the principle of virtual works $w_I - w_E = 0$ when any one of the free parameters a_k is changed from equilibrium.

As an example of implementation, we may consider the special choice

$$\varphi_k(x) = \sin \left(\frac{k\pi x}{l} \right)$$

The application of (8.2.17) then results in the standard solutions as a Fourier series:

$$a_k = -\frac{\int_0^l F \sin (k\pi x/l)\, dx}{(l/2)\tau(k\pi/l)^2}$$

Details are left to the reader.

8.3   THE GALERKIN METHOD

We have seen that the solution U of a partial differential equation may be approximated by an expression of the form

$$U \simeq u = \sum_k \varphi_k a_k \tag{8.3.1}$$

where the $\varphi_k(x)$ are a basis of prescribed function of x.

The "variational" method was predicated on the fact that for many partial differential equations which originate from the physical sciences, the exact solution minimizes some functional, and forcing (8.3.1) to minimize the same functional provides a "natural" principle which is used in the family of approximations known as the Rayleigh–Ritz methods.

One may generalize these ideas and seek approximations of the form (8.3.1) by the use of principles other than those based on the calculus of variations. Let

$$LU = F(x) \tag{8.3.2}$$

be the general form of a partial differential equation in some domain D in x, to the solution of which we seek an approximation of the form (8.3.1)

The degree to which u fails to satisfy (8.3.2) is expressed by the *equation residual*, thus defined:

$$\Re = Lu - F \tag{8.3.3}$$

The smaller $\Re$, the better the approximation.

The essence of Galerkin's method is to require that this residual be orthogonal to the set of basis functions φ_k, that is, that

$$\langle \varphi_k, \Re \rangle \equiv \int_D \varphi_k \Re \, dD = 0; \qquad k = 1, 2, \ldots, N \tag{8.3.4}$$

using the free parameters in u to reach this objective.

EXAMPLE

By way of example, consider again the problem of finding an approximation to the steady-state displacement of a loaded string:

$$\tau \frac{d^2U}{dx^2} + F(x) = 0 \qquad (8.3.5)$$

with the homogeneous boundary conditions

$$U(0) = U(l) = 0 \qquad (8.3.6)$$

With a set of basis functions $\{\varphi_k(x)\}$ that satisfy those boundary conditions:

$$\varphi_k(0) = \varphi_k(l) = 0 \qquad \text{for all } k \qquad (8.3.7)$$

we form the approximation and equation residual

$$u = \sum_{k=1}^{N} \varphi_k a_k$$
$$\Re = \tau \sum_k \frac{d^2\varphi_k}{dx^2} a_k + F(x) \qquad (8.3.8)$$

Galerkin's conditions of orthogonality of this residual with respect to the basis functions read

$$\langle \varphi_k, \Re \rangle = \int_0^l \varphi_k \left[\tau \sum_k \frac{d^2\varphi_k}{dx^2} a_k + F(x) \right] dx = 0 \qquad (8.3.9)$$

Note that invoking the rule of integration by parts allows us to rewrite this expression as

$$\int_0^l \left(-\frac{d\varphi_k}{dx} \tau \sum_{k'=1}^N \frac{d\varphi_{k'}}{dx} a_k + \varphi_k F \right) dx \qquad (8.3.10)$$

This expression may be seen to be identical with (8.2.17) from the variational, Rayleigh–Ritz approximation to the same problem. It is this analogy that led Galerkin (1915) to express the general method expressed by (8.3.4). But whereas the Rayleigh–Ritz method is restricted to equations that satisfy a variational principle, the Galerkin method does not have that limitation.

Another important implication of (8.3.10) is this: whereas consideration of (8.3.9) would lead us to believe that the basis functions $\{\varphi_k(x)\}$ must be twice-differentiable (because of the presence of the $d^2\varphi_k/dx^2$ terms), (8.3.10) shows that *simple* differentiability is sufficient. More on this follows.

8.3.1 WEAK SOLUTIONS
OF A DIFFERENTIAL EQUATION

The concept of *weak solution* is central in attempts to analyze mathematically the class of approximation methods based on expressing the solution as (8.3.1). The distinction between strong and weak solution deserves clarification.

A *strong solution* of

$$LU = F$$

is a function $U(x)$ that satisfies this equation in every point of D. If L is a differential operator of order m, then U *must* belong to $C^{m-1}(D)$, the class of functions with continuous partial derivatives of order $(m - 1)$ everywhere in D. Now, we can expand the class of functions in which we seek solutions (or approximate solutions) to $LU = F$ by considering test functions $v(x)$ which belong to some space $\mathcal{V}$. The equation is multiplied by those test functions and integrated over D to yield

$$\langle v(x), Lu - F \rangle = 0 \tag{8.3.11}$$

A function $u(x)$ that satisfies this equation for every function $v(x)$ in $\mathcal{V}$ is a *weak solution* of our equation. Analytically, everything depends on the choice of $\mathcal{V}$. If $\mathcal{V}$ contains only $C^{m-1}(D)$ with appropriate boundary conditions, then a formal integration by parts shifts all the derivatives of u onto v, leading to the equation

$$\langle L^*v, u \rangle - \langle v, F \rangle = 0 \tag{8.3.12}$$

where L^* is the adjoint of L. In this weakest form, u is not required to have any continuity at all.

On the other extreme, if $\mathcal{V}$ includes all Dirac delta functions, then no derivatives can be shifted from u onto v and u will have to belong to $C^{m-1}(D)$ and satisfy the equation $Lu = F$ in the classical or "*strong*" sense.

What emerges from this is that to every test space $\mathcal{V}$ corresponds a solution space $\mathcal{U}$ to which functions $u(x)$ must belong to qualify as *weak* solutions.

In Galerkin's method, $\mathcal{U}$ and $\mathcal{V}$ are the same space, that spanned by the set of basis function $\{\varphi_k(x)\}$. Whereas strong solutions of $LU = F$ must belong to $C^{m-1}(D)$, it is sufficient for Galerkin (weak) solutions to belong to $C^{(m/2)-1}(D)$ only.

The excellent book by Strang and Fix (1973) is recommended reading for those interested in pursuing this analysis in more detail.

8.4 WEIGHTED RESIDUAL METHODS

For an equation expressed in the general form

$$LU = F \qquad (8.4.1)$$

the essence of Galerkin's method is to first express the equation residual

$$\Re \equiv Lu - F \qquad (8.4.2)$$

where u is a trial solution expressed as a linear combination of basis functions:

$$U \simeq u = \sum_k \varphi_k a_k \qquad (8.4.3)$$

and then to require that $\Re$ be orthogonal to those basis functions:

$$\langle \varphi_k, \Re \rangle \equiv \int_D \varphi_k \Re \, dD = 0; \qquad k = 1, 2, \ldots \qquad (8.4.4)$$

where D is the domain of the equation. A generalization of this procedure is obtained when the Galerkin conditions (8.4.4) are replaced by the more general condition

$$\langle w_k, \Re \rangle \equiv \int_D w_k \Re \, dD = 0 \qquad (8.4.5)$$

where the $\{w_k(x)\}$ are a set of prescribed "weight" functions. The name *method of weighted residual* was coined by Crandall (1956) to describe this general procedure, which is summarized by (8.4.3)–(8.4.5). Galerkin's procedure is thus a particular case of the method of weighted residuals with $w_k(x) = \varphi_k(x)$.

8.4.1 COLLOCATION METHOD[5]

This is a special case of the weighted residual method. Let

$$\{x_n; n = 1, 2, \ldots, N\} \qquad (8.4.6)$$

be points in D, called *collocation points*. The weight functions are then chosen as

$$w_n = \delta(x - x_n) \qquad (8.4.7)$$

---

[5]Also called *method of selected points* (Lanczos, 1956) and *pseudospectral approximation* (Orszag, 1971, 1972).

where the $\delta(x - x_n)$ are Dirac delta functions. The simple consequence of this choice is that the equation (8.4.5) becomes

$$\Re(x_n) = 0; \qquad n = 1, 2, \ldots, N \tag{8.4.8}$$

that is, it is required that the residual vanish in the N collocation points. For the equation

$$LU(x) = F(x) \tag{8.4.9}$$

the residual is

$$\Re = L(\sum_k \varphi_k(x)a_k) - F(x) \tag{8.4.10}$$

and the collocation conditions (8.4.8) here are

$$L(\sum_k \varphi_k(x)a_k)_n - F(x_n) = 0; \qquad n = 1, 2, \ldots, N \tag{8.4.11}$$

If the problem is linear, then these equations become

$$\sum_k L\varphi_k(x_n)a_k - F(x_n) = 0 \tag{8.4.12}$$

which form a linear system of algebraic equations with matrix

$$\{L\varphi_k(x_n)\}$$

The conditions under which this system of equations has a unique solution $\{a_k\}$ is that the matrix be regular, that is,

$$\det\{L\varphi_k(x_n)\} \neq 0 \tag{8.4.13}$$

The question of where the collocation points should be located to give the best results is worth pursuing (see Lanczos, 1956, pp. 504–507). An interesting version is the "orthogonal collocation method," in which the collocation points are chosen as the zeros of orthogonal polynomials on D. The underlying theoretical justification is to be found with the principles of Gaussian quadrature (see Section 12.4). Certain aspects will be described in Section 11.5.

Solution Space for the Collocation Method. Because the weight functions $w = \delta(x - x_n)\delta(y - y_n)\ldots$ are not differentiable, *no* derivatives of u in the expression of the weak conditions (8.4.5), which we may rewrite as

$$\langle w, Lu - F \rangle = 0$$

may be shifted from u onto w. Therefore, if L is a differential operator of order m, u must belong to $C^{m-1}(D)$, the class of functions with continuous derivatives up to order $(m - 1)$ in D. This stands in contrast with Galerkin's method, where u had to belong to $C^{(m/2)-1}(D)$ only.

Fourier Sine Series as an Application of Collocation.   We may show that the Fourier sine series method of Section 4.4 can be interpreted as a special case of collocation. For the problem

$$LU \equiv -\frac{d^2U}{dx^2} = F(x)$$

$$U(0) = U(l) = 0$$

(8.4.14)

let

$$U \simeq u = \sum_{k=1}^{N} \varphi_k(x)b_k$$

$$\varphi_k(x) = \sin\left(\frac{k\pi x}{l}\right)$$

(8.4.15)

and

$$x_n = nh = n\frac{l}{N+1}; \qquad n = 1, 2, \ldots N$$

The residual is

$$\Re = \sum_{k=1}^{N} b_k \left(\frac{k\pi}{l}\right)^2 \sin\left(\frac{k\pi x}{l}\right) - F(x)$$

and the collocation conditions are

$$\sum_{k=1}^{N} b_k \left(\frac{k\pi}{l}\right)^2 \sin\left(\frac{k\pi x_n}{l}\right) - F(x_n) = 0; \qquad n = 1, 2, \ldots, N \qquad (8.4.16)$$

This system of equations can be solved formally for the $\{b_k\}$ by using the orthogonality relations

$$\sum_{n=1}^{N} \sin\left(\frac{k_1 \pi x_n}{l}\right) \sin\left(\frac{k_2 \pi x_n}{l}\right) = \begin{cases} 0 & \text{when } k_1 \neq k_2 \\ \dfrac{N+1}{2} & \text{when } k_1 = k_2 \neq 0 \end{cases}$$

to derive, for $k = 1, 2, \ldots, N$,

$$b_k = \left(\frac{l}{k\pi}\right)^2 \frac{2}{N+1} \sum_{n=1}^{N} F(x_n) \sin\left(\frac{k\pi x_n}{l}\right) \qquad (8.4.17)$$

These are precisely the equations found in Section 4.4 as an application of discrete Fourier series.

8.4.2 SUBDOMAIN METHOD

This is another special case of weighted residuals. In this case, the domain D is covered with N subdomains,

$$D_1, D_2, \ldots, D_N$$

which are not necessarily covering the whole of D and which may have overlaps. The weighting functions $\{w_n\}$ are chosen as

$$w_n = \begin{cases} 1 & \text{in } D_n \\ 0 & \text{outside } D_n \end{cases} \tag{8.4.18}$$

This method was first described by Biezeno (1923–1924).

8.5 SPECTRAL METHODS

8.5.1 THE HERMITIAN PROPERTY

The eigenfunctions of a differential operator L in D associated to given homogeneous boundary conditions on the boundary ∂D are, by definition, those functions φ that satisfy the boundary conditions and for which

$$L\varphi = \lambda\varphi$$

where λ is a scalar. Certain elliptic operators have the property that their eigenvalues λ are a discrete set of *real* numbers, and that the corresponding eigenfunctions form an orthogonal set. Such operators are called *Hermitian*.

When the operator L and boundary conditions of an elliptic problem are *Hermitian*, then using the eigenfunctions of L as the basis functions of a Galerkin-type approximation of the solution results in a particularly simple formulation. (We shall see that the Hermitian property holds for *self-adjoint* operators.)

Before entering in this, we shall review some of the relevant mathematics.

8.5.2 SELF-ADJOINT OPERATORS

Let L be a linear operator on functions $U(x)$ in D. Also, let

$$BU = 0 \tag{8.5.1}$$

be boundary conditions to be satisfied by U on ∂D where B is a linear homogeneous operator on U.

A (nonzero) function $\varphi(x)$ that satisfies the boundary conditions (8.5.1) is called an eigenfunction of the operator L if it satisfies

$$L\varphi(x) = \lambda\varphi(x) \qquad (8.5.2)$$

where λ is a scalar, called the corresponding eigenvalue. This equation is called the characteristic equation for the operator L.

There are operators L for which any number λ is an eigenvalue. For instance, if

$$L = \frac{\partial}{\partial x}$$

$$D \equiv x \geq 0$$

and

$$BU \equiv U(0) - 1 = 0$$

then

$$\varphi(x) = e^{\lambda x}$$

is an eigenfunction for any real or complex value of λ.

But operators L which occur in elliptic problems often have the interesting property that:

1. Only for an infinite sequence of *discrete, real* values $\{\lambda_k\}$ of λ does (8.5.2) have a solution.

2. The corresponding eigenfunctions $\{\varphi_k(x)\}$ form an orthogonal set of functions on D:

$$\langle \varphi_{k_1}, \varphi_{k_2} \rangle \equiv \int_D \varphi_{k_1}\varphi_{k_2}\, dD = 0; \qquad k_1 \neq k_2 \qquad (8.5.3)$$

This turns out to have far-reaching consequences, and has been used profusely in the applied mathematics of the last two centuries.

Also interesting are the possible computational uses of this property. The set of all eigenvalues λ of an operator is called the *spectrum* of that operator, and computational methods that make use of the eigen functions of an operator as a *basis* in which the solution is to be expressed are sometimes called *spectral methods*.

The Self-Adjoint Property.   The operator L associated to the boundary conditions (8.5.2) is called self-adjoint if, for any two functions $U(x)$ and $V(x)$ that satisfy the boundary conditions, the relation

$$\langle U, LV \rangle = \langle LU, V \rangle \qquad (8.5.4)$$

holds, where

$$\langle f, g \rangle = \int_D fg \, dD \tag{8.5.5}$$

THEOREM 1 *The eigenvalues of a self-adjoint operator are real.*

Proof. Let λ_k be an eigenvalue of a self-adjoint operator and $\varphi_k(x)$ be the associated eigenfunction. If λ_k were complex, then $\lambda_{k'} = \bar{\lambda}_k$ would also be an eigenvalue of L with eigenfunction[6] $\bar{\varphi}_k$. By the self-adjoint property of L we would then have

$$\langle \bar{\varphi}_k, L\varphi_k \rangle = \langle L\bar{\varphi}_k, \varphi_k \rangle$$

or

$$\langle \bar{\varphi}_k, \lambda_k\varphi_k \rangle = \lambda_k\langle \bar{\varphi}_k, \varphi_k \rangle = \bar{\lambda}_k\langle \bar{\varphi}_k, \varphi_k \rangle = \bar{\lambda}_k\langle \bar{\varphi}_k, \varphi_k \rangle \tag{8.5.6}$$

which implies that $\lambda_k = \bar{\lambda}_k$, thus that λ_k is real. $\square$

THEOREM 2 *The eigenfunctions of a self-adjoint operator are orthogonal.*

Proof. Let φ_1 and φ_2 be the eigenfunctions corresponding to two distinct eigenvalues λ_1 and λ_2. By the self-adjoint property of L we have

$$\langle \varphi_1, L\varphi_2 \rangle = \langle L\varphi_1, \varphi_2 \rangle$$

or

$$\lambda_2\langle \varphi_1, \varphi_2 \rangle = \lambda_1\langle \varphi_1, \varphi_2 \rangle \tag{8.5.7}$$

Since $\lambda_1 \neq \lambda_2$, this implies that

$$\langle \varphi_1, \varphi_2 \rangle = 0 \tag{8.5.8}$$

(i.e., orthogonality). $\square$

8 5.3 APPLICATIONS

Now consider the elliptic problem,

$$LU = F \qquad \text{in } D \tag{8.5.9}$$

$$BU = 0 \qquad \text{on } \partial D \tag{8.5.10}$$

---

[6] It is assumed that all coefficients in L are *real*.

where L associated to the boundary conditions (8.5.10) is self-adjoint (i.e., Hermitian). We seek a solution in the form of a function expansion:

$$U \simeq u = \sum_{k=1}^{N} \varphi_k(x)a_k \qquad (8.5.11)$$

where the $\varphi_k(x)$ are eigenfunctions of L. Applying the formal steps of Galerkin's method, we form the residual

$$\Re \equiv L(\sum_k \varphi_k a_k) - F \qquad (8.5.12)$$

The equations

$$\langle \varphi_k, \Re \rangle = 0; \qquad k = 1, 2, \ldots, N \qquad (8.5.13)$$

become, owing to the orthogonality of the $\{\varphi_k(x)\}$,

$$a_k = \frac{\langle \varphi_k, F \rangle}{\langle \varphi_k, \varphi_k \rangle} \qquad (8.5.14)$$

which can be computed one a_k at a time. Therein lies the attractiveness of the method.

The first Poisson's equation solver of Section 7.1 was in fact an application of this method. There are also many special techniques used to compute neutron flux shapes inside nuclear reactors which make use of the self-adjoint property of the neutron diffusion operator in the manner described here (they are sometimes called synthesis techniques in this context), see Stacey (1967) and references cited.

8.6   INTRODUCTION TO THE FINITE-ELEMENT METHOD

The finite-element method is somewhat of a crossbreed between function-expansion methods and finite-difference methods. What is conserved from finite-difference methods is that the discrete parameters which are to represent the numerical approximation with finite elements are still *local values* of the solution. The domain D of the problem is divided into subregions D_e. The shape of the finite elements is generally simple polygons such as triangles and quadrilaterals in two dimensions; pyramids and triangular and rectangular prisms in three dimensions;[7] and so on. In each element, the approximate solution assumes the form of a linear combination of prescribed functions

$$U(x, y, \ldots) \simeq u(x, y, \ldots) = \sum_n \varphi_n(x, y, \ldots)u_n \qquad (8.6.1)$$

where the $\varphi_n(x, y, \ldots)$ are analytically simple, such as low-order polynomials and the like, and the $\{u_n\}$ are, as in the case of finite-difference methods, local or nodal values of the solution[8]:

We have a representation of U as a function expansion of the kind discussed in Chapter 7, which is different in each element D_e and whose coefficients u_n are approximations of the solution in discrete points[9] which are either contained in or directly adjacent to the element D_e (rather than *generalized* parameters a_k as was the case in the function-expansion methods described in the preceding section).

Having chosen a *form of representation* such as (8.6.1) for the numerical solution, one must proceed to derive equations to be satisfied by the u_n for (8.6.1) to be indeed an approximation of the problem at hand. As in the case of the function-expansion methods described in the preceding section, several principles may be used to derive these equations. Specifically:

1.   Variational principles may be invoked when they apply.

2.   Galerkin's or weighted residual types of minimization may be used.

3.   In some cases, methods invoking physical principles such as *conservation* in calculations related to transport processes, or the principle of *virtual works* of classical mechanics may be used directly to derive equations for the $\{u_n\}$.

EXAMPLE

As a simple example, consider our model second-order equation

$$\frac{d^2U}{dx^2} + F(x) = 0 \qquad (8.6.2)$$

with boundary conditions

$$U(0) = U(l) = 0 \qquad (8.6.3)$$

---

[7]The sides of those elements may be curved; see Chapter 12.

[8]There are occasional exceptions to this statement. Although rare, one finds in the literature a few examples of the use of "nodeless" variables in finite elements.

[9]Partial derivatives at the nodes—$(\partial u/\partial x)_n$, $(\partial u/\partial y)_n$, . . .—are also sometimes carried. Such representations are called Hermitian.

Let the domain $x \in [0, l]$ be divided in $N + 1$ subdomains of equal length by the points

$$x_n = nh; \qquad h = \frac{l}{N + 1}$$

See Figure 8.6.1. In each interval, we choose to approximate U by the linear interpolant between the nodal points as illustrated in Figure 8.6.2. Such a piecewise-linear function may be expressed analytically as

$$u(x) = \sum_n \varphi_n(x)u_n = \{\varphi_n(x)\}^T\{u_n\} \qquad (8.6.4)$$

where the $\{\varphi_n(x)\}$ are the linear "roof" or "chapeau" functions[10]:

$$\varphi_n(x) = \begin{cases} 1 - \dfrac{|x - x_n|}{h} & \text{for } |x - x_n| \leq h \\ 0 & \text{elsewhere} \end{cases} \qquad (8.6.5)$$

See Figure 8.6.3. We may choose to approximate $F(x)$ in the same basis of linear functions:

$$F(x) \simeq f(x) = \sum_n \varphi_n(x)f_n = \{\varphi_n(x)\}^T\{f_n\}; \qquad f_n = F(x_n) \qquad (8.6.6)$$

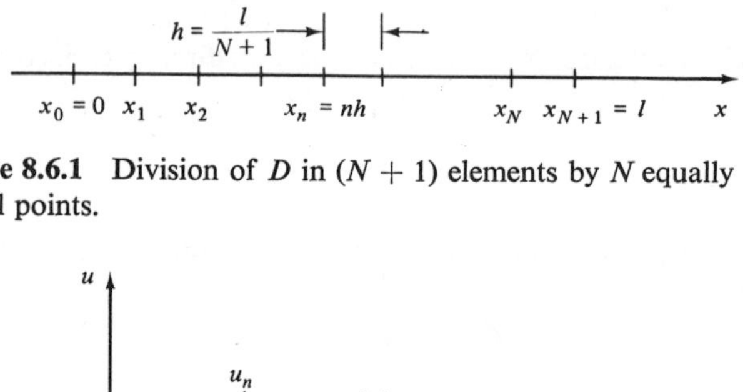

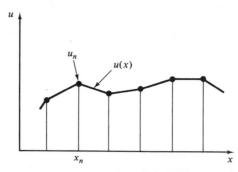

Figure 8.6.1   Division of D in $(N + 1)$ elements by N equally spaced nodal points.

Figure 8.6.2   Linear finite-element representation of the function $U(x)$.

---

[10]"Chapeau" means "hat" in French.

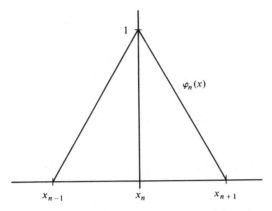

Figure 8.6.3 Linear basis or "chapeau" function.

In the spirit of Galerkin's method, we form the equation residual

$$\Re = \frac{d^2u}{dx^2} + f = \sum_n \left(\frac{d^2\varphi_n}{dx^2} u_n + \varphi_n f_n \right) \tag{8.6.7}$$

and impose that it be orthogonal to the basis functions (8.6.5) for all interior points: that is,

$$0 = \langle \varphi_n, \Re \rangle$$
$$= \int_0^l \varphi_n \sum_m \frac{d^2\varphi_m}{dx^2} u_m \, dx + \int_0^l \varphi_n \sum_m \varphi_m f_m \, dx; \qquad n = 1, 2, \ldots, N \tag{8.6.8}$$

The first integral term would appear to introduce a difficulty, since the second derivative of the linear basis functions does not exist in the usual sense. But this difficulty is removed when the rule of integration by parts is invoked:

$$\int \varphi_n \frac{d^2\varphi_m}{dx^2} \, dx = \left(\varphi_n \frac{d\varphi_m}{dx} \right) - \int \frac{d\varphi_n}{dx} \frac{d\varphi_m}{dx} \, dx$$
$$= 0 - \int \frac{d\varphi_n}{dx} \frac{d\varphi_m}{dx} \, dx$$
$$= \begin{cases} \dfrac{-2}{h} & \text{when } n = m \\[2mm] \dfrac{1}{h} & \text{when } |n - m| = 1 \\[2mm] 0 & \text{otherwise} \end{cases} \tag{8.6.9}$$

The other integral in (8.6.8) is

$$\int \varphi_n \varphi_m \, dx = \begin{cases} \dfrac{2h}{3} & \text{when } n = m \\[2mm] \dfrac{h}{6} & \text{when } |n - m| = 1 \\[2mm] 0 & \text{otherwise} \end{cases} \qquad (8.6.10)$$

Whereupon collecting results gives us the system of algebraic equations

$$\frac{u_{n-1} - 2u_n + u_{n+1}}{h^2} + \frac{1}{6}(f_{n-1} + 4f_n + f_{n+1}) = 0;$$

$$n = 1, 2, \ldots, N \qquad (8.6.11)$$

which approximates the differential equation (8.6.2). We note that this expression is a discrete approximation. It *resembles* the familiar expression obtained by classical finite differences

$$\frac{u_{n-1} + u_{n+1} - 2u_n}{h^2} + f_n = 0 \qquad (8.6.12)$$

The second term of (8.6.11) amounts to an approximation of the mean value of $F(x)$:

$$\overline{F(x)} \text{ in } [x_{n-1}, x_{n+1}] = \frac{1}{2h} \int_{x_{n-1}}^{x_{n+1}} F(x) \, dx \simeq \frac{1}{6}(f_{n-1} + 4f_n + f_{n+1}) \qquad (8.6.13)$$

obtained by Simpson's rule of quadrature. The approximation may be expressed in matrix notations as

$$\mathbf{A}_1\{u_n\} = \mathbf{A}_2\{f_n\} \qquad (8.6.11a)$$

where $\mathbf{A}_1$ and $\mathbf{A}_2$ are the matrices

$$\mathbf{A}_1 = \frac{1}{h^2} \begin{bmatrix} -2 & 1 & & & & & \\ 1 & -2 & 1 & & & \mathbf{0} & \\ & 1 & -2 & 1 & & & \\ & & \cdot & \cdot & \cdot & & \\ & & & \cdot & \cdot & \cdot & \\ \mathbf{0} & & & & \cdot & \cdot & 1 \\ & & & & & 1 & -2 \end{bmatrix} \qquad (8.6.14)$$

$$
\mathbf{A}_2 = \frac{1}{6}
\begin{bmatrix}
1 & 4 & 1 & & & & \\
& 1 & 4 & 1 & & \mathbf{0} & \\
& & \cdot & \cdot & \cdot & & \\
& & & \cdot & \cdot & \cdot & \\
& & & & \cdot & \cdot & \cdot \\
& \mathbf{0} & & & \cdot & \cdot & \cdot \\
& & & & 1 & 4 & 1
\end{bmatrix}
\tag{8.6.15}
$$

and

$$
\{u_n\} =
\begin{Bmatrix}
u_1 \\ u_2 \\ \cdot \\ \cdot \\ \cdot \\ u_n \\ \cdot \\ \cdot \\ \cdot \\ u_N
\end{Bmatrix};
\qquad
\{f_n\} =
\begin{Bmatrix}
f_0 \\ f_1 \\ \cdot \\ \cdot \\ \cdot \\ f_n \\ \cdot \\ \cdot \\ \cdot \\ f_{N+1}
\end{Bmatrix}
\tag{8.6.16}
$$

8.6.1 SEVERAL-DIMENSIONAL PROBLEMS

The greatest success of the finite-element method has been with problems in several spatial dimensions. As a typical problem, consider Poisson's equation

$$
-\nabla^2 U \equiv -\left(\frac{\partial^2 U}{\partial x^2} + \frac{\partial^2 U}{\partial y^2}\right) = F(x, y)
\tag{8.6.17}
$$

in a domain D (Figure 8.6.4a) associated with Dirichlet boundary conditions

$$
U = G \qquad \text{on } \partial D
\tag{8.6.18}
$$

The steps involved in solving this problem with the aid of the finite-element method are, as explained before:

1. The domain D is divided in two-dimensional subregions D_e. For example, Figure 8.6.4b shows a division of D in triangular subdomains.

2. Nodes are selected at the vertices (and often additional points) of each subdomain D_e. The discrete parameters describing the approximate solution are the values at the nodes or "nodal" values:

$$
u_n \simeq U(x_n, y_n)
$$

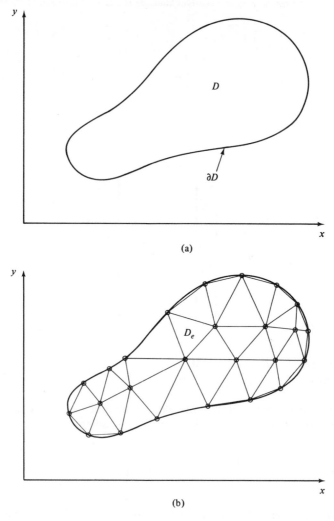

Figure 8.6.4 (*a*) Two-dimensional domain *D* and (*b*) its division in triangular elements D_e.

For example, the nodes in Figure 8.6.4b are at the vertices of the triangular elements.

3. A simple polynomial shape that interpolates between the nodal values is assumed for the approximate solution in each element. In our example, we are more-or-less forced by the number of nodal points per element (three) to choose linear interpolation (i.e., planes) to represent the approximate solution in each element D_e. We show in Section 10.2 how such a piecewise-linear function can be expressed in the form (8.6.1) of a weighted sum of basis functions.

4. Some principle, leading to integrals of the type (8.6.8), is used to obtain discrete equations from which the $\{u_n\}$ can be computed. The manner in which this can be done for our example of Poisson's equation approximated by linear elements over triangles is described in Section 10.2.

The field in which the finite-element method has been most successful is that of the analysis of stresses and deformation in continuous structures such as airplane fuselages, pressure vessels, turbine blades, and the like. Figure 8.6.5 shows the division of a pressure vessel and of an aerodynamic structure in finite elements which were used to compute deformations and internal stresses.

8.6.2 IMPORTANCE OF THE FINITE-ELEMENT METHOD

Mathematicians often claim that the remarkable success of the finite-element method over finite-difference methods is due to the fact that the former is more accurate. Although it may be true that the finite-element method is indeed sometimes more accurate, there is another factor which, in our opinion, is as important in explaining its success: The computer solution of an engineering problem involving partial differential equations with either the finite-difference or finite-element method consists essentially of two steps (Figure 8.6.6):

1. Derivation of a discrete approximation of the problem consisting in a finite set of algebraic equations between the unknown nodal values $\{u_n\}$ of a numerical solution.

2. Solution of the set of algebraic equations derived in step 1, in order to obtain the numerical solution.

The form of the discrete equations that are solved in step 2 is not very different between the two methods, and the algorithms and computer programs used for their solution are similar in complexity, in fact, often almost identical. But with finite-difference methods, except for relatively simple problems and simple geometries, the formulation of the discrete equations must be done by hand, with pencil and paper, which uses up a great deal of time and energy on the part of the user, the engineer.

By contrast, the finite-element method is eminently suitable to the synthesis of algorithms that "automate" the derivation of the discrete equations, reducing this step to a set of basic rules which can be incorporated into computer codes. The only work required from the user is to supply the code with his or her choice of the location of the nodes, the shape of the elements, and so on. The discrete equations are derived and then *assembled* by finite-element codes

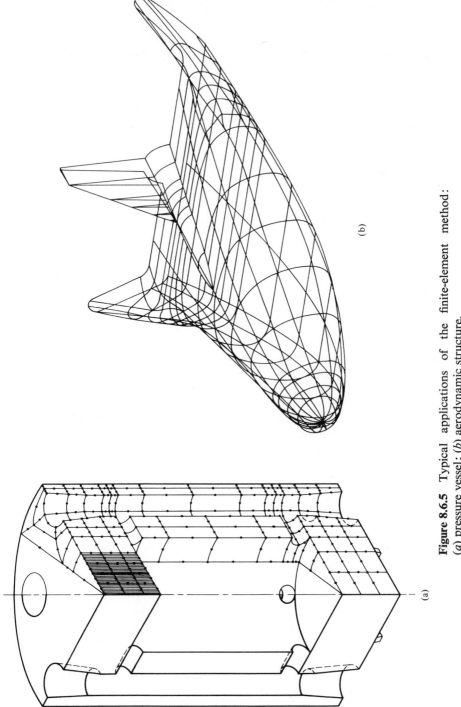

Figure 8.6.5 Typical applications of the finite-element method: (a) pressure vessel; (b) aerodynamic structure.

(b)

(a)

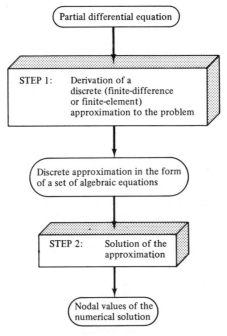

Figure 8.6.6

prior to the execution of step 2. The finite-element method may thus be viewed as a labor-saving device, delegating to the computer the tedious task of calculations and bookkeeping involved in the derivation of the discrete equations in step 1. The computer implementation of these derivations requires large computers (for the large engineering problems we are accustomed to). It is thus by no coincidence that the spectacular rise of the finite-element method (in the 1960s) came at the same time that large, fast scientific computers were becoming available.

8.6.3 MORE HISTORY

Although the name "finite-element method" was coined in 1960 (by Clough, in a paper dealing with elasticity problems), the concepts of the method were reasonably familiar at the time. Finite differences had been used by engineers for decades, as well as methods in which the solution of differential equations is approximated by simple shapes such as polynomials. The idea of putting the two together by dividing the domain into finite sections in which the equation is approximated locally by such simple shapes was a natural extension, but it presented technical difficulties. An important paper in this path of development of ideas was published by Hrenikoff in 1941.

He was concerned with the problem of deformation in elastic solids, and he assumed that a continuous structure could be divided into structural sections (or "elements," although Hrenikoff did not use this word) interconnected *only* at a finite number of interface "nodal points." This results in great simplification, since dividing a continuum in sections with a continuous (rather than discrete) interface results in much more complicated mathematics. Hrenikoff's "framework method" of approximation was found to work, and became the starting point of what was to be the finite-element method in the engineering community.

It is not until the mid-1950s, though (rekindled by the emerging availability of electronic computers), that the method became systematically developed as a working tool. A series of papers were published on this subject by Argyris and collaborators (starting in 1954). At about the same time, a landmark paper was published by Turner et al. (1956) in which triangular elements are used to solve plane-stress problems. It is then that the engineering community started to recognize that the finite-element method was an efficient tool to be used in elasticity problems, in particular with the help of automatic computers, to which it is well suited.

Mathematicians quote Courant (1943) as the origin of the finite-element method. In that paper, Courant used triangular elements to generate trial functions to which he applied the Rayleigh–Ritz method to approximate solutions of column torsion problems.

It is not until the works of Besseling (1963), Fraeijs de Veubeke (1964a, 1964b) and Jones (1964) that it became fully recognized that the engineers' finite-element method, where the equations are developed by what is known as the "stiffness method," is equivalent to Courant's Ritz method.

That the general ideas leading to the finite-element method were in the air is also illustrated by a paper published by MacNeal in 1953, quite unrelated to those other developments. What this paper deals with is the problem of synthesizing an asymmetric network of electrical resistances for the analog solution of Laplace's equation (see Section 7.2). While standard methods work well to generate resistance values to be used for symmetric networks (these are networks corresponding to *standard finite differences* on regular rectangular or triangular meshes), they fail in the asymmetric case. The method used by MacNeal in that case is in fact a finite-element method, in which continuity over subdomains is used to generate the discrete equations.

9

Two-Point
Boundary-Value Problems II

9.1 LINEAR FINITE ELEMENTS
FOR ONE-DIMENSIONAL PROBLEMS

In Section 1.4 we introduced linear finite elements for the solution of a one-dimensional boundary-value problem. Further aspects of the application of linear elements will be considered here. We shall rely heavily on examples to introduce new concepts. Although the examples and the element shapes are obviously simple, most concepts which they introduce remain applicable to more complex problems (e.g., problems in several dimensions) and more complex finite-element shapes.

9.1.1 FINITE-ELEMENT-GALERKIN
SOLUTION OF A NONHOMOGENEOUS
DIFFUSION PROBLEM

The steady-state distribution of temperature in a heated slab is described by the equation

$$\frac{d}{dx}\left(P(x)\frac{dU}{dx}\right) + F(x) = 0 \quad \text{in } x \in (0, l) \tag{9.1.1}$$

where $U(x)$ = temperature distribution

$P(x) =$ a given "conductivity" function of the material of the slab
$F(x) =$ distributed heat source in the slab
$x =$ linear distance perpendicular to the surfaces of the slab

It is assumed that the temperature is imposed on both surfaces:

$$U(0) = U_0 \quad \text{(prescribed)}$$
$$U(l) = U_l \quad \text{(prescribed)}$$

(9.1.2)

which constitute the boundary conditions of the equation (Figure 9.1.1).

We seek an approximation of the solution $U(x)$ using linear finite elements on a regular grid: that is,

$$U(x) \simeq u(x) = \sum_{n=0}^{N+1} \varphi_n(x)u_n$$

(9.1.3)

where $u_n \simeq U(x_n)$ and the $\varphi_n(x)$ are the linear finite elements or chapeau functions illustrated in Figure 9.1.2. The application of Galerkin's method asks that we first form the equation residual:

$$\Re \equiv \frac{d}{dx}\left(P(x)\frac{du(x)}{dx}\right) + F(x)$$

(9.1.4)

and then equate to zero its scalar product with each basis function $\varphi_n(x)$ corresponding to a "free" point; that is, for the points $n = 1, 2, \ldots, N$ where U is not otherwise specified (in $n = 0$ and $N + 1$, U is a given boundary condition),

$$0 = \langle \varphi_n, \Re \rangle$$

$$= \left\langle \varphi_n, \frac{d}{dx}\left(P(x)\sum_{m=0}^{N+1}\frac{d\varphi_m}{dx}u_m\right) + F(x) \right\rangle$$

$$= \sum_m \left\langle \varphi_n, \frac{d}{dx}\left(P(x)\frac{d\varphi_m}{dx}\right) \right\rangle u_m + \langle \varphi_n, F(x) \rangle; \quad n = 1, 2, \ldots, N \quad (9.1.5)$$

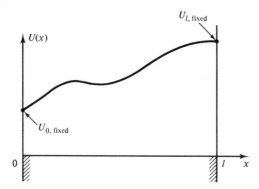

Figure 9.1.1

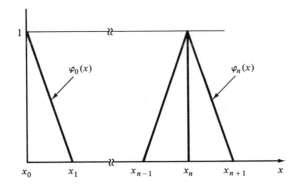

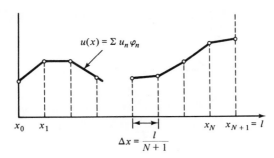

Figure 9.1.2

Or, after integration by parts of the first term,

$$0 = -\sum_m \left\langle \frac{d\varphi_n}{dx}, P(x)\frac{d\varphi_m}{dx}\right\rangle u_m + \langle \varphi_n, F(x)\rangle; \qquad n = 1, 2, \ldots, N \qquad (9.1.6)$$

The coefficients of this system of equations contain the integrals

$$\left\langle \frac{d\varphi_n}{dx}, P(x)\frac{d\varphi_m}{dx}\right\rangle = \int_0^l \frac{d\varphi_n}{dx}P(x)\frac{d\varphi_m}{dx}\,dx \qquad (9.1.7)$$

and

$$\langle \varphi_n, F(x)\rangle = \int_0^l \varphi_n F(x)\,dx \qquad (9.1.8)$$

The value of these integrals could be obtained by:

1. Analytic evaluation, in those occurrences where such analytic evaluation is feasible.

2. Numerical quadrature, which may be warranted when the variation of either dP/dx or dF/dx is likely to be large within one interval Δx.

3. Expressing $P(x)$ and $F(x)$ also by a linear finite-element approximation.

We may show that the third alternative leads to a simple treatment. We choose to express P and F as a weighted sum of the same linear basis functions $\varphi_n(x)$:

$$P(x) \simeq p(x) = \sum_{n=0}^{N+1} \varphi_n(x) p_n$$
$$F(x) \simeq f(x) = \sum_{n=0}^{N+1} \varphi_n(x) f_n \tag{9.1.9}$$

where p_n and f_n are the local values of P and F at the collocation points:

$$p_n = P(x_n); \qquad f_n = F(x_n) \tag{9.1.10}$$

Then the integrals (9.1.7)–(9.1.8) may be evaluated[1]:

$$\left\langle \frac{d\varphi_n}{dx}, P\frac{d\varphi_m}{dx} \right\rangle \simeq \left\langle \frac{d\varphi_n}{dx}, p\frac{d\varphi_m}{dx} \right\rangle$$

$$= \begin{cases} \dfrac{p_{n-1}+p_n}{2\,\Delta x} + \dfrac{p_n+p_{n+1}}{2\,\Delta x} & \text{when } m = n \\[2ex] \dfrac{p_{n-1}+p_n}{2\,\Delta x} & \text{when } m = n-1 \\[2ex] \dfrac{p_n+p_{n+1}}{2\,\Delta x} & \text{when } m = n+1 \\[2ex] 0 & \text{otherwise} \end{cases} \tag{9.1.11}$$

$$\langle \varphi_n, F \rangle \simeq \langle \varphi_n, f \rangle = \left\langle \varphi_n, \sum_m \varphi_m f_m \right\rangle$$
$$= \frac{\Delta x}{6}(f_{n-1} + 4f_n + f_{n+1}) \tag{9.1.12}$$

Inserting these results into (9.1.6) gives

$$\frac{p_{n-1}+p_n}{2}\frac{u_{n-1}-u_n}{\Delta x} + \frac{p_{n+1}+p_n}{2}\frac{u_{n+1}-u_n}{\Delta x}$$
$$+ \frac{\Delta x}{6}(f_{n-1} + 4f_n + f_{n+1}) = 0; \qquad n = 1, 2, \ldots, N \tag{9.1.13}$$

---

[1]Details are left to the reader. Note that $d\varphi_n/dx$ and $d\varphi_m/dx$ are piecewise-constant $(= \pm 1/\Delta x$ or zero), thus leaving almost trivial expressions to be integrated.

We have thus been able to derive an expression [equation (9.1.13)] that strongly resembles a finite-difference approximation of our problem. At no time, though, did we need to have recourse to the use of difference approximations to local values of derivatives in the original equation (9.1.1).

In matrix form this may be written as

$$\frac{1}{2\,\Delta x}\begin{bmatrix} \cdot & & \cdot & & \cdot & 0 \\ & \cdot & & \cdot & & \cdot \\ & & \cdot & & \cdot & \\ p_{n-1}+p_n & -(p_{n-1}+2p_n+p_{n+1}) & p_n+p_{n+1} & \\ & \cdot & & \cdot & & \\ & & \cdot & & \cdot & \\ 0 & & \cdot & & \cdot & \cdot \end{bmatrix}\begin{Bmatrix} u_1 \\ \cdot \\ \cdot \\ u_n \\ \cdot \\ \cdot \\ u_N \end{Bmatrix}$$

$$= -\frac{1}{2\,\Delta x}\begin{Bmatrix} (p_0+p_1)U_0 \\ 0 \\ \cdot \\ \cdot \\ \cdot \\ 0 \\ (p_N+p_{N+1})U_{N+1} \end{Bmatrix} - \frac{\Delta x}{6}\begin{bmatrix} 1 & 4 & 1 & & & 0 \\ \cdot & \cdot & \cdot & & & \\ & \cdot & \cdot & \cdot & & \\ & & \cdot & \cdot & \cdot & \\ & & & \cdot & \cdot & \cdot \\ 0 & & & 1 & 4 & 1 \end{bmatrix}\begin{Bmatrix} f_0 \\ \cdot \\ \cdot \\ \cdot \\ \cdot \\ \cdot \\ f_{N+1} \end{Bmatrix}$$

$$(9.1.14)$$

Or, with obvious notations:

$$\mathbf{Ku} = \mathbf{b} + \mathbf{s} \qquad (9.1.14a)$$

where $\mathbf{K}$ is an N by N matrix[2] and $\mathbf{b}$ and $\mathbf{s}$ are N-vectors containing the contributions of the given boundary terms (U_0 and U_{N+1}) and of the source terms $\{f_n\}$, respectively.

9.1.2 A NONLINEAR PROBLEM: QUASILINEARIZATION

Consider next a nonlinear version of the preceding problem, by having P and/or F become functions of the solution U:

$$\frac{d}{dx}\left(P(x,U)\frac{dU}{dx}\right) + F(x,U) = 0 \qquad (9.1.15)$$

---

[2]Which we may call a "conductivity matrix"; the traditional name, in the lore of the finite-element method, is "stiffness matrix."

All the steps leading to equation (9.1.14) remain applicable, but the p_n and f_n are now to be interpreted as functions of u_n (and, of course, x_n):

$$\frac{p(u_{n-1}) + p(u_n)}{2}\left(\frac{u_{n-1} - u_n}{\Delta x}\right) + \frac{p(u_n) + p(u_{n+1})}{2}\left(\frac{u_{n+1} - u_n}{\Delta x}\right)$$

$$+ \frac{\Delta x}{6}[f(u_{n-1}) + 4f(u_n) + f(u_{n+1})] = 0 \qquad (9.1.16)$$

Or, in notations analogous to (9.1.14a),

$$\mathbf{K(u)u = b(u) + s\,(u)} \qquad (9.1.17)$$

If the variation of P and F with U can be assumed to be small (i.e., if $\partial P/\partial U$ and $\partial F/\partial U$ are small in some sense), then (9.1.17) may be considered as a set of equations written in a natural *quasi-linear* form. This suggests the following sequence of calculation (see also Section 6.4):

1. Assume an initial guess for the solution:

$$\{u_n^0; n = 1, 2, \ldots, N\} \equiv \{\mathbf{u}^0\} \qquad (9.1.18)$$

2. Solve for $\{u_n^1\}$ the *linear* system

$$\mathbf{K(u^0)u^1 = b(u^0) + s(u^0)} \qquad (9.1.19)$$

3. Repeat step 2 iteratively, as

$$\mathbf{K(u^k)u^{k+1} = b(u^k) + s(u^k)} \qquad (9.1.20)$$

where $\mathbf{u}^k \equiv \{u_n^k\}$ is the estimate of $\mathbf{u}$ after the kth iteration. If convergent, this procedure will produce the solution of the *nonlinear* problem by the solution of an iterative sequence of *linear* problems.

9.1.3 A CONCENTRATED SOURCE

Suppose, in the preceding problem, that $F(x)$ is a concentrated source:

$$F(x) = f_s\,\delta(x - x_s) \qquad (9.1.21)$$

where δ is the usual Dirac delta function and x_s is the location of the source

$$0 < x_s < l \qquad (9.1.22)$$

It is not correct to assume that $F(x)$ may be represented as a combination of linear chapeau functions.

This case can be evaluated analytically: we simply have here (referring to Figure 9.1.3)

$$\langle \varphi_{n-1}, F \rangle = f_s \left(\frac{x_n - x_s}{\Delta x} \right)$$

$$\langle \varphi_n, F \rangle = f_s \left(\frac{x_s - x_{n-1}}{\Delta x} \right)$$

(9.1.23)

where x_{n-1} and x_n are the grid points adjacent to x_s.

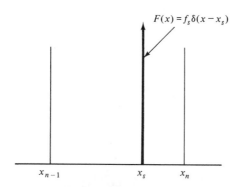

Figure 9.1.3   Concentrated source.

9.1.4 NONHOMOGENEOUS DIFFUSION OVER AN IRREGULAR MESH

We consider again the nonhomogeneous diffusion equation of Section 9.1.1:

$$\frac{d}{dx} \left(P(x) \frac{dU}{dx} \right) + F(x) = 0$$

(9.1.24)

to be approximated this time by linear finite elements over the irregular mesh defined by

$$x_{n+1} - x_n = \Delta x_n; \qquad n = 0, 1, 2, \ldots, N$$

(9.1.25)

As before, we express an approximation of the solution as

$$U(x) \simeq u(x) = \sum_n \varphi_n(x) u_n$$

(9.1.26)

and similarly for $P(x)$ and $F(x)$. After having expressed the Galerkin conditions of orthogonality of the residual with respect to the basis functions, we are led to evaluating the integrals analogous to (9.1.11) and (9.1.12):

$$\left\langle \frac{d\varphi_n}{dx}, p\frac{d\varphi_m}{dx} \right\rangle = \begin{cases} \dfrac{p_{n-1}+p_n}{2\,\Delta x_{n-1}} + \dfrac{p_n+p_{n+1}}{2\,\Delta x_n} & \text{when } m = n \\[2mm] -\dfrac{p_{n-1}+p_n}{2\,\Delta x_{n-1}} & \text{when } m = n-1 \\[2mm] -\dfrac{p_n+p_{n+1}}{2\,\Delta x_n} & \text{when } m = n+1 \\[2mm] 0 & \text{otherwise} \end{cases} \qquad (9.1.27)$$

and

$$\langle \varphi_n, f(x) \rangle = \langle \varphi_n, \sum_m \varphi_m f_m \rangle$$

$$= \frac{\Delta x_{n-1}}{6}(f_{n-1}+2f_n) + \frac{\Delta x_n}{6}(2f_n+f_{n+1}) \qquad (9.1.28)$$

Inserting these results in (9.1.5) produces

$$\frac{p_{n-1}+p_n}{2}\left(\frac{u_{n-1}-u_n}{\Delta x_{n-1}}\right) + \frac{p_n+p_{n+1}}{2}\left(\frac{u_{n+1}-u_n}{\Delta x_n}\right)$$

$$+ \frac{\Delta x_{n-1}}{6}(f_{n-1}+2f_n) + \frac{\Delta x_n}{6}(2f_n+f_{n+1}) = 0;$$

$$n = 1, 2, \ldots, N \qquad (9.1.29)$$

9.1.5 BOUNDARY CONDITIONS OF THE SECOND KIND

Consider the problem

$$\frac{d^2U}{dx^2} + F(x) = 0 \qquad \text{in } x \in [0, l] \qquad (9.1.30)$$

$$U(0) = 0 \qquad (9.1.31)$$

$$\left(\frac{\partial U}{\partial x}\right)_l = g(U_l) \qquad (9.1.32)$$

This may describe the steady-state distribution of temperature inside a slab, with distributed heat source $F(x)$ and temperature-dependent heat flux $g(U)$ at the surface $x = l$. Consider the finite-element representation of $U(x)$ with the same linear basis as before. For all *interior* points, the mathematics of the preceding examples apply, thus:

$$\frac{u_{n-1} - 2u_n + u_{n+1}}{\Delta x} + \frac{\Delta x}{6}(f_{n-1} + 4f_n + f_{n+1}) = 0;$$

$$n = 1, 2, \ldots, N \qquad (9.1.33)$$

The situation is somewhat different in the last point, since the boundary condition (9.1.32) must be taken into account.

One may proceed as follows. We apply in $n = N + 1$ the standard Galerkin condition:

$$\langle \varphi_{N+1}, \mathfrak{R} \rangle = \int \varphi_{N+1}(x)\left(\frac{d^2u}{dx^2} + f\right) dx$$

$$= 0 \qquad (9.1.34)$$

This expression becomes, after integration by parts,

$$\int_{x_N}^{l} \left[-\frac{d\varphi_{N+1}}{dx}\left(\frac{du}{dx}\right) + \varphi_{N+1}f \right] dx + \left(\frac{du}{dx}\right)_l = 0 \qquad (9.1.35)$$

We now replace $(du/dx)_l$ by its imposed value (9.1.32) and obtain

$$-\frac{u_{N+1} - u_N}{\Delta x} + \frac{\Delta x}{6}(f_N + 2f_{N+1}) = -g(u_{N+1}) \qquad (9.1.36)$$

which, together with the N equations (9.1.34), constitutes the $(N + 1)$ equations to be used for the determination of the $\{u_n\}$.

This result was obtained by using Galerkin's method, somewhat freely interpreted in the point $x_{N+1} = l$. That this procedure was correct may be shown by invoking the calculus of variations. The functional

$$J(U) = \int_0^l \Phi(U, U') \, dx + G(U_l) \qquad (9.1.37)$$

where

$$\Phi = -\tfrac{1}{2}(U')^2 + FU$$

$$G(U) = \int^U g(v) \, dv + \text{constant} \qquad (9.1.38)$$

may be verified to be stationary for functions $u(x)$ that satisfy (9.1.30)–

(9.1.32). Indeed [from (9.1.31)],

$$\delta J = \int_0^l \left[-\frac{d}{dx}\left(\frac{\partial \Phi}{\partial U'}\right) + \frac{\partial \Phi}{\partial U} \right] \delta U \, dx + \left[\frac{\partial \Phi}{\partial U'} \delta U \right]_0^l + \left(\frac{\partial G}{\partial U_l}\right) \delta U_l$$

$$= \int_0^l \left(\frac{d^2U}{dx^2} + F\right) \delta U \, dx - \left[\left(\frac{dU}{dx}\right) \delta U \right]_0^l + g(U_l) \, \delta U_l \qquad (9.1.39)$$

The integral equals zero because its integrand does. Moreover,

$$\left(\frac{dU}{dx}\right)_l - g(U_l) = 0 \qquad (9.1.40)$$

since this is the imposed boundary condition (9.1.32). We may thus derive the discrete finite-element equations by writing, for $u(x)$ given by

$$u = \sum_n \varphi_n(x) u_n$$

the expression of $J(u)$ and letting, in the spirit of the Rayleigh–Ritz method,

$$\frac{\partial J(u)}{\partial u_n} = 0; \qquad n = 1, 2, \ldots, N + 1$$

For $n = N + 1$, this equation becomes

$$\frac{\partial J}{\partial u_{N+1}} = \int \left[-\frac{d\varphi_{N+1}}{dx}\left(\frac{du}{dx}\right) + \varphi_{N+1}F \right] dx + g(u_{N+1})$$

$$= 0$$

which, after evaluation of the integral, becomes identical to (9.1.36).

9.2 QUADRATIC FINITE ELEMENTS

Quadratic finite elements in one dimension are obtained by constructing an approximation that consists of pieces of parabolas. Since a parabola

$$u(x) = a_0 + a_1 x + a_2 x^2 \qquad (9.2.1)$$

has *three* free coefficients, each element must have *three* nodes to uniquely define u. These are normally the two end nodes and one inside node.

As a simple application we consider here the case of equally spaced points:

$$x_n = n \, \Delta x$$

The domain of one element consists of two intervals (Figure 9.2.1)

$$x_{n-1} \leq x \leq x_{n+1} \qquad (9.2.2)$$

The nodes which belong to that element are the two end nodes, x_{n-1} and x_{n+1}, and the center node, x_n.

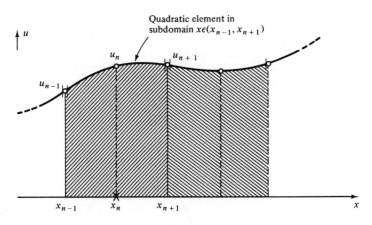

Figure 9.2.1

9.2.1 QUADRATIC BASIS FUNCTIONS

To proceed as before, we express $u(x)$ in the interval (x_{n-1}, x_{n+1}) as the sum

$$u(x) = u_{n-1}\varphi_{n-1}(x) + u_n\varphi_n(x) + u_{n+1}\varphi_{n+1}(x) \qquad (9.2.3)$$

where φ_{n-1}, φ_n, and φ_{n+1} are the quadratic basis functions illustrated in Figure 9.2.2. These basis function are the fundamental Lagrange interpolation polynomials:

$$\varphi_m(x) = \frac{(x - x_{n_1})(x - x_{n_2})}{(x_m - x_{n_1})(x_m - x_{n_2})} \qquad (9.2.4)$$

where n_1 and n_2 are

$$(n - 1) \text{ and } (n + 1) \qquad \text{for } m = n$$
$$(n - 1) \text{ and } \quad n \qquad \text{for } m = n + 1$$
$$n \quad \text{ and } (n + 1) \qquad \text{for } m = n - 1$$

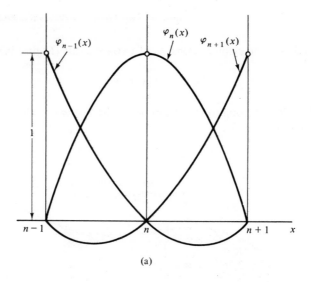

(a)

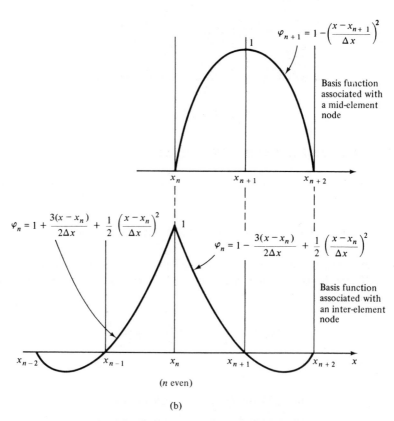

$$\varphi_{n+1} = 1 - \left(\frac{x - x_{n+1}}{\Delta x}\right)^2$$

Basis function associated with a mid-element node

$$\varphi_n = 1 + \frac{3(x - x_n)}{2\Delta x} + \frac{1}{2}\left(\frac{x - x_n}{\Delta x}\right)^2$$

$$\varphi_n = 1 - \frac{3(x - x_n)}{2\Delta x} + \frac{1}{2}\left(\frac{x - x_n}{\Delta x}\right)^2$$

Basis function associated with an inter-element node

(n even)

(b)

Figure 9.2.2 (a) Quadratic basis functions in the interval (x_{n-1}, x_{n+1}); (b) quadratic basis functions associated to individual nodes.

The same process is repeated over all the subdomains or elements of width $2\,\Delta x$, and the overall function so obtained may be expressed as the sum

$$u(x) = \sum_n \varphi_n(x)u_n \qquad (9.2.3a)$$

where the $\varphi_n(x)$ are of two kinds, depending on the parity of n (Figure 9.2.2b).

EXAMPLE

Consider again our simple two-point boundary-value problem

$$\frac{d^2U}{dx^2} + F(x) = 0 \qquad \text{in } x \in (0, l) \qquad (9.2.5)$$

$$U(0) = U(l) = 0 \qquad (9.2.6)$$

A quadratic finite-element approximation of this problem is obtained by dividing $[0, l]$ is an even number of intervals:

$$\Delta x = \frac{l}{N+1}; \qquad N \text{ odd}$$

$$x_n = n\,\Delta x \qquad\qquad\qquad (9.2.7)$$

over which U is approximated by quadratic finite elements as (9.2.3). We form the residual

$$\Re(x) \equiv \frac{d^2u}{dx^2} + F$$

$$= \sum_n \frac{d^2\varphi_n}{dx^2}u_n + F \qquad (9.2.8)$$

which is, following Galerkin's procedure, made to be orthogonal to the basis functions $\{\varphi_n(x)\}$:

$$0 = \langle \varphi_n, \Re \rangle \equiv \int_0^l \varphi_n(x)\Re(x)\,dx \qquad (9.2.9)$$

or

$$\sum_m \left\langle \varphi_n, \frac{d^2\varphi_m}{dx^2} \right\rangle u_m + \langle \varphi_n, F \rangle = 0; \qquad n = 1, 2, \dots, N \qquad (9.2.10)$$

This brings about the following integrals, which are easily evaluated analytically (details are left to the reader):

n even:

$$\left\langle \varphi_n, \frac{d^2\varphi_n}{dx^2} \right\rangle = -\left\langle \frac{d\varphi_n}{dx}, \frac{d\varphi_m}{dx} \right\rangle$$

$$= \begin{cases} -\dfrac{7}{3\,\Delta x} & \text{if } m = n \\[2ex] \dfrac{4}{3\,\Delta x} & \text{if } m = n \pm 1 \\[2ex] -\dfrac{1}{6\,\Delta x} & \text{if } m = n \pm 2 \end{cases}$$

n odd:

$$\left\langle \varphi_n, \frac{d^2\varphi_m}{dx^2} \right\rangle = \begin{cases} -\dfrac{8}{3\,\Delta x} & \text{if } m = n \\[2ex] \dfrac{4}{3\,\Delta x} & \text{if } m = n \pm 1 \end{cases}$$

The integrals $\langle \varphi_n, F(x) \rangle$ remain to be evaluated. Except in special cases where they can be computed analytically, approximation will be required. A form that is consistent with the representation of U is to approximate $F(x)$ in the same basis of quadratic functions:

$$F(x) \simeq f(x) = \sum_n \varphi_n(x) f_n; \qquad f_n = F(x_n)$$

whence the approximation

$$\langle \varphi_n, F \rangle \simeq \sum_n \langle \varphi_n, \varphi_m \rangle f_n$$

The individual integrals are easily evaluated analytically:

n even:

$$\langle \varphi_n, \varphi_m \rangle = \begin{cases} \dfrac{8\,\Delta x}{15} & \text{if } m = n \\[2ex] \dfrac{2\,\Delta x}{15} & \text{if } m = n \pm 1 \\[2ex] -\dfrac{\Delta x}{15} & \text{if } m = n \pm 2 \end{cases}$$

n odd:

$$\langle \varphi_n, \varphi_m \rangle = \begin{cases} \dfrac{16 \, \Delta x}{15} & \text{if } m = n \\[2mm] \dfrac{2 \, \Delta x}{15} & \text{if } m = n \pm 1 \end{cases}$$

Collecting these results, we obtain the equations (*n* odd)

$$\left. \begin{aligned} -\frac{-u_{n-2} + 8u_{n-1} - 14u_n + 8u_{n+1} - u_{n+2}}{4 \, \Delta x^2} \\[2mm] = \frac{1}{10}(-f_{n-2} + 2f_{n-1} + 8f_n + 2f_{n+1} - f_{n+2}) \\[2mm] -\frac{u_n - 2u_{n+1} + u_{n+2}}{\Delta x^2} = \frac{1}{10}(f_n + 8f_{n+1} + f_{n+2}) \end{aligned} \right\} \quad (9.2.11)$$

This constitutes a system of linear equations with a pentadiagonal matrix.

9.3 HERMITE CUBIC FINITE ELEMENTS

A function $u(x)$ may be interpolated by a polynomial that matches not only u_n at the collocation points x_n but also matches the slopes $(\partial u/\partial x)_n$ (and possibly higher derivatives) in those points; this is referred to as *Hermite*[3] *interpolation.*

The application of this concept to finite-element approximation leads to using as discrete parameters not only the function values u_n, but also the derivatives at the mesh points,

$$u_n' = \left(\frac{\partial u}{\partial x} \right)_n$$

For the one-dimensional problems considered here, the approximation in the element domain (x_n, x_{n+1}) is then defined by the *four* parameters $(u_n, u_n', u_{n+1}, u_{n+1}')$. The simplest form of polynomial interpolation using this data is a piecewise *cubic polynomial in each interval* (x_n, x_{n+1}) which is determined by the matching of these four conditions at the boundaries (Figure 9.3.1).

---

[3]Charles Hermite (1822–1901), French mathematician.

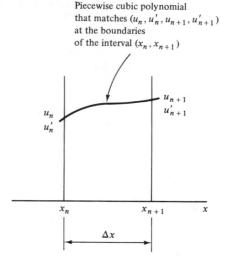

Piecewise cubic polynomial
that matches $(u_n, u'_n, u_{n+1}, u'_{n+1})$
at the boundaries
of the interval (x_n, x_{n+1})

Figure 9.3.1 Hermite cubic interpolation in the element (x_n, x_{n+1}).

9.3.1 HERMITE CUBIC BASIS FUNCTIONS

Without further justification, it may be observed that the elementary basis functions (Figure 9.3.2)

$$\varphi_n(x) = 1 - 3\left(\frac{x - x_n}{\Delta x}\right)^2 + 2\left|\frac{x - x_n}{\Delta x}\right|^3$$

$$\psi_n(x) = \begin{cases} (x - x_n)\left(\dfrac{x - x_{n+1}}{\Delta x}\right)^2 & \text{in } (x_n, x_{n+1}) \\ (x - x_n)\left(\dfrac{x - x_{n-1}}{\Delta x}\right)^2 & \text{in } (x_{n-1}, x_n) \end{cases} \qquad (9.3.1)$$

allow the direct writing of the Hermite cubic interpolation in the element domain (x_n, x_{n+1}) as the sum

$$u(x) = u_n\varphi_n(x) + u'_n\psi_n(x) + u_{n+1}\varphi_{n+1}(x) + u'_{n+1}\psi_{n+1}(x) \qquad (9.3.2)$$

This results from the relations that are satisfied by the φ_n and ψ_n: namely,

$$\varphi_n(x_{n+1}) = \left(\frac{d\varphi_n}{dx}\right)_n = \left(\frac{d\varphi_n}{dx}\right)_{n+1} = 0$$

$$\psi_n(x_n) = \psi_n(x_{n+1}) = \left(\frac{d\psi_n}{dx}\right)_{n+1} = 0$$

$$\varphi_n(x_n) = \varphi_{n+1}(x_{n+1}) = 1 \qquad (9.3.3)$$

$$\left(\frac{d\psi_n}{dx}\right)_n = \left(\frac{d\psi_{n+1}}{dx}\right)_{n+1} = 1$$

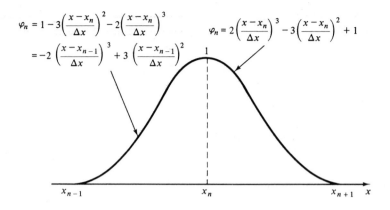

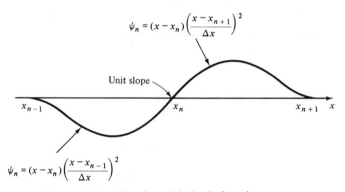

Figure 9.3.2 Hermite cubic basis functions.

The complete Hermite cubic finite-element representation of $U(x)$ on the x axis is thus given by the sum over all n:

$$u(x) = \sum_n \varphi_n(x)u_n + \psi_n(x)u'_n$$

$$= \{\varphi_n\}^T\{u_n\} + \{\psi_n\}^T\{u'_n\} \tag{9.3.4}$$

Note that, by contrast with linear and quadratic elements, $u(x)$ with Hermite cubic elements belongs to C^1 (both u and u' are continuous across mesh points).

EXAMPLE 1

We treat as an example the standard second-order equation

$$\frac{d^2U}{dx^2} + F(x) = 0$$

$$U(0) = U(l) = 0 \tag{9.3.5}$$

whose solution is to be approximated by Hermite cubic elements [expression (9.3.4)] with

$$x_n = n \, \Delta x; \qquad \Delta x = \frac{l}{N+1} \tag{9.3.6}$$

Similarly, we approximate $F(x)$ as

$$F(x) \simeq f(x) = \sum_n [\varphi_n(x)f_n + \psi_n(x)f'_n];$$
$$f_n \simeq F(x_n); \qquad f'_n \simeq F'(x_n) \tag{9.3.7}$$

The equation residual is

$$\mathfrak{R} \equiv \sum_n (\varphi'_n u_n + \psi''_n u'_n) + \sum_n (\varphi_n f_n + \psi_n f'_n) \tag{9.3.8}$$

and the Galerkin orthogonality conditions read

$$0 = \langle \varphi_n, \mathfrak{R} \rangle$$
$$= \int \varphi_n \mathfrak{R} \, dx \tag{9.3.9}$$

and

$$0 = \langle \psi_n, \mathfrak{R} \rangle$$
$$= \int \psi_n \mathfrak{R} \, dx \tag{9.3.10}$$

This brings about the fundamental integrals, which are easily evaluated (the details are left to the reader as an exercise)

$$\langle \varphi_n, \varphi_m \rangle = \begin{cases} \dfrac{312}{420} \Delta x & \text{when } m = n \\[2mm] \dfrac{54}{420} \Delta x & \text{when } m = n \pm 1 \end{cases}$$

$$\langle \varphi_n, \psi_m \rangle = \begin{cases} 0 & \text{when } m = n \\[2mm] \dfrac{13}{420} \Delta x^2 & \text{when } m = n - 1 \\[2mm] -\dfrac{13}{420} \Delta x^2 & \text{when } m = n + 1 \end{cases}$$

$$\langle \psi_n, \psi_m \rangle = \begin{cases} \dfrac{8}{420} \Delta x^3 & \text{when } m = n \\[2mm] -\dfrac{3}{420} \Delta x^3 & \text{when } m = n \pm 1 \end{cases}$$

$$\left\langle \varphi_n, \frac{d^2\varphi_m}{dx^2} \right\rangle = -\left\langle \frac{d\varphi_n}{dx}, \frac{d\varphi_m}{dx} \right\rangle$$

$$= \begin{cases} -\dfrac{72}{30\,\Delta x} & \text{when } m = n \\[2ex] \dfrac{36}{30\,\Delta x} & \text{when } m = n \pm 1 \end{cases}$$

$$\left\langle \varphi_n, \frac{d^2\psi_m}{dx^2} \right\rangle = -\left\langle \frac{d\varphi_n}{dx}, \frac{d\psi_m}{dx} \right\rangle = \left\langle \frac{d^2\varphi_n}{dx^2}, \psi_m \right\rangle$$

$$= \begin{cases} 0 & \text{when } m = n \\[2ex] \dfrac{3}{30} & \text{when } m = n - 1 \\[2ex] -\dfrac{3}{30} & \text{when } m = n + 1 \end{cases}$$

$$\left\langle \psi_n, \frac{d^2\psi_m}{dx^2} \right\rangle = -\left\langle \frac{d\psi_n}{dx}, \frac{d\psi_m}{dx} \right\rangle$$

$$= \begin{cases} -\dfrac{8\,\Delta x}{30} & \text{when } m = n \\[2ex] \dfrac{\Delta x}{30} & \text{when } m = n \pm 1 \end{cases}$$

Bringing these results into (9.3.9) and (9.3.10) results, for each n, in the pair of equations

$$36\left(\frac{u_{n-1} - 2u_n + u_{n+1}}{\Delta x^2}\right) - 6\left(\frac{u'_{n+1} - u'_{n-1}}{2\,\Delta x}\right)$$

$$= -\frac{1}{14}[54f_{n-1} + 312f_n + 54f_{n+1} + 13\,\Delta x(f'_{n+1} - f'_{n-1})] \quad (9.3.11)$$

$$3\left(\frac{u_{n+1} - u_{n-1}}{\Delta x^2}\right) + \frac{u'_{n-1} - 8u'_n + u'_{n+1}}{\Delta x}$$

$$= -\frac{1}{14}[-13f_{n-1} + 13f_{n+1} - \Delta x(3f'_{n-1} - 8f'_n + 3f'_{n+1})] \quad (9.3.12)$$

EXAMPLE 2

As a second example, we consider the steady-state bending of a thin, uniform beam described by the equation

$$\frac{d^2}{dx^2}\left(EI\frac{d^2U}{dx^2}\right) = F(x); \qquad x \in [0, l] \tag{9.3.13}$$

to be approximated with the Hermite–Galerkin finite-element procedure. We let

$$U(x) \simeq u(x) = \sum_n \varphi_n(x)u_n + \psi_n(x)u'_n \tag{9.3.14}$$

where the φ_n and ψ_n are the Hermite cubic basis functions (9.3.1). The length l of the beam has been divided into $(N + 1)$ finite elements of length Δx by N interior points:

$$\Delta x = \frac{l}{N+1}; \qquad x_n = n\,\Delta x$$

We also assume that the load $F(x)$ is concentrated at the nodes[4]:

$$F(x) = \sum_n f_n\, \delta(x - x_n) \tag{9.3.15}$$

The residual orthogonality conditions are

$$\langle \varphi_n, \mathfrak{R} \rangle = 0 \quad \text{and} \quad \langle \psi_n, \mathfrak{R} \rangle = 0 \tag{9.3.16}$$

Since φ_n and ψ_n are piecewise third-degree polynomials and since $\mathfrak{R}$ contains fourth derivatives, *integration by parts* is to be invoked to evaluate the above. Specifically, for all interior points we find

$$
\begin{aligned}
\langle \varphi_n, \mathfrak{R} \rangle &\equiv \int \varphi_n \left[\frac{d^2}{dx^2} EI \left(\sum_m \frac{d^2\varphi_m}{dx^2} u_m + \frac{d^2\psi_m}{dx^2} u'_m \right) - \sum_m f_m\, \delta(x - x_m) \right] dx \\
&= \int \frac{d^2\varphi_n}{dx^2} \left(EI \sum_m \frac{d^2\varphi_m}{dx^2} u_m + \frac{d^2\psi_m}{dx^2} u'_m \right) dx - f_n = 0
\end{aligned}
$$

and

$$
\begin{aligned}
\langle \psi_n, \mathfrak{R} \rangle &\equiv \int \psi_n \left[\frac{d^2}{dx^2} EI \left(\sum_m \frac{d^2\varphi_m}{dx^2} u_m + \frac{d^2\psi_m}{dx^2} u'_m \right) - \sum_m f_m\, \delta(x - x_m) \right] dx \\
&= \int \frac{d^2\psi_n}{dx^2} \left(EI \sum_m \frac{d^2\varphi_m}{dx^2} u_m + \frac{d^2\psi_m}{dx^2} u'_m \right) dx = 0
\end{aligned} \tag{9.3.17}
$$

---

[4]Assumptions of this kind are common procedure in civil and mechanical engineering, whence this problem originates.

The elementary integrals to be evaluated are[5]

$$
\left.
\begin{aligned}
\left\langle \frac{d^2\varphi_n}{dx^2}, \frac{d^2\varphi_n}{dx^2} \right\rangle &= \frac{24}{\Delta x^3} \\[6pt]
\left\langle \frac{d^2\varphi_n}{dx^2}, \frac{d^2\varphi_{n-1}}{dx^2} \right\rangle &= \left\langle \frac{d^2\varphi_n}{dx^2}, \frac{d^2\varphi_{n-1}}{dx^2} \right\rangle = -\frac{12}{\Delta x^3} \\[6pt]
\left\langle \frac{d^2\varphi_n}{dx^2}, \frac{d^2\psi_n}{dx^2} \right\rangle &= 0 \\[6pt]
\left\langle \frac{d^2\varphi_n}{dx^2}, \frac{d^2\psi_{n+1}}{dx^2} \right\rangle &= -\left\langle \frac{d^2\varphi_n}{dx^2}, \frac{d^2\psi_{n-1}}{dx^2} \right\rangle = \frac{6}{\Delta x^2} \\[6pt]
\left\langle \frac{d^2\psi_n}{dx^2}, \frac{d^2\psi_n}{dx^2} \right\rangle &= \frac{8}{\Delta x} \\[6pt]
\left\langle \frac{d^2\psi_n}{dx^2}, \frac{d^2\psi_{n+1}}{dx^2} \right\rangle &= \left\langle \frac{d^2\psi_n}{dx^2}, \frac{d^2\psi_{n-1}}{dx^2} \right\rangle = \frac{2}{\Delta x}
\end{aligned}
\right\} \qquad (9.3.18)
$$

Collecting these results into (9.3.17), we finally obtain the pair of equations for $n = 1, 2, \ldots, N$

$$
\boxed{
\begin{aligned}
6\mathbb{E}\mathbb{I}\left[2\left(\frac{-u_{n-1} + 2u_n - u_{n+1}}{\Delta x^3} \right) + \frac{u'_{n+1} - u'_{n-1}}{\Delta x^2} \right] &= f_n \\[8pt]
2\mathbb{E}\mathbb{I}\left[3\left(\frac{u_{n-1} - u_{n+1}}{\Delta x} \right) + (u'_{n-1} + 4u'_n + u'_{n+1}) \right] &= 0
\end{aligned}
} \qquad (9.3.19)
$$

9.4 THE ENGINEERS' DISPLACEMENT METHOD

What we have described so far is, with few exceptions, the mathematicians' version of the finite-element method. The "displacement method," which we are about to describe in this section, illustrates the engineers' approach. We are using the problem of the bending of a beam as an example, and will thus arrive through an altogether different formalism at a set of discrete equations which are identical to those obtained in the preceding section.

The displacement method approaches the problem of deriving finite-element discrete equations by imposing the condition that the approximate solution must satisfy the principle of virtual works (from classical mechanics). This can be done without invoking explicitly variational principles of physics

---

[5]We assume that $\mathbb{E}\mathbb{I}$ is independent of x.

(although the two are equivalent) and without invoking some residual minimization or projection technique, as in the application of Galerkin's method.

The displacement method is not restricted to continuous media (or continua). In fact, the method had long been used by engineers for the calculation of deformations of bar structures. The method was extended to continua by idealizing them with finite pieces or "finite elements," whence the name given to the method.

9.4.1 THE PRINCIPLE OF VIRTUAL WORKS

When a mechanical (or thermodynamical) system is at equilibrium, the total energy (internal energy + external energy) is stationary (generally a minimum). That is, a small (virtual) displacement of the state of the system from equilibrium brings about no change in the total energy. Since the change of energy is work, this principle may be stated:

> The internal work and external work corresponding to the (small) displacement of a system away from a state of equilibrium are equal.

We may recognize that this principle is that formulated in Section 8.1 [equation (8.1.24)]. But the formulation there relied on the expression of total energy (which is a *functional*) and then expressing work as the *variation* of that functional, which required the use of the calculus of variations.

In the displacement method we are concerned primarily with the solution of deformation problems of elastic continua. Internal geometrical deformations are called *strains*, and the corresponding internal elastic forces are called *stresses*. One relies on fundamental elasticity laws to express the internal work w_I from first principles, and then express

$$w_I = w_E \qquad (9.4.1)$$

as the fundamental relation from which approximate relations are derived (without ever having to express the variation of a functional $\delta J = 0$, as in the variational method). In (9.4.1) the external work takes the form

$$w_E = \int_{\text{volume}} (\text{external forces})(\delta \cdot \text{displacements}) \, d(\text{vol}) \qquad (9.4.2)$$

and the internal work

$$w_I = \int_{\text{volume}} (\text{stress})(\delta \cdot \text{strain}) \, d(\text{vol}) \qquad (9.4.3)$$

The internal deformations of continua are approximated by finite-element shapes with a finite number of nodal values, and expressing the equality of w_E and w_I provides the conditions from which the nodal values may be computed. The specific steps may be summarized as follows:

1. *Definition of the elements.* The continua are divided into *finite elements.* The elements are interconnected at a finite number of nodal points situated on their boundaries. The displacement of these nodal points (and possibly that of additional "interior" nodal points not on the boundaries) will be the set of discrete parameters of the problem.

2. *Definition of displacements.* A set of *displacement* or *shape functions* is chosen to define uniquely the displacement within each finite element in terms of its nodal displacements.

3. *Strains.* The *internal strains* are expressed in terms of partial derivatives of displacements with respect to distance. Thus, the shape functions uniquely define the state of strain within each element in terms of its nodal displacements.

4. *Stresses.* The strains (together with any initial strains), and the laws expressing the elastic properties of the material, define the state of *stress* throughout each element.

5. *Relating external loads to nodal displacements.* It is assumed (at least in the simpler applications of the method) that external forces or loads are applied at the nodes only. The principle of virtual works is then used to express the fact that for an arbitrary, small displacement $\{\delta u_n\}$ of the nodal variables $\{u_n\}$, the work of the discrete approximation of the external forces

$$w_E = \sum_n f_n \, \delta u_n = \{f_n\}^T \{\delta u_n\} \qquad (9.4.4)$$

must be equal to the internal work of the finite-element shape that approximates the continuum,

$$w_I = \sum_{\text{elements}} \int_{\text{volume}} (\text{stress})(\delta \cdot \text{strain}) \, d(\text{vol}) \qquad (9.4.5)$$

EXAMPLE

Consider the problem of finding the steady-state bending of a beam. The governing equation is [see (9.3.13)]

$$\frac{d^2}{dx^2}\left(\mathbb{EI}\,\frac{d^2U}{dx^2}\right) = F(x) \tag{9.4.6}$$

This equation is given for reference only, since it is not used in the displacement method, which proceeds from more fundamental principles. As was said before, it is based on an application of the principle of virtual works (which applies to mechanical systems at equilibrium) and on the use of the local stress/local strain relationship, which is, in this case,

$$\mathbb{M} = -\mathbb{EI}\,\frac{d^2U}{dx^2} \tag{9.4.7}$$

where $\mathbb{M}$ = internal moment of forces in the beam = *stress*
   $\mathbb{EI}$ = flexural rigidity of the beam
   $\dfrac{d^2U}{dx^2}$ = curvature of the beam = *strain*

Step 1:   Definition of elements and variables.   To obtain a finite-elements solution, we divide the length of the beam into finite elements of length Δx by the N interior nodes

$$x_n = n\,\Delta x; \qquad \Delta x = \frac{l}{N+1} \tag{9.4.8}$$

Nodal values that are sufficient to express the solution are the beam's *position* and *slope*:

$$
\begin{aligned}
u_n &\simeq U(x_n) \\
u'_n &\simeq \frac{dU}{dx}(x_n)
\end{aligned}
\tag{9.4.9}
$$

We assume that the external force F is applied at the nodes only, as concentrated forces f_n:

$$F(x) = \sum_n f_n\,\delta(x - x_n) \tag{9.4.10}$$

The specific form in which we seek to formulate approximate equations is as the linear system

$$\mathbf{K}\{u_n\} = \{f_n\} \tag{9.4.11}$$

where[6]

$$\{u_n\} = \begin{Bmatrix} \cdot \\ \cdot \\ \cdot \\ u_n \\ u'_n \\ u_{n+1} \\ u'_{n+1} \\ u_{n+2} \\ \cdot \\ \cdot \\ \cdot \end{Bmatrix} \quad \text{and} \quad \{f_n\} = \begin{Bmatrix} \cdot \\ \cdot \\ \cdot \\ f_n \\ 0 \\ f_{n+1} \\ 0 \\ f_{n+2} \\ \cdot \\ \cdot \\ \cdot \end{Bmatrix} \tag{9.4.12}$$

are vectors of the nodal values of the solution and of the external force, $\mathbf{K}$ is the stiffness matrix, and the objective is to determine the numerical value of its elements.

Step 2: **Choice of the shape functions.**   The solution *in each element* (x_n, x_{n+1}) is determined by *four* parameters: the positions u_n and u_{n+1} and slopes u'_n and u'_{n+1} at both extremities. With four parameters, a simple shape function is the cubic polynomial (Figure 9.4.1)

$$u(y) = a_0 + a_1 y + a_2 y^2 + a_3 y^3 \tag{9.4.13}$$

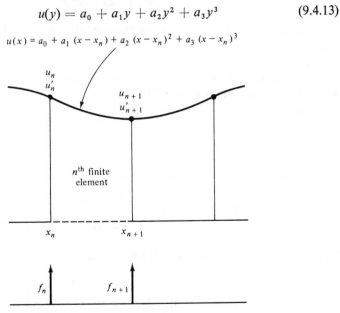

$$u(x) = a_0 + a_1 (x - x_n) + a_2 (x - x_n)^2 + a_3 (x - x_n)^3$$

Figure 9.4.1   Finite-element approximation of a thin beam.

---

[6]Zeros in $\{f_n\}$ correspond to the absence of external torque, which is assumed in this simple example.

where y is the local variable

$$y = x - x_n \tag{9.4.14}$$

and the coefficients a_0, a_1, a_2, and a_3 are different from element to element. [Note that (9.4.13) is identical to the Hermite cubic polynomials described in Section 9.3.] The slope of the element solution is obtained by derivation of (9.4.13):

$$u'(x) \equiv \frac{du}{dx} = \frac{du}{dy} = a_1 + 2a_2 y + 3a_3 y^2 \tag{9.4.15}$$

These relations may be expressed in matrix form as

$$\{u^e(y)\} = \begin{Bmatrix} u(y) \\ u'(y) \end{Bmatrix} = \begin{bmatrix} 1 & y & y^2 & y^3 \\ 0 & 1 & 2y & 3y^2 \end{bmatrix} \begin{Bmatrix} a_0 \\ a_1 \\ a_2 \\ a_3 \end{Bmatrix}$$

$$\equiv [\Phi^e]^T \{a_k^e\} \tag{9.4.16}$$

where

$$[\Phi^e] \equiv \begin{bmatrix} 1 & y & y^2 & y^3 \\ 0 & 1 & 2y & 3y^2 \end{bmatrix}^T$$

is called the *shape-function matrix* of the approximation. This relation gives the internal displacements in relation to the shape parameters a_0, a_1, a_2, and a_3.

The internal displacements are now related to the nodal displacements. At the node x_n, $y = 0$ and therefore

$$\begin{aligned} u_n &= a_0 \\ u'_n &= a_1 \end{aligned} \tag{9.4.17}$$

At the node x_{n+1}, $y = \Delta x$ and therefore

$$\begin{aligned} u_{n+1} &= a_0 + a_1 \Delta x + a_2 \Delta x^2 + a_3 \Delta x^3 \\ u'_{n+1} &= \qquad\; a_1 \qquad + 2a_2 \Delta x + 3a_3 \Delta x^2 \end{aligned} \tag{9.4.18}$$

Recasting these results in matrix form gives us

$$\begin{Bmatrix} u_n \\ u'_n \\ u_{n+1} \\ u'_{n+1} \end{Bmatrix} = \begin{bmatrix} 1 & 0 & 0 & 0 \\ 0 & 1 & 0 & 0 \\ 1 & \Delta x & \Delta x^2 & \Delta x^3 \\ 0 & 1 & 2\Delta x & 3\Delta x^2 \end{bmatrix} \begin{Bmatrix} a_0 \\ a_1 \\ a_2 \\ a_3 \end{Bmatrix} \tag{9.4.19}$$

or, defining

$$\{u_n^e\} = \begin{Bmatrix} u_n \\ u_n' \\ u_{n+1} \\ u_{n+1}' \end{Bmatrix}; \qquad C = \begin{bmatrix} 1 & 0 & 0 & 0 \\ 0 & 1 & 0 & 0 \\ 1 & \Delta x & \Delta x^2 & \Delta x^3 \\ 0 & 1 & 2\,\Delta x & 3\,\Delta x^2 \end{bmatrix}$$

we may write

$$\{u_n^e\} = C\{a_k^e\} \tag{9.4.20}$$

The inverse is easily found:

$$\begin{Bmatrix} a_0 \\ a_1 \\ a_2 \\ a_3 \end{Bmatrix} = \begin{bmatrix} 1 & 0 & 0 & 0 \\ 0 & 1 & 0 & 0 \\ -\dfrac{3}{\Delta x^2} & -\dfrac{2}{\Delta x} & \dfrac{3}{\Delta x} & -\dfrac{2}{\Delta x} \\ \dfrac{2}{\Delta x^3} & \dfrac{1}{\Delta x^2} & -\dfrac{2}{\Delta x^3} & \dfrac{1}{\Delta x^2} \end{bmatrix} \begin{Bmatrix} u_n \\ u_n' \\ u_{n+1} \\ u_{n+1}' \end{Bmatrix}$$

or

$$\{a_k^e\} = C^{-1}\{u_n^e\} \tag{9.4.21}$$

Inserting this relation into the expression (9.4.16) of the internal displacements gives

$$\{u^e(y)\} = [\Phi^e(y)]^T C^{-1}\{u_n^e\} = \{\varphi^e(y)\}^T\{u_n^e\} \tag{9.4.22}$$

The notation $\{\varphi^e\}^T$ used to represent the product $[\Phi^e]^T[c]^{-1}$ is introduced to draw our attention to the fact that the $\{\varphi^e\}$ are identical to the basis functions used in the conventional finite-element Galerkin-type methods.

Step 3:   Strain/displacement relationship. The strains (in this case the curvatures d^2u/dx^2) within each element are now related to the displacements $\{u^e(y)\}$ and hence [through (9.4.22)] to the nodal displacements $\{u_n^e\}$. This is obtained by simple differentiation:

$$strain = \epsilon = -\frac{d^2u}{dy^2} = -2a_2 - 6a_3 y \tag{9.4.23}$$

which can be written in matrix form as

$$\epsilon(y) = [0 \quad 0 \quad -2 \quad -6y]\{a_k^e\}$$

or, using (9.4.21), as

$$\epsilon(y) = [0 \quad 0 \quad -2 \quad -6y]\mathbf{C}^{-1}\{u_n^e\}$$
$$= \mathbf{B}\{u_n^e\} \tag{9.4.24}$$

where $\mathbf{B}$ is the row vector of functions

$$\mathbf{B} = [0 \quad 0 \quad -2 \quad -6y]\mathbf{C}^{-1}$$

$$= \left[\frac{6}{\Delta x^2} - \frac{12y}{\Delta x^3} \quad \frac{4}{\Delta x} - \frac{6y}{\Delta x^2} \quad -\frac{6}{\Delta x^2} + \frac{12y}{\Delta x^3} \quad \frac{2}{\Delta x} - \frac{6y}{\Delta x^2}\right] \tag{9.4.25}$$

Step 4: Stress/strain relationship. The fundamental relation (9.4.7) becomes, in its finite-element form:

$$m(y) = -\mathbb{EI}\frac{d^2u}{dy^2} = \mathbb{EI}\epsilon(y)$$

$$= \mathbb{EI}\mathbf{B}\{u_n^e\} \tag{9.4.26}$$

(m is the approximation of M.)

Step 5: Relating external loads to nodal displacements. This is where the essence of the method applies. "Closing the loop" requires that the fundamental *equilibrium law of mechanics expressing the equality of virtual works* be invoked. The fundamental relation that holds at equilibrium is:

$$\begin{bmatrix}\text{work of external}\\ \text{forces}\end{bmatrix} = \begin{bmatrix}\text{work of internal forces}\\ \text{(or change in internal energy)}\end{bmatrix} \tag{9.4.27}$$

both expressed when a "*virtual*" *displacement* of the system is effected. We denote such a virtual displacement by

$$\{\delta u_n\} \equiv \begin{Bmatrix} \vdots \\ \vdots \\ \delta u_n \\ \delta u_n' \\ \delta u_{n+1} \\ \delta u_{n+1}' \\ \vdots \\ \vdots \end{Bmatrix} \tag{9.4.28}$$

In the finite-element model we have

$$\begin{pmatrix} \text{work of external} \\ \text{forces} \end{pmatrix} = w_E = \sum_n f_n \, \delta u_n = \{f_n\}^T \{\delta u_n\} \qquad (9.4.29)$$

and

$$\begin{pmatrix} \text{work of internal} \\ \text{forces} \end{pmatrix} = w_I = \int_0^l \text{stress}(\delta \cdot \text{strain}) \, dx$$

$$= \sum_n \int_0^{\Delta x} (\text{stress})_n (\delta \cdot \text{strain})_n \, dy$$

$$= \sum_n w_{I,n}^e \qquad (9.4.30)$$

where each term under the $\sum_n$ sign is the interior work in one finite element. These terms may be computed individually:

$$w_{I,n}^e = \int_0^{\Delta x} (\text{stress})_n (\delta \cdot \text{strain})_n \, dy$$

$$= \int_0^{\Delta x} m(y) \, \delta\epsilon(y) \, dy$$

$$= \int_0^{\Delta x} \{u_n^e\}^T \mathbf{B}^T \mathbb{E} \mathbf{B} \{\delta u_n^e\} \, dy$$

$$= \int_0^{\Delta x} \{u_n^e\}^T \begin{bmatrix} \dfrac{6}{\Delta x^2} - \dfrac{12y}{\Delta x^3} \\[2mm] \dfrac{4}{\Delta x} - \dfrac{6y}{\Delta x^2} \\[2mm] -\dfrac{6}{\Delta x^2} + \dfrac{12y}{\Delta x^3} \\[2mm] \dfrac{2}{\Delta x} - \dfrac{6y}{\Delta x^3} \end{bmatrix} \mathbb{E}$$

$$\times \begin{bmatrix} \dfrac{6}{\Delta x^2} - \dfrac{12y}{\Delta x^3} & \dfrac{4}{\Delta x} - \dfrac{6y}{\Delta x^2} & -\dfrac{6}{\Delta x^2} + \dfrac{12y}{\Delta x^3} & \dfrac{2}{\Delta x} - \dfrac{6}{\Delta x^3} \end{bmatrix} dy \quad (9.4.31)$$

Upon integration we find that

$$w_{I,n}^e = \frac{\mathbb{E}}{\Delta x^3} \{u_n^e\}^T \begin{bmatrix} 12 & 6\,\Delta x & -12 & 6\,\Delta x \\ 6\,\Delta x & 4\,\Delta x^2 & -6\,\Delta x & 2\,\Delta x^2 \\ -12 & -6\,\Delta x & 12 & -6\,\Delta x \\ 6\,\Delta x & 2\,\Delta x^2 & -6\,\Delta x & 4\,\Delta x^2 \end{bmatrix} \{\delta u_n^e\}^T \qquad (9.4.32)$$

Pulling these results together gives

$$\{\delta u_n\}^T\{f_n) = \{\delta u_n\}^T\mathbf{K}\{u_n\} \tag{9.4.33}$$

where

$$\mathbf{K} = \frac{EI}{\Delta x^3}
\begin{bmatrix}
 & -6\,\Delta x & 2\,\Delta x^2 & & 0 & & & 0 \\
-6\,\Delta x & 24 & 0 & -12 & 6\,\Delta x & & & \\
2\,\Delta x^2 & 0 & 8\,\Delta x^2 & -6\,\Delta x & 2\,\Delta x^2 & & 0 & \\
& -12 & -6\,\Delta x & 24 & 0 & -12 & 6\,\Delta x & 0 \\
0 & 6\,\Delta x & 2\,\Delta x^2 & 0 & 8\,\Delta x^2 & -6\,\Delta x & 2\,\Delta x^2 & \\
& & & -12 & -6\,\Delta x & 24 & 0 & -12 \\
& 0 & & 6\,\Delta x & 2\,\Delta x^2 & 0 & 8\,\Delta x^2 & -6\,\Delta x \\
0 & & & & 0 & -12 & 6\,\Delta x &
\end{bmatrix}
\tag{9.4.34}$$

is the stiffness matrix. The principle of virtual works states that this relation must be true for all virtual displacements $\{\delta u_n\}$. This entails

$$\mathbf{K}\{u_n\} = \{f_n\} \tag{9.4.35}$$

which is the set of equations that allow the displacements $\{u_n\}$ to be computed from the forces $\{f_n\}$. As we may easily verify, this result is identical to (9.3.19), obtained with the "mathematical" finite-element Galerkin method.

10

Poisson's Equation

10.1 FINITE-ELEMENT APPROXIMATION

10.1.1 POISSON'S EQUATION
WITH TRIANGULAR AND RECTANGULAR
ELEMENTS

Before analyzing in some detail the geometry of finite elements and the theory of the method in several dimensions, we shall consider in this chapter the treatment of *Poisson's equation with simple triangular and rectangular elements*. This shall have the advantage of providing a simple setting in which questions arising in more complex applications of the finite-element method are more easily resolved.

The simplest form of Poisson's equation in the plane is

$$-\nabla^2 U = F \qquad (10.1.1)$$

where $F(x, y)$ is a given forcing function. In *Cartesian coordinates*, this equation becomes

$$-\left(\frac{\partial^2 U}{\partial x^2} + \frac{\partial^2 U}{\partial y^2}\right) = F(x, y) \qquad (10.1.2)$$

However, the notation (10.1.1) is meant to be independent of the coordinates system. A less simple equation that we shall also consider is

$$-\nabla(P(x, y)\,\nabla U) + Q(x, y)U = F(x, y) \qquad (10.1.3)$$

where P and Q are given functions of x and y. (Semantically, this equation is closer to Helmholz's equation than to Poisson's.) Equations (10.1.1) or (10.1.3) apply in a finite closed domain D in the (x, y) plane (Figure 10.1.1) Different kinds of boundary conditions may be assigned;

  1.   *Dirichlet boundary conditions, also called boundary conditions of the first kind.* The value of U is prescribed on the boundary

$$U = U_B(x, y) \qquad \text{on } \partial D \qquad (10.1.4)$$

  2.   *Neumann boundary condition also called boundary conditions of the second kind.* The value of the outward normal derivative to the boundary is prescribed

$$\frac{\partial U}{\partial n} = G(x, y) \qquad \text{on } \partial D \qquad (10.1.5)$$

  3.   *Boundary conditions of the third kind* (*sometimes called Robbin's boundary conditions*). These are a combination of the preceding two;

$$\frac{\partial U}{\partial n} = G(x, y) + H(x, y)U \qquad \text{on } \partial D \qquad (10.1.6)$$

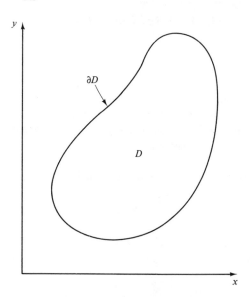

Figure 10.1.1

Boundary conditions of the first, second, and third kinds may apply simultaneously on different segments of ∂D.

10.1.2 FINITE-ELEMENT SOLUTION

The first step in a finite-element solution is in the choice of the type of finite elements that shall be used; that is, one must choose:

1. The geometric shape of the subdomains D_e in which D shall be divided or "tesselated."

2. The number and location of nodes that will be associated with this division of D into the $\{D_e\}$.

3. The family of analytic functions that will be used to express the finite-element solution $u(x, y)$ in each subdomain D_e (with very few exceptions this choice is *polynomials*).

The finite-element solution is then expressed as the function expansion

$$U(x, y) \simeq u(x, y) = \sum_n \varphi_n(x, y) u_n \qquad (10.1.7)$$

where the $\{\varphi_n(x, y)\}$ are basis function chosen in 3, and the $\{u_n\}$ are the corresponding nodal variables.

10.1.3 DERIVATION
OF THE DISCRETE EQUATIONS

We restrict ourselves at first to boundary conditions of the first kind.[1] A distinction is to be made between *interior points* $\{n\}$, where the solution is to be found, and *boundary points* $\{n_B\}$, where the solution is given at the onset:

$$u_{n_B} = U(x_{n_B}, y_{n_B}); \qquad n_B = 1, 2, \ldots, N_B \qquad (10.1.8)$$

The $\{u_n\}$ in all interior points

$$n = 1, 2, \ldots, N$$

are to be found as the solution of corresponding discrete equations. These equations may be derived in the present case either by invoking the variational method or by the more direct Galerkin method. Both lead to an identical result.

---

[1]See Section 10.4 for the mathematics corresponding to other boundary conditions.

Derivation of the Equations by a Variational Principle (the Rayleigh-Ritz Method).   The solution of equation (10.1.1) associated to boundary conditions of the first kind (10.1.4) minimizes the functional

$$J \equiv \int_D \left\{ \frac{1}{2}\left[\left(\frac{\partial U}{\partial x}\right)^2 + \left(\frac{\partial U}{\partial y}\right)^2 \right] - UF \right\} dD$$

$$= \int_D \left[\tfrac{1}{2}(\nabla U)^2 - UF \right] dD \tag{10.1.9}$$

where ∇U is the gradient

$$\nabla U \equiv \begin{bmatrix} \dfrac{\partial U}{\partial x} \\[2mm] \dfrac{\partial U}{\partial y} \end{bmatrix} \tag{10.1.10}$$

Indeed, the Euler–Lagrange equation for a minimum of J is

$$\nabla\left(\frac{\partial \Phi}{\partial(\nabla U)}\right) - \frac{\partial \Phi}{\partial U} = 0; \qquad \Phi = \tfrac{1}{2}(\nabla U)^2 - FU \tag{10.1.11}$$

which is precisely (10.1.1). Following the Rayleigh–Ritz principle, we substitute $u(x, y)$ for U in J and minimize $J(u)$ with respect to the parameters $\{u_n\}$. We have

$$J(u) = \int_D \left[\tfrac{1}{2}(\textstyle\sum_n \nabla\varphi_n u_n)^2 - \sum_n \varphi_n u_n F \right] dD \tag{10.1.12}$$

The necessary conditions for a minimum are, for all interior points $\{n\}$,

$$0 = \frac{\partial J(u)}{\partial u_n} = \int_D (\nabla\varphi_n^T \nabla u - \varphi_n F)\, dD; \qquad n = 1, 2, \ldots, N \tag{10.1.13}$$

After evaluation, this becomes a system of linear algebraic equations in the $\{u_n\}$ which is to be solved to produce the approximate solution. In this formulation, boundary values $\{u_{n_B}\}$ appear in the evaluation of (10.1.13) in those points that are one element away from the boundary. Those given boundary values are placed in the right-hand side of the system of algebraic equations to be solved, together with the contribution of the forcing function $F(x, y)$.

Derivation of the Equations by Galerkin's Method.   To apply Galerkin's method to the same problem, we form the equation residual

$$-\Re \equiv \nabla^2 u + F(x, y) \tag{10.1.14}$$

and express the N orthogonality conditions

$$0 = -\langle \varphi_n, \Re \rangle \equiv \int_D \varphi_n(\nabla^2 u + F)\, dD; \qquad u = 1, 2, \ldots, N \qquad (10.1.15)$$

The first term in this equation may be integrated by parts [note that all corresponding $\varphi_n(x, y)$ are zero on the boundary ∂D]:

$$0 = \int_D (-\nabla \varphi_n^T \nabla u + \varphi_n F)\, dD$$
$$= -\langle \nabla \varphi_n, \nabla u \rangle + \langle \varphi_n, F \rangle \qquad (10.1.16)$$

We see that this equation is identical to (10.1.13) and the computational procedure is thus the same. We should not, however, conclude that the variational and Galerkin methods are always identical. There are many problems whose solution does not obey some form of variational principle; Galerkin's method (and other weighted residual methods) remains applicable, whereas the variational method does not.

10.1.4 THE GENERALIZED POISSON EQUATION

A somewhat less simple case is that of the generalized Poisson's equation

$$-\nabla(P(x, y) \nabla U) + Q(x, y)U = F(x, y) \qquad (10.1.17)$$

where $P(x, y)$ represents a space-dependent diffusivity coefficient, $Q(x, y)U$ is a linear source term, and $F(x, y)$ is the usual forcing function. This is the Euler–Lagrange equation for the functional

$$J = \int \left[\tfrac{1}{2}P(\nabla U)^2 + Q\frac{U^2}{2} - FU \right] dD \qquad (10.1.18)$$

In the finite-element approximation of this equation, the Galerkin conditions (or the variational equations) are

$$\langle \varphi_n, \Re \rangle \equiv \langle \varphi_n, -\nabla(p \nabla u) + qu - f \rangle = 0 \qquad (10.1.19)$$

where $u, p, q,$ and f are approximations to $U, P, Q,$ and F in the finite-element representation. In this formulation, the $\{\varphi_n\}$ are the basis functions used to approximate U; that is,

$$U \simeq u = \sum_n \varphi_n u_n$$

But it is *not* necessarily the case that P, Q, and F be approximated in the same form (i.e., be approximated in the same subspace spanned by the φ_n). We *may, when convenient,* use the approximations

$$P \simeq p = \sum_n \varphi_n p_n$$

$$Q \simeq q = \sum_n \varphi_n q_n$$

$$F \simeq f = \sum_n \varphi_n f_n$$

but there are cases where it is more convenient or necessary to do otherwise.

10.2   LINEAR ELEMENTS ON TRIANGLES

10.2.1   *TRIANGULAR ELEMENTS IN TWO DIMENSIONS*

Possibly the *simplest* finite-element geometry that is acceptable[2] for elliptic equations of the second order in two dimensions is that obtained by covering D with triangles having nodes (or solution points) at the vertices (Figure 10.2.1) and in which the finite-element solution $u(x, y)$ consists of plane triangular "tiles." Such element shapes were used by Courant (1943)

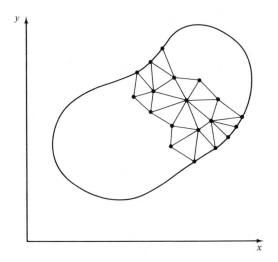

Figure 10.2.1

---

[2]See Section 8.3 for a discussion of what is "acceptable."

in one of the first papers describing the finite-element method. The triangular finite element in two spatial dimensions holds an important place because it is very flexible in covering geometrically an arbitrary domain D. To quote from G. Birkhoff (1972a): "It can be considered as a humble but invaluable workhorse," and from O. C. Zienkiewicz (1977, p. 164): "The advantage of an arbitrary triangular shape in approximating to any boundary shape has been amply demonstrated."

In the process of triangulation, exterior nodes are chosen on *or near* the boundary line such that the surrounding polygonal line provides a reasonable approximation to ∂D, as illustrated in Figure 10.2.2. In each

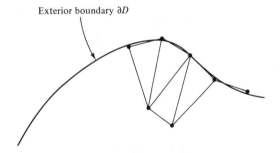

Exterior boundary ∂D

Figure 10.2.2   Choice of the exterior nodal points to provide a polynomial approximation of the curved boundary ∂D.

subdomain or element, we assume the approximation $u(x, y)$ to be a plane. This is equivalent to expressing u as $\sum_{n} \varphi_n(x, y)u_n$, where the basis functions $\{\varphi_n\}$ are the pyramidal functions shown in Figure 10.2.3a.

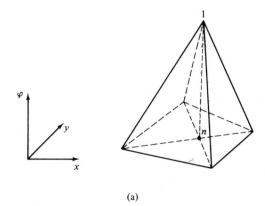

(a)

Figure 10.2.3   (*a*) Pyramidal basis function $\varphi_n(x, y)$; (*b*) piecewise linear approximation of $U(x, y)$ by $u(x, y)$ over triangular elements.

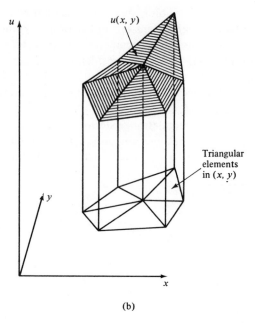

u(x, y)

Triangular
elements
in (x, y)

(b)

Figure 10.2.3   (Cont.)

10.2.2   ORGANIZATION OF THE CALCULATION ELEMENT BY ELEMENT

We shall limit ourselves at first to the simple form (10.1.1) of Poisson's equation. The generalized form (10.1.3) of Poisson's equation is treated later, in Section 10.3. According to (10.1.13) or (10.1.15), we are to evaluate the coefficients that appear in

$$\frac{\partial J(u)}{\partial u_n} = \langle \varphi_n, \Re \rangle$$

$$= \int_D (\nabla \varphi_n^T \nabla u - \varphi_n F) dD; \qquad n = 1, 2, \ldots, N \qquad (10.2.1)$$

to obtain the discrete form of the equation (a system of algebraic equations) prior to its solution for the $\{u_n\}$. Evaluation of the individual terms is more conveniently achieved when those calculations are carried out *element by element* rather than node by node. This will be shown here. We begin by expressing the integral (10.2.1) as the sum

$$0 = \sum_e \frac{\partial J_e(u)}{\partial u_n} \qquad (10.2.2)$$

where $J_e(u)$ is the integral (10.1.12) over the eth element:

$$J_e(u) \equiv \int_{D_e} [\tfrac{1}{2}(\nabla u)^2 - uF] dD_e \qquad (10.2.3)$$

The summation in (10.2.2) has nonzero contributions only from those elements that are adjacent to the node n. But since $\partial J_e(u)/\partial u_n$ will eventually be needed for *all three vertices of the eth element*, it is more efficient to develop these three expressions simultaneously for one element and to take care of their appropriate use in the process called *assembling*, to be later defined. We thus define

$$\left\{\frac{\partial J_e}{\partial u_{n_e}}\right\} = \int_{D_e} (\{\nabla \varphi_{n_e}^T\} \nabla u_e - \{\varphi_{n_e}\}F) \, dD_e \qquad (10.2.4)$$

as the 3-vector of elementary parts of (10.2.1) which are to be evaluated *in* the element D_e. Here $\{\varphi_{n_e}\}$ stands for

$$\{\varphi_{n_e}(x, y)\} = \begin{Bmatrix} \varphi_{1_e}(x, y) \\ \varphi_{2_e}(x, y) \\ \varphi_{3_e}(x, y) \end{Bmatrix} \qquad (10.2.5)$$

which is the set of shape or basis functions limited to the element D_e; 1_e, 2_e, and 3_e are a local labeling of the nodes corresponding to that element (Figure 10.2.4) and $\{\nabla \varphi_{n_e}^T\}$ are the corresponding gradients in D_e:

$$\{\nabla \varphi_{n_e}^T\} = \begin{Bmatrix} \nabla \varphi_{1_e}^T \\ \nabla \varphi_{2_e}^T \\ \nabla \varphi_{3_e}^T \end{Bmatrix} \qquad (10.2.6)$$

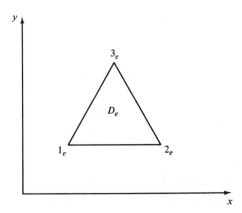

Figure 10.2.4   Local labeling of nodes in the triangular element D_e.

The equivalent of (10.2.2) in the Galerkin formulation is obtained by expressing

$$\mathcal{R}_e = \begin{cases} \mathcal{R} & \text{in } D_e \\ 0 & \text{elsewhere} \end{cases} \tag{10.2.7}$$

Whereupon the Galerkin conditions (10.1.15)–(10.1.16) become

$$0 = \langle \varphi_n, \mathcal{R} \rangle = \sum_e \langle \varphi_n, \mathcal{R}_e \rangle$$

$$= \sum_e \int_{D_e} \varphi_n (-\nabla^2 u_e - F)\, dD_e \tag{10.2.8}$$

The equivalent of (10.2.4) is (taking integration by part into account)

$$\langle \{\varphi_{n_e}\}, \mathcal{R} \rangle = \int_{D_e} \{\varphi_{n_e}\} \mathcal{R}\, dD_e$$

$$= \int_{D_e} (\{\nabla \varphi_{n_e}^T\} \nabla u_e - \{\varphi_{n_e}\} F)\, dD_e \tag{10.2.9}$$

which is the row vector of three scalars $\langle \varphi_{n_e}, \mathcal{R} \rangle$ identical to (10.2.5):

$$\langle \{\varphi_{n_e}\}, \mathcal{R} \rangle = \left\{ \frac{\partial J_e}{\partial u_{n_e}} \right\} \tag{10.2.10}$$

We may also express the approximate solution *in* D_e as

$$u(x, y) = u_e(x, y) = \{\varphi_{n_e}\}^T \{u_{n_e}\}$$

where

$$\{u_{n_e}\} = \begin{Bmatrix} u_{1_e} \\ u_{2_e} \\ u_{3_e} \end{Bmatrix}$$

is the 3-vector of nodal values belonging to D_e.

Another form in which u_e may be expressed in D_e, and which we shall use for (and only for) intermediate calculations, is

$$u(x, y) = u_e(x, y) = a_0 + a_1 x + a_2 y$$

where the parameters a_1, a_2, and a_3 are to be determined from the collocation conditions at the nodes:

$$\begin{rcases} (u_e)_{1_e} = u_{1_e} \\ (u_e)_{2_e} = u_{2_e} \\ (u_e)_{3_e} = u_{3_e} \end{rcases} \quad \text{or} \quad \{(u_e)_{n_e}\} = \{u_{n_e}\}$$

These conditions are (subscript e omitted)

$$\begin{bmatrix} 1 & x_1 & y_1 \\ 1 & x_2 & y_2 \\ 1 & x_3 & y_3 \end{bmatrix} \begin{Bmatrix} a_0 \\ a_1 \\ a_2 \end{Bmatrix} = \begin{Bmatrix} u_1 \\ u_2 \\ u_3 \end{Bmatrix} \tag{10.2.11}$$

Upon solving for the a_i, we find

$$\begin{Bmatrix} a_0 \\ a_1 \\ a_2 \end{Bmatrix} = \frac{1}{2S_e} \begin{bmatrix} x_2 y_3 - x_3 y_2 & x_3 y_1 - x_1 y_3 & x_1 y_2 - x_2 y_1 \\ y_2 - y_3 & y_3 - y_1 & y_1 - y_2 \\ x_3 - x_2 & x_1 - x_3 & x_2 - x_1 \end{bmatrix} \begin{Bmatrix} u_1 \\ u_2 \\ u_3 \end{Bmatrix} \tag{10.2.12}$$

where[3]

$$|2S_e| = \left| \det \begin{bmatrix} 1 & x_1 & y_1 \\ 1 & x_2 & y_2 \\ 1 & x_3 & y_3 \end{bmatrix} \right| = 2(\text{area of element } D_e) \tag{10.2.13}$$

The usefulness of this result is illustrated by the following. To compute the gradient of $u(x, y)$ in D_e [as needed in (10.2.9)], we find by simple observation of (10.2.12) that

$$\nabla u_e = \begin{bmatrix} \dfrac{\partial u_e}{\partial x} \\ \dfrac{\partial u_e}{\partial y} \end{bmatrix} = \begin{bmatrix} a_1 \\ a_2 \end{bmatrix} = \frac{1}{2S_e} \begin{bmatrix} y_2 - y_3 & y_3 - y_1 & y_1 - y_2 \\ x_3 - x_2 & x_1 - x_3 & x_2 - x_1 \end{bmatrix} \begin{Bmatrix} u_1 \\ u_2 \\ u_3 \end{Bmatrix} \tag{10.2.14}$$

Upon defining the matrix

$$\mathbf{B}_e = \frac{1}{2S_e} \begin{bmatrix} y_2 - y_3 & y_3 - y_1 & y_1 - y_2 \\ x_3 - x_2 & x_1 - x_3 & x_2 - x_1 \end{bmatrix} \tag{10.2.15}$$

the expression (10.2.14) may be rewritten simply as

$$\nabla u_e = \mathbf{B}_e \begin{Bmatrix} u_1 \\ u_2 \\ u_3 \end{Bmatrix} = \mathbf{B}_e \{u_{n_e}\} \tag{10.2.14a}$$

The expression of the gradients $\nabla \varphi_{n_e}$ in D_e may be obtained by applying

---

[3] S_e is positive when the ordering of the nodes in D_e is counterclockwise, and negative in the other case.

this result. With $\{u_{n_e}\}$ replaced in (10.2.14) by

$$\begin{Bmatrix} 1 \\ 0 \\ 0 \end{Bmatrix}, \quad \begin{Bmatrix} 0 \\ 1 \\ 0 \end{Bmatrix}, \quad \text{and} \quad \begin{Bmatrix} 0 \\ 0 \\ 1 \end{Bmatrix}$$

successively, we obtain in matrix form:

$$\{\nabla\varphi_{n_e}\} \equiv [\nabla\varphi_{1_e} \quad \nabla\varphi_{2_e} \quad \nabla\varphi_{3_e}] = \mathbf{B}_e \begin{bmatrix} 1 & 0 & 0 \\ 0 & 1 & 0 \\ 0 & 0 & 1 \end{bmatrix} = \mathbf{B}_e \qquad (10.2.15a)$$

(Note that this result was not unexpected: *the columns of* $\mathbf{B}_e$ *are simply the gradients* $\nabla\varphi_{n_e}$ *in* D_e.) This also reads

$$\nabla\varphi_{1_e} = \frac{1}{2S_e}\begin{vmatrix} y_2 - y_3 \\ x_3 - x_2 \end{vmatrix}$$

$$\nabla\varphi_{2_e} = \frac{1}{2S_e}\begin{vmatrix} y_3 - y_1 \\ x_1 - x_3 \end{vmatrix}$$

$$\nabla\varphi_{3_e} = \frac{1}{2S_e}\begin{vmatrix} y_1 - y_2 \\ x_2 - x_1 \end{vmatrix}$$

We thus have a simple expression of the gradients of the basis functions in D_e in terms of the coordinates of the three vertices of that element.

10.2.3 EVALUATION OF THE INTEGRALS

The two components appearing in the integral (10.2.9) may be evaluated separately. We first note that both gradients appearing in

$$\int_{D_e} \{\nabla\varphi_{n_e}\}^T \nabla u_e \, dx \, dy$$

are, because of the linearity of the shape functions, independent of x and y (the expressions $\{\nabla\varphi_{n_e}\}$ and $\{\nabla u_e\}$ are constants in D_e). The integration is thus trivial:

$$\int_{D_e} \{\nabla\varphi_{n_e}\}^T \nabla u_n \, dx \, dy = S_e \mathbf{B}_e^T \mathbf{B}_e \{u_{n_e}\} \qquad (10.2.16)$$

If (as appears logical) we choose to approximate F in D_e by the same linear shape functions, that is,

$$F \simeq f_e = \sum_{n_e} \varphi_{n_e} f_{n_e} = \{\varphi_{n_e}\}^T \{f_{n_e}\}; \qquad f_{n_e} = F(x_{n_e}, y_{n_e}) \qquad (10.2.17)$$

then the last term of (10.2.9) becomes

$$\int_{D_e} \{\varphi_{n_e}\} F \, dx \, dy \simeq \int_{D_e} \{\varphi_{n_e}\} \{\varphi_{n_e}\}^T \{f_{n_e}\} \, dx \, dy \qquad (10.2.18)$$

in which the individual integrals are easily evaluated:

$$\int_{D_e} \varphi_{n_e} \varphi_{n'_e} \, dx \, dy = \begin{cases} \dfrac{S_e}{6} & \text{when } n_e = n'_e \\[2mm] \dfrac{S_e}{12} & \text{when } n_e \neq n'_e \end{cases}$$

or

$$\int_{D_e} \{\varphi_{n_e}\} \{\varphi_{n_e}\}^T \, dx \, dy = \frac{S_e}{12} \begin{bmatrix} 2 & 1 & 1 \\ 1 & 2 & 1 \\ 1 & 1 & 2 \end{bmatrix} \qquad (10.2.19)$$

Whence, collecting results into (10.2.9), we find

$$\langle \{\varphi_{n_e}\}, \Re \rangle = S_e \left[\mathbf{B}_e^T \mathbf{B}_e \{u_{n_e}\} - \frac{1}{12} \begin{bmatrix} 2 & 1 & 1 \\ 1 & 2 & 1 \\ 1 & 1 & 2 \end{bmatrix} \{f_{n_e}\} \right] \qquad (10.2.20)$$

An important remark to be made about the result expressed by equation (10.2.20) is that a different (local) coordinate system may be chosen for each element. Prior to "assembling," equation (10.2.20) has become *independent of this local coordinate system* for each element (see also Example 1 in the next section).

10.2.4 ASSEMBLING

The element equations (10.2.20) are to be summed up, or *assembled*, according to (10.2.2) or (10.2.8). This assembling consists in adding to *each equation corresponding to one node* the contributions coming from all those elements that are adjacent to that node (Figure 10.2.5).

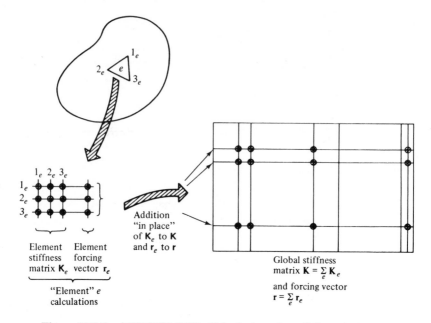

Figure 10.2.5 ASSEMBLING: Calculation of coefficients is done *one element at a time*. When all contributions have been added to the global system of equations, *then* **Ku** = **r** is solved for **u**.

The final result is generally expressed in matrix form as

$$\mathbf{K}\{u_n\} = \{r_n\} \qquad (10.2.21)$$

where **K** is often referred to as the stiffness matrix.[4] An interesting property of the system (10.2.21) is that it is *symmetric*. Indeed, we may observe that the element submatrices (10.2.20) are symmetric, and that addition (during assembling) preserves this property. This has important consequences, since *storing* and *solving* a system that has a symmetric matrix is more economical than when symmetry is not present. As we shall see later, the same procedure applied to the general elliptic equation

$$-\nabla(P(x, y) \nabla U) + Q(x, y)U = F$$

with boundary conditions of the first, second, and third kinds also results in symmetric systems. Symmetry is a genuine property of certain elliptic equations which is preserved by the finite-element method.

---

[4]The term "stiffness matrix" originates from structural engineering, where the finite-element method was first used and where **K** does indeed represent the stiffness of the structure.

EXAMPLE 1

Consider the problem of determining the deflection of a stretched, thin circular membrane uniformly loaded:

$$-\nabla^2 U = F = \text{constant} \qquad \text{in } D$$
$$U = 0 \qquad \text{on } \partial D$$

(10.2.22)

To keep things simple, we assume a single interior node located at the center and M nodes equally spaced on the boundary ∂D. This results in a division of D into M isosceles triangular elements D_e (Figure 10.2.6).

Albeit a natural set of coordinates for this problem is circular (r, α), the calculation in each element may be based on a set of *local* rectangular coordinates (x, y).

One element with its own coordinates is illustrated in Figure 10.2.7. The node coordinates are

$$(0, 0); \qquad \left(R \cos \frac{\alpha}{2}, -R \sin \frac{\alpha}{2} \right); \qquad \left(R \cos \frac{\alpha}{2}, R \sin \frac{\alpha}{2} \right)$$

The element's area is

$$S_e = R^2 \cos \left(\frac{\alpha}{2} \right) \sin \left(\frac{\alpha}{2} \right) = \frac{R^2 \sin \alpha}{2}$$

(10.2.23)

The gradients matrix $\mathbf{B}_e$ is evaluated from (10.2.15)

$$\mathbf{B}_e = \frac{R}{2S_e} \begin{bmatrix} -2 \sin \left(\frac{\alpha}{2} \right) & \sin \left(\frac{\alpha}{2} \right) & \sin \left(\frac{\alpha}{2} \right) \\ 0 & -\cos \left(\frac{\alpha}{2} \right) & \cos \left(\frac{\alpha}{2} \right) \end{bmatrix}$$

(10.2.24)

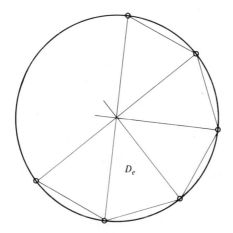

D_e

Figure 10.2.6

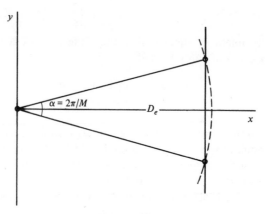

Figure 10.2.7

whence

$$\mathbf{B}_e^T\mathbf{B}_e = \left[\frac{\sin\,(\alpha/2)}{R\,\sin\,\alpha}\right]^2 \begin{bmatrix} 4 & - & - \\ - & - & - \\ - & - & - \end{bmatrix} \qquad (10.2.25)$$

where (— —) indicates those elements that are not needed in further calculations and have therefore not been computed. Applying (10.2.20) and being reminded that

$$\{u_{n_e}\} = \begin{Bmatrix} u_{n_e} \\ 0 \\ 0 \end{Bmatrix}; \qquad \{f_{n_e}\} = \begin{Bmatrix} F \\ F \\ F \end{Bmatrix}$$

we find

$$\frac{\partial J_e}{\partial u_{n_1}} = S_e\left\{-4\left[\frac{\sin\,(\alpha/2)}{R\,\sin\,\alpha}\right]^2 u_{n_{1e}} + \tfrac{1}{3}F\right\} \qquad (10.2.26)$$

Since $N = 1$ (the node at the center), the assembled system consists of a single equation. Assembling consists [per equation (10.2.8)] in adding the M equations identical to (10.2.26) for the M elements, resulting in the equation for the center node:

$$\sum_e \frac{\partial J_e}{\partial u_{n_1}} = MS_e\left\{-4\left[\frac{\sin\,(\alpha/2)}{R\,\sin\,\alpha}\right]^2 u_1 + \tfrac{1}{3}F\right\} = 0 \qquad (10.2.27)$$

and

$$u_1 = \frac{FR^2}{12}\left[\frac{\sin\,(2\pi/M)}{\sin\,(\pi/M)}\right]^2 \qquad (10.2.28)$$

To provide a means of comparison, the exact solution of (10.2.22) is easily derived. We obtain analytically:

$$U(r) = \frac{F}{4}(R^2 - r^2)$$

which for the center point $r = 0$ is

$$u_1 \simeq U(0) = \frac{FR^2}{4} \tag{10.2.29}$$

By contrast (10.2.28) gives

$$\text{For } M = 4: \quad u_1 = \frac{FR^2}{6} \tag{10.2.30a}$$

$$\text{For } M \rightarrow \infty: \quad u_1 = \frac{FR^2}{3} \tag{10.2.30b}$$

A treatment of the boundary for the case $M = 4$ that would appear to be heuristically more correct is to approximate D by the square of an area equal to that of the membrane shown in Figure 10.2.8.

The corresponding corrected radius is

$$R' = \sqrt{\frac{\pi}{2}}R \tag{10.2.31}$$

Then (10.2.30a) becomes

$$\text{For } M = 4: \quad u_1 = \frac{F(R')^2}{6} = \frac{FR^2}{6}\left(\frac{\pi}{2}\right)$$

$$= \frac{FR^2}{3.82} \tag{10.2.32}$$

which is much closer to the exact answer (10.2.29)

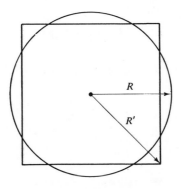

Figure 10.2.8

It is worthwhile noting again that at no time has it been necessary to use other than local rectangular coordinates in each triangular element.

EXAMPLE 2

Pursuing the mathematics of Example 1, we seek a finite-element approximation to the first eigenfunction of the Laplacian operator in a circle, which is the problem of seeking the lowest value of λ for which

$$\nabla^2 U + \lambda U = 0$$
$$U(R) = 0 \qquad (10.2.33)$$

has a nontrivial solution. Proceeding as before, the analog of (10.2.26) now becomes

$$\frac{\partial J_e}{\partial u_{1_e}} = S_e \left\{ 4 \left[\frac{\sin (\pi/M)}{R \sin (2\pi/M)} \right]^2 u_{1_e} + \frac{\lambda}{12} [2 \quad 1 \quad 1] \begin{bmatrix} u_{1_e} \\ 0 \\ 0 \end{bmatrix} \right. \qquad (10.2.34)$$

whence, as before,

$$\left\{ 4 \left[\frac{\sin (\pi/M)}{R \sin (2\pi/M)} \right]^2 + \frac{\lambda}{6} \right\} u_{1_e} = 0 \qquad (10.2.35)$$

which admits a nonzero solution u_{1_e} only if

$$\lambda = -\frac{24}{R_2} \left[\frac{\sin (\pi/M)}{\sin (2\pi/M)} \right]^2 \qquad (10.2.36)$$

The *exact* first eigenfunction of this problem is, for comparison,

$$U(r) = J_0 \left(\gamma \frac{r}{R} \right)$$

where J_0 is the Bessel function of order zero, and for the first mode

$$\frac{\gamma}{R} = \frac{1}{2.404}$$

The corresponding eigenvalue is

$$\lambda_{\text{exact}} = \frac{\nabla^2 J_0(\gamma r/R)}{J_0(\gamma r/R)} = -\left(\frac{2.404}{R} \right)^2 = -\frac{5.78}{R^2}$$

By contrast, the approximation (10.2.36) gives

For $M = 4$: $[R'$ given by (10.2.31)$] = \lambda_{\text{app}} = -\dfrac{24}{\pi R^2} = -\dfrac{7.64}{R^2}$

For $M \rightarrow \infty$: $\qquad\qquad\qquad\qquad \lambda_{\text{app}} = -\dfrac{6}{R^2}$

Problem

Repeat the calculations of Examples 1 and 2 with a division of the circular membrane into the $3M$ triangles illustrated in Figure 10.2.9.

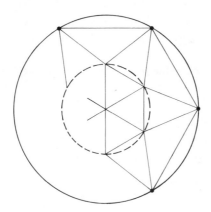

Figure 10.2.9

10.2.5 INSERTING BOUNDARY CONDITIONS OF THE FIRST KIND

There are several ways in which boundary conditions of the first kind may be inserted. A simple way consists in ignoring, during the element calculation and assembling process, which nodes belong to the boundary. That is, the matrix **K** in (10.2.21) is formed as an $(N + N_B)$ by $(N + N_B)$ matrix (where, as we remember, N is the number of interior nodes and N_B is the number of boundary nodes). Then the equation for each boundary node $n = n_B$ is replaced by

$$1 \cdot u_n = u_{n_B,\,\text{imposed}}$$

and all the terms on the left-hand side of (10.2.21) that contain u_{n_B} are displaced to the right-hand side.

To illustrate this procedure, suppose that our system is 4 by 4 and that u_3 is a given boundary point. At the end of the assembling process, we have obtained

$$\begin{bmatrix} a_{11} & a_{12} & a_{13} & a_{14} \\ a_{21} & a_{22} & a_{23} & a_{24} \\ a_{31} & a_{32} & a_{33} & a_{34} \\ a_{41} & a_{42} & a_{43} & a_{44} \end{bmatrix} \begin{bmatrix} u_1 \\ u_2 \\ u_3 \\ u_4 \end{bmatrix} = \begin{bmatrix} r_1 \\ r_2 \\ r_3 \\ r_4 \end{bmatrix} \tag{10.2.37}$$

After proceeding as indicated, we transform this system into

$$\begin{bmatrix} a_{11} & a_{12} & 0 & a_{14} \\ a_{21} & a_{22} & 0 & a_{24} \\ 0 & 0 & 1 & 0 \\ a_{41} & a_{42} & 0 & a_{44} \end{bmatrix} \begin{bmatrix} u_1 \\ u_2 \\ u_3 \\ u_4 \end{bmatrix} = \begin{bmatrix} r_1 - a_{13}u_{3B} \\ r_2 - a_{23}u_{3B} \\ u_{3B} \\ r_4 - a_{43}u_{3B} \end{bmatrix} \tag{10.2.38}$$

where u_{3B} is the assigned value for u_3 (on the boundary). Note that it would be sufficient to modify only the third line of (10.2.37): that is, the system

$$\begin{bmatrix} a_{11} & a_{12} & a_{13} & a_{14} \\ a_{21} & a_{22} & a_{23} & a_{24} \\ 0 & 0 & 1 & 0 \\ a_{41} & a_{42} & a_{43} & a_{44} \end{bmatrix} \begin{bmatrix} u_1 \\ u_2 \\ u_3 \\ u_4 \end{bmatrix} = \begin{bmatrix} r_1 \\ r_2 \\ u_{3B} \\ r_4 \end{bmatrix} \tag{10.2.39}$$

is equivalent to (10.2.38). But $\{a_{ij}\}$ is a symmetric matrix and (10.2.38) is also symmetric and, as we have indicated, both *storage* in computer memory and *solution* techniques may take advantage of this property. By contrast, the system (10.2.39) has lost its symmetry and becomes more expensive to handle.

A second way to handle boundary conditions of the first kind is given in Section 10.4.

10.3 SPECIAL CASES

In this section we describe the mathematics of the extension of linear elements on triangles to the treatment of slightly more complicated cases.

10.3.1 SPACE-DEPENDENT DIFFUSIVITY

Consider the case of Poisson's equation with parameters that depend on location. Example of a physical situation where this occurs is that of steady-state diffusion in a nonhomogeneous material. The correct form of equation (10.1.1) is then

$$-\nabla(P(x, y)\, \nabla U) = F(x, y) \qquad (10.3.1)$$

where U is the quantity being diffused and $P(x, y)$ is the corresponding diffusivity. If P is a smoothly variable function, then we may approximate it in the same basis as U:

$$P(x, y) \simeq p(x, y) = \sum_n \varphi_n(x, y) p_n \qquad (10.3.2)$$

where

$$p_n = P(x_n, y_n)$$

and the $\varphi_n(x)$ are the linear basis functions of Figure 10.2.3. Equations (10.2.3), (10.2.4), and (10.2.20) become, respectively,

$$J_e = \int_{D_e} [\tfrac{1}{2} p (\nabla u)^2 - uF]\, dx\, dy \qquad (10.3.3)$$

$$\langle \{\varphi_{n_e}\}, \mathcal{R}_e \rangle = \left\{ \frac{\partial J_e}{\partial u_{n_e}} \right\} = \int_{D_e} (\{\nabla \varphi_{n_e}\}^T p \, \nabla u_e - \{\varphi_{n_e}\}^T F)\, dD_e$$

$$= \int_{D_e} (\mathbf{B}_e^T p \mathbf{B}_e \{u_{n_e}\} - \{\varphi_{n_e}\}^T \{\varphi_{n_e}\}\{f_{n_e}\})\, dD_e$$

$$= S_e \left[\tfrac{1}{3}(p_{1_e} + p_{2_e} + p_{3_e}) \mathbf{B}_e^T \mathbf{B}_e \{u_{n_e}\} - \frac{1}{12}\begin{bmatrix} 2 & 1 & 1 \\ 1 & 2 & 1 \\ 1 & 1 & 2 \end{bmatrix} \{f_{n_e}\} \right] \qquad (10.3.4)$$

Extending triangular elements for Poisson's equation to the case of space-dependent properties is thus easy to handle, since it consists merely in multiplying the matrix $[\mathbf{B}_e^T \mathbf{B}_e]$ by the mean value of the diffusivity (from the value at the three vertices) in each element prior to assembling.

10.3.2 CONCENTRATED SOURCE

The approximation of the source term $F(x, y)$ in the same basis of linear functions (10.2.17) was not mandatory. To illustrate this, consider the case of a concentrated source:

$$F(x, y) = S(t)\, \delta(x - x_s)\, \delta(y - y_s) \qquad (10.3.5)$$

where (x_s, y_s) is the location of the source. The last term in (10.3.4) then becomes simply

$$\int_{D_e} \{\varphi_{n_e}\}^T F\, dD = \{\varphi_{n_e}(x_s, y_s)\} S(t) \qquad (10.3.6)$$

In the particular case where (x_s, y_s) is one of the nodes, say $(x_s = x_n, y_s = y_n)$, then

$$\int \varphi_m(x, y)F\, dD = \begin{cases} S(t) & \text{when } m = n \\ 0 & \text{when } m \neq n \end{cases} \qquad (10.3.7)$$

10.3.3 AXISYMMETRIC SOLUTION OF POISSON'S EQUATION IN THREE DIMENSIONS

The preceding mathematics lend themselves well to an extension to the semi-two-dimensional case derived from axisymmetry in three dimensions. Poisson's equation expressed in cylindrical coordinates is

$$-\nabla^2 U \equiv -\left(\frac{\partial^2}{\partial r^2} + \frac{1}{r}\frac{\partial}{\partial r} + \frac{1}{r^2}\frac{\partial^2}{\partial \alpha^2} + \frac{\partial^2}{\partial z^2}\right)U$$
$$= -\left[\frac{1}{r}\frac{\partial}{\partial r}\left(r\frac{\partial}{\partial r}\right) + \frac{1}{r^2}\frac{\partial^2}{\partial \alpha^2} + \frac{\partial^2}{\partial z^2}\right]U = F \qquad (10.3.8)$$

where r, α, and z are the radial, angular, and axial coordinates, respectively (Figure 10.3.1). *If both F and the boundary conditions are axisymmetric*, then $\partial^2 U/\partial \alpha^2 = 0$ and (10.3.8) becomes the axisymmetric form of Poisson's equation in cylindrical coordinates:

$$-\nabla^2 U \equiv -\left[\frac{1}{r}\frac{\partial}{\partial r}\left(r\frac{\partial}{\partial r}\right) + \frac{\partial^2}{\partial z^2}\right]U(r, z) = F(r, z) \qquad (10.3.9)$$

To reduce this equation to the preceding case, we multiply it by r to obtain

$$-\left[\frac{\partial}{\partial r}\left(r\frac{\partial}{\partial r}\right) + r\frac{\partial^2}{\partial z^2}\right]U = -\left[\frac{\partial}{\partial r}\left(r\frac{\partial}{\partial r}\right) + \frac{\partial}{\partial z}\left(r\frac{\partial}{\partial z}\right)\right]U = rF \qquad (10.3.10)$$

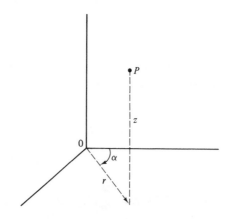

Figure 10.3.1 Cylindrical coordinates (r, α, z).

We note that this equation is identical to Poisson's equation in a nonisotropic medium (10.3.1):

1. With r and z considered as Cartesian coordinates.
2. With a space-dependent P which is equal to r.
3. With a modified source term F' equal to rF.

That is (10.3.10) is identical to

$$-\nabla(P \nabla U) = F' = rF \qquad (10.3.11)$$

where ∇ is the *Cartesian* operator

$$\nabla \equiv \begin{bmatrix} \dfrac{\partial}{\partial r} \\ \dfrac{\partial}{\partial z} \end{bmatrix} \qquad (10.3.12)$$

Equation (10.3.10) may thus be approximated by linear shape functions on triangles, and the mathematics of equations (10.3.1) through (10.3.4) apply exactly. The shape in three dimensions of one element is that of a triangular ring (Figure 10.3.2).

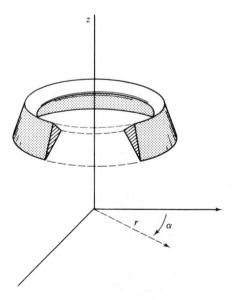

Figure 10.3.2 Triangular ring generated by a triangular element in (r, z).

Note that if the equation were to contain a space-dependent axisymmetric P to begin with:

$$-\left[\frac{1}{r}\frac{\partial}{\partial r}\left(P(r, z)r\frac{\partial}{\partial r}\right) + \frac{\partial}{\partial z}\left(P(r, z)\frac{\partial}{\partial z}\right)\right]U = F(r, z) \qquad (10.3.13)$$

then multiplying this equation by r still results, as before, in

$$-\left[\frac{\partial}{\partial r}\left(P'\frac{\partial}{\partial r}\right) + \frac{\partial}{\partial z}\left(P'\frac{\partial}{\partial z}\right)\right]U = F' \qquad (10.3.14)$$

with the modified terms

$$P' = rP$$
$$F' = rF$$

As before, a finite-element approximation based on an approximation of Poisson's equation (10.3.11) treating r and z as Cartesian coordinates remains applicable.

10.3.4 LINEAR TERM QU

A convenient way to treat the linear term QU is to assume that in each element Q is constant, equal to the average value:

$$q_e = \tfrac{1}{3}(q_{1_e} + q_{2_e} + q_{3_e})$$

Then [see (10.2.20) for the mathematics]

$$\langle\{\varphi_{n_e}\}, q_e u\rangle = \frac{S_e q_e}{12}\begin{bmatrix} 2 & 1 & 1 \\ 1 & 2 & 1 \\ 1 & 1 & 2 \end{bmatrix}\{u_{n_e}\} \qquad (10.3.15)$$

A second way in which this term may be handled is by assuming linear interpolation in each element between the values $\{q_{n_e}u_{n_e}\}$ at the nodes. That is, we approximate QU as

$$QU \simeq \sum_n \varphi_n(x, y)q_n u_n$$

where the φ_n are the linear basis functions on triangles. Then, instead of (10.3.15), we obtain

$$\langle\{\varphi_{n_e}\}, \sum_n \varphi_n q_n u_n\rangle = \frac{S_e}{12}\begin{bmatrix} 2 & 1 & 1 \\ 1 & 2 & 1 \\ 1 & 1 & 2 \end{bmatrix}\begin{Bmatrix} q_{1_e}u_{1_e} \\ q_{2_e}u_{2_e} \\ q_{3_e}u_{3_e} \end{Bmatrix} \qquad (10.3.16)$$

10.3.5 FIRST-ORDER TERM (ADVECTION)

The addition of a first-order term to the standard Poisson equation results in

$$-\nabla^2 U + \mathbb{c}^T \nabla U = F \tag{10.3.17}$$

where $\mathbb{c}$ is a vector that has the dimensions of a displacement in the plane. In Cartesian coordinates,

$$\mathbb{c} = \begin{bmatrix} c_x \\ c_y \end{bmatrix}$$

Equation (10.3.12) might describe, for instance, the steady-state distribution $U(x, y)$ of a polluting substance in soil where $\nabla^2 U$ represents diffusion, F is a distributed source or sink, and

$$\mathbb{c}^T \nabla U$$

represents entrainment or *advection* by subsurface water flow. With the presence of a first-order term, we cannot find a function J which is minimized by solutions of the equation. Thus we cannot apply the variational method. But Galerkin's method applies with no difficulty. The equation residual is

$$\mathfrak{R} = -\sum_n \nabla^2 \varphi_n u_n + \mathbb{c}^T \sum_n \nabla \varphi_n u_n - F \tag{10.3.18}$$

where, with triangular elements, we have *in D_e*,

$$\sum_n \nabla \varphi_n u_n \equiv \nabla u_e = \mathbf{B}_e \{u_{n_e}\}$$

We shall assume at first that $\mathbb{c}$ is independent of location. We find as an extension of (10.2.20),

$$\langle \{\varphi_{n_e}\}, \mathfrak{R} \rangle = S_e \left[\mathbf{B}_e^T \mathbf{B}_e \{u_{n_e}\} + \tfrac{1}{3}\mathbb{c}^T \mathbf{B}_e\{u_{n_e}\} - \frac{1}{12}\begin{bmatrix} 2 & 1 & 1 \\ 1 & 2 & 1 \\ 1 & 1 & 2 \end{bmatrix}\{f_{n_e}\} \right] \tag{10.3.19}$$

where the new term is

$$\langle \{\varphi_{n_e}\}, \mathbb{c}^T \sum_n \nabla \varphi_n u_n \rangle = \frac{S_e}{3}\mathbb{c}^T \mathbf{B}_e\{u_{n_e}\} \tag{10.3.20}$$

If $\mathbb{c}$ is not constant,

$$\{\mathbb{c}_n\} \equiv \begin{bmatrix} \{c_{x,\,n}\} \\ \{c_{y,\,n}\} \end{bmatrix}$$

then we may assume that the variation of $\mathbb{c}$ in each element is linear:

$$\mathbb{c}(x, y) = \sum_n \varphi_n \mathbb{c}_n$$

We then find for the corresponding term in (10.3.19):

$$\langle \{\varphi_{n_e}\}, \mathbb{c}^T \, \nabla u \rangle = \langle \{\varphi_{n_e}\}, \{\varphi_{1_e} \varphi_{2_e} \varphi_{3_e}\} \{\mathbb{c}_{n_e}^T\} \mathbf{B}_e \{u_{n_e}\} \rangle$$

$$= \frac{S_e}{12} \begin{bmatrix} 2 & 1 & 1 \\ 1 & 2 & 1 \\ 1 & 1 & 2 \end{bmatrix} \{\mathbb{c}_{n_e}^T\} \mathbf{B}_e \{u_{n_e}\} \qquad (10.3.21)$$

In this expression, $\{\mathbb{c}_{n_e}^T\}$ stands for the matrix

$$\{\mathbb{c}_{n_e}^T\} \equiv \begin{bmatrix} c_{x,\,1_e} & c_{y,\,1_e} \\ c_{x,\,2_e} & c_{y,\,2_e} \\ c_{x,\,3_e} & c_{y,\,3_e} \end{bmatrix} \qquad (10.3.22)$$

so that

$$\{\mathbb{c}_{n_e}^T\} \mathbf{B}_e \{u_{n_e}\}$$

is a 3-vector. *Note that with the presence of a first-order term, the stiffness matrix is no longer symmetric.*

10.4   BOUNDARY CONDITIONS

We have so far considered only boundary conditions of the first kind for Poisson's equation; that is,

$$U = U_B(x, y) \qquad \text{on the boundary} \qquad (10.4.1)$$

We will consider in this section the other two kinds of boundary conditions (see Figure 10.4.1):

1. Boundary condition of the second kind (Neumann): *imposed "flux":*

$$\frac{\partial U}{\partial n} = g(x, y) \qquad \text{on the boundary} \qquad (10.4.2)$$

   where $\partial U / \partial n$ is the normal (outward) derivative on the boundary.

2. Boundary conditions of the third kind:

$$\frac{\partial U}{\partial n} = g + hU \qquad (10.4.3)$$

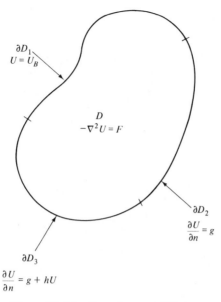

Figure 10.4.1 Boundary conditions.

10.4.1 VARIATIONAL EQUATIONS

Before considering the finite-element treatment of boundary conditions, we return to the calculus of variations and ask: *Which functional (if any) is "extremized" by solutions of Poisson's equations*

$$-\nabla^2 U \equiv -\left(\frac{\partial^2 U}{\partial x^2} + \frac{\partial^2 U}{\partial y^2}\right) = F(x, y) \qquad (10.4.4)$$

in D, when boundary conditions of the type (10.4.2) *or* (10.4.3) *apply?*

To be more specific, we shall assume that the boundary line ∂D is divided in three segments:

∂D_1: segment of the boundary ∂D where conditions (10.4.1) hold
∂D_2: segment of the boundary ∂D where conditions (10.4.2) hold
∂D_3: segment of the boundary ∂D where conditions (10.4.3) hold

Thus (Figure 10.4.1)

$$\partial D = \partial D_1 + \partial D_2 + \partial D_3 \qquad (10.4.5)$$

In practice, we may delete ∂D_2, since this is merely a special case of ∂D_3 with $h(x, y) = 0$. Repeating the derivation of Section 9.1 for the functional

$$J(U) = \int_D \Phi(U, \nabla U) \, dD \qquad (10.4.6)$$

we find

$$\delta J = \int_D \left[\frac{\partial \Phi}{\partial U} \delta U + \frac{\partial \Phi}{\partial (\nabla U)} \delta \nabla U \right] dD$$

$$= \int_D \left[\frac{\partial \Phi}{\partial U} \delta U + \frac{\partial \Phi}{\partial (\nabla U)} \nabla \delta U \right] dD \qquad (10.4.7)$$

and, upon integrating the second term by parts,

$$\delta J = \int_D \frac{\partial \Phi}{\partial U} - \nabla \left(\frac{\partial \Phi}{\partial (\nabla U)} \right) \right] \delta U \, dD$$

$$+ \int_{\partial D} \left(\frac{\partial \Phi}{\partial (\nabla U)} \right)_{\text{normal}} \delta U \, ds \qquad (10.4.8)$$

Applying this to the case we are interested in (from Section 10.3),

$$J = \int_D \Phi \, dx \, dy; \qquad \Phi = \tfrac{1}{2} (\nabla U)^2 - UF \qquad (10.4.9)$$

equation (10.4.8) becomes

$$\delta J = - \int_D (F + \nabla^2 U) \delta U \, dD + \int_{\partial D} \frac{\partial U}{\partial n} \delta U \, ds \qquad (10.4.10)$$

We then ask what functional [other than (10.4.9)] is stationary for functions $U(x, y)$ which satisfy both Poisson's equation,

$$\nabla^2 U + F = 0 \qquad \text{in } D$$

and the specified boundary conditions,

$$U = U_B \qquad \text{on } \partial D_1$$

$$\frac{\partial U}{\partial n} = g + hU \qquad \text{on } \partial D_3$$

$$(\partial D = \partial D_1 + \partial D_3)$$

Without further theory, we may observe that the *augmented functional*

$$J' = \int_D [\tfrac{1}{2} (\nabla U)^2 - FU] \, dD - \int_{\partial D_3} (gU + \tfrac{1}{2} hU^2) \, ds \qquad (10.4.11)$$

is the answer. Indeed, we find

$$\delta J' = \delta J - \delta \int_{\partial D_3} (gU + \tfrac{1}{2}hU^2)\, ds$$

$$= -\int_D (F + \nabla^2 U)\, \delta U\, dD + \int_{\partial D} \frac{\partial U}{\partial n}\, \delta U\, ds$$

$$- \int_{\partial D_3} (g + hU)\, \delta U\, ds \qquad (10.4.12)$$

The integrand of the first integral is identically zero in D, since it is a solution of (10.4.4) and the sum of the second and third integrals also vanishes since

$$\delta U = 0 \qquad \text{on } \partial D_1 \qquad (10.4.13)$$

and

$$\frac{\partial U}{\partial n} - (g + hU) = 0 \qquad \text{on } \partial D_3$$

Thus,

$$\delta J'(U) = 0 \qquad (10.4.14)$$

for the function $U(x, y)$, which satisfies equation (10.4.4) *and* the boundary conditions (10.4.1) and (10.4.3).

Generalization.   For any boundary condition of the form

$$\frac{\partial U}{\partial n} = b(U) \qquad \text{on } \partial D_3 \qquad (10.4.15)$$

the corresponding functional which is stationary is

$$J'(U) = J(U) - \int_{\partial D_3} B(U)\, ds \qquad (10.4.16)$$

where

$$B(U) = \int^U b(v)\, dv + \text{constant} \qquad (10.4.17)$$

Natural Boundary Conditions.   Note that if the boundary condition along $\partial D_2 + \partial D_3$ were

$$\frac{\partial U}{\partial n} = 0 \qquad \text{on } \partial D_2 + \partial D_3 \qquad (10.4.18)$$

then no addition to J would be required to make it stationary. That is, the variation of the functional (10.4.6) is zero for the function $U(x, y)$ solution of (10.4.4) with the boundary condition (10.4.1) on ∂D_1 and (10.4.18) on $\partial D_2 + \partial D_3$. The boundary condition (10.4.18) is called a *natural boundary condition* for Poisson's equation.

10.4.2 FINITE-ELEMENT EQUATIONS

With these results in hand, we now return to the problem of deriving the corresponding finite-element discrete equations. We shall restrict discussion to the case of linear shape functions on triangles, but the results extend to other types of finite elements with minimal modifications.

We shall consider both the variational and Galerkin approaches to the derivation of these equations. It will be shown that if properly applied, the Galerkin approach yields results identical to those obtained by the variational method.

Derivation of Finite-Element Equations by the Variational Method.   We form the approximate solution

$$u(x, y) = \sum_n \varphi_n(x, y) u_n = \{\varphi_n\}^T \{u_n\} \tag{10.4.19}$$

where $\{u_n\}$ are the nodal values of the approximate solution in all interior and boundary nodes, and $\{\varphi_n\}$ are the linear basis functions defined in Section 10.3. To (10.4.19) corresponds the functional [from (10.4.11)]

$$J'(u) = J'(\{u_n\})$$
$$= \int_D [\tfrac{1}{2}(\nabla u)^2 - uF]\, dD - \int_{\partial D_3} (gu + \tfrac{1}{2}hu^2)\, ds \tag{10.4.20}$$

whose variation is, following the Rayleigh–Ritz method, to be equated to zero.

For each nodal point n that is interior to D and each point n_B that is on the boundary ∂D_3, we write

$$\frac{\partial J'(u)}{\partial u_n} = 0 \tag{10.4.21}$$

and follow exactly the procedure of Section 10.3.

Points that *are on* ∂D_1 have solution values u_{n_B} that are given, and thus (10.4.21) *is not* to be expressed in those points.

The expression of (10.4.21) in *points that are on the boundary* ∂D_3 now contains an additional term:

$$\frac{\partial J'(u)}{\partial u_{n_B}} = \int_D (\nabla \varphi_{n_B}^T \nabla u - \varphi_{n_B} F)\, dD - \int_{\partial D_3} \varphi_{n_B}(g + hu)\, ds = 0 \tag{10.4.22}$$

The second term is a line integral that is to be evaluated along boundary sides (those which approximate ∂D) of those elements having one vertex in n_B and one side on the boundary. The only new term that need be evaluated (in addition to those derived in Section 10.3) is

$$\int_{\partial D_3} \varphi_{n_B}(g + hu) \, ds \qquad (10.4.23)$$

This is decomposed into

$$\int_{\partial D_{left}} + \int_{\partial D_{right}} \qquad (10.4.24)$$

where the two integration paths are as shown in Figure 10.4.2. We assume at first for simplicity that g and h are constants. Then, we find

$$\int_{\partial D_3} \varphi_{n_B}(g + hu) \, ds = l_{\text{left}}\left[\frac{g}{2} + \frac{h}{6}(u_{n_B-1} + 2u_{n_B})\right]$$

$$+ l_{\text{right}}\left[\frac{g}{2} + \frac{h}{6}(2u_{n_B} + u_{n_B+1})\right] \qquad (10.4.25)$$

where l_{left} and l_{right} are the *side length* to the left and right of n_B, respectively. Extension to nonconstant g and h is easily achieved by using common concepts of numerical integration. (See Section 10.4.3.)

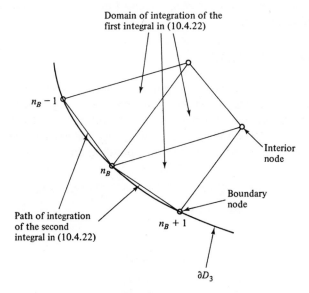

Domain of integration of the first integral in (10.4.22)

$n_B - 1$

Interior node

n_B

Boundary node

Path of integration of the second integral in (10.4.22)

$n_B + 1$

∂D_3

Figure 10.4.2

Derivation of the Finite-Element Equations by Galerkin's Method.   The same results can be obtained by application of Galerkin's method. We simply form the equation residual

$$\mathfrak{R} \equiv -\nabla^2 u - F \tag{10.4.26}$$

and require, following the recipe, that it be orthogonal to the basis functions corresponding to all nodal points where u_n is not specified:

$$0 = \langle \varphi_n, \mathfrak{R} \rangle$$
$$= -\int_D \varphi_n (\nabla^2 u + F) \, dD \tag{10.4.27}$$

Next, consider this relation for a point n_B on the boundary ∂D_3. Integration by parts yields (sign changed)

$$0 = \int_D (\nabla \varphi_{n_B}^T \nabla u - \varphi_{n_B} F) \, dD - \int_{\partial D_3} \varphi_{n_B} \left(\frac{\partial u}{\partial n} \right) ds \tag{10.4.28}$$

If we now replace $(\partial u / \partial n)$ by its imposed value (10.4.3), then we obtain

$$0 = \int_D (\nabla \varphi_{n_B}^T \nabla u - \varphi_{n_B} F) \, dD - \int_{\partial D_3} \varphi_{n_B} (g + hu) \, ds \tag{10.4.29}$$

which is identical to (10.4.22). Note that this result was obtained with Galerkin's method without having at any time to invoke results from the calculus of variations.

10.4.3 VARIABLE CASE

Consider next the case where g in the expression (10.4.3) of the boundary condition is dependent on location. Previously developed mathematics apply easily to this case (in particular, see Section 9.1). We let $g(x, y)$ on boundary sides of the third kind be the linear interpolant between nodal values where it is given. Then the corresponding term in (10.4.25) becomes

$$\int \varphi_{n_B} g \, ds = \frac{l_{\text{left}}}{6} (g_{n_B - 1} + 2g_{n_B}) + \frac{l_{\text{right}}}{6} (2g_{n_B} + g_{n_B + 1}) \tag{10.4.30}$$

instead of

$$(l_{\text{left}} + l_{\text{right}}) \frac{g}{2}$$

Note that, as expected, the two become identical when g is constant.

10.4.4 A DIFFERENT TREATMENT
OF BOUNDARY CONDITIONS
OF THE FIRST KIND

We have shown at the end of Section 10.2 one way to handle boundary conditions of the first kind. We show another way here.

Consider fictitious boundary conditions of the third kind on ∂D_3 expressed as

$$\left(\frac{\partial U}{\partial n}\right)_B = \alpha(U_B - U)$$

where α is (so far) an arbitrary number. As prescribed, we use for those points the equations (10.4.22)–(10.4.25), which result in appropriate terms in the left- and right-hand sides of $\mathbf{Ku} = \mathbf{r}$. If we want to impose $u_{n_B} = u_{n_B,\text{given}}$, we may simply let $\alpha \longrightarrow \infty$ (in practice, $\alpha \simeq 10^{15}$ to 10^{20}).

To illustrate this procedure we refer to equation (10.2.37) that is, we assume that the total number of points is 4, and that

$$\begin{bmatrix} a_{11} & a_{12} & a_{13} & a_{14} \\ a_{21} & a_{22} & a_{23} & a_{24} \\ a_{31} & a_{32} & a_{33} & a_{34} \\ a_{41} & a_{42} & a_{43} & a_{44} \end{bmatrix} \begin{bmatrix} u_1 \\ u_2 \\ u_3 \\ u_4 \end{bmatrix} = \begin{bmatrix} r_1 \\ r_2 \\ r_3 \\ r_4 \end{bmatrix}$$

is the system obtained at the end of the assembly process, ignoring which points are on the boundary. According to (10.4.22)–(10.4.25), we then augment this system as

$$\begin{bmatrix} a_{11} & a_{12} & a_{13} & & a_{14} \\ a_{21} & a_{22} & a_{23} & & a_{24} \\ a_{31} & a_{32} & a_{33}+c & & a_{34} \\ a_{41} & a_{42} & a_{43} & & a_{44} \end{bmatrix} \begin{bmatrix} u_1 \\ u_2 \\ u_3 \\ u_4 \end{bmatrix} = \begin{bmatrix} r_1 \\ r_2 \\ r_3 + cu_{3_B} \\ r_4 \end{bmatrix} \qquad (10.4.31)$$

(where $c \simeq 10^{15}$), which has the desired result.

10.4.5 BOUNDARY CONDITIONS
FOR THE GENERALIZED FORM
OF POISSON'S EQUATION

We consider next the generalized form of Poisson's equation,

$$-\nabla(P(x, y) \nabla U) + Q(x, y)U = F(x, y) \qquad (10.4.32)$$

It was noted previously that this is the Euler–Lagrange equation of the functional

$$J = \int_D \left[\tfrac{1}{2}P(\nabla U)^2 + Q\frac{U^2}{2} - FU \right] dD \qquad (10.4.33)$$

To determine the way of treating boundary conditions of the second and third kinds, let us express the variation of J,

$$\delta J \equiv J(U + \delta U) - J(U)$$
$$= \int_D \left[(P\,\nabla U)^T\,\delta\nabla U + (QU - F)\delta U \right] dD \qquad (10.4.34)$$

which becomes, after integration by parts of the first term,

$$\delta J = \int_D \left[(-\nabla P\,\nabla U + QU - F)\,\delta U \right] dD + \int_{\partial D} \left[(P\,\nabla U)_n\,\delta U \right] ds \qquad (10.4.35)$$

where $(\cdot)_n$ denotes the component of $(\cdot)$ normal to the boundary:

$$(P\,\nabla U)_n \equiv P\frac{\partial U}{\partial n} \qquad (10.4.36)$$

Natural boundary conditions (i.e., boundary conditions that equate δJ to zero) are thus

$$P\frac{\partial U}{\partial n} = 0 \quad \text{or} \quad \frac{\partial U}{\partial n} = 0 \qquad (10.4.37)$$

as in the case of the simple form of Poisson's equation.

10.4.6 BOUNDARY CONDITIONS OF THE THIRD KIND

In view of (10.4.35), the "natural" way to express boundary conditions of the third kind for equation (10.4.32) is as

$$(P\,\nabla U)_n \equiv P\frac{\partial U}{\partial n} = g + hU \qquad (10.4.38)$$

[rather than (10.4.3)]. We may then observe that the extended functional

$$J' = J - \int_{\partial D_3} \left(gU + h\frac{U^2}{2} \right) ds$$

is stationary for solutions of the problem. Indeed, we find

$$\delta J' = \delta J - \int_{\partial D_3} [(g + hU)\,\delta U]\,ds = \delta J - \int_{\partial D_3} P\frac{\partial U}{\partial n}\,\delta U\,ds$$

which is zero for functions U that satisfy both the Euler–Lagrange equation (10.4.32) in D and the boundary conditions (10.4.38) on ∂D_3.

10.5   THE MERMAID COMPUTER CODE

We give in this section the listing of a computer code (called MERMAID) for the solution of the extended Poisson equation

$$-\nabla(P(x, y)\,\nabla U) + Q(x, y)U = F$$

in the plane. The code is written in FORTRAN and uses linear finite elements on triangles. The mathematics are those developed in preceding sections of this chapter.

Boundary conditions accepted by the code may be of the first, second, and third kinds along different portions of ∂D.

10.5.1   SUBROUTINES

The code consists of the following subroutines:

> FORMK
> ELEMNT
> B3KIND
> SIDEB3
> B1KIND
> SOLVE

It has, in addition, a block DATA, where the data for the problem are entered.

Subroutine FORMK.   This subroutine assembles the conductivity or stiffness matrix CONDCT and right-hand-side vector $R1$ of the system of equations

$$\text{CONDCT} \cdot \{u_n\} = R1 \tag{10.5.1}$$

per equations (10.2.8), (10.2.20), and (10.3.4)–(10.3.13). The calculation of the coefficients is done element by element in subroutine ELEMNT, which is called by FORMK once for each element.

Subroutine ELEMNT.   Does the calculation (10.2.8) one element at a time and returns those results to FORMK.

Subroutine B3KIND.   Adds the appropriate terms to the stiffness matrix and right-hand side for those nodes that are on a boundary side of the third kind. Calculations are done one side at a time, in SIDEB3, which is called by B3KIND.

Subroutine SIDEB3.   Does the calculation (10.4.24)–(10.4.30) one side at a time and returns those results to B3KIND.

Subroutine B1KIND.   Inserts boundary conditions of the first kind using the method illustrated by equation (10.2.38).

Subroutine SOLVE.   Solves (10.5.1) to produce the solution $\{u_n\}$.

10.5.2 THE MERMAID COMPUTER CODE

```
      CONTROL MAIN PROGRAM
C****************************************************************************
C*                                                                         *
C*   *  *  *****  ****     *   *      *      *  ***                         *
C*   ** **  *      *   *   **  **    * *     *  *  *                        *
C*   * * *  *****  ****    * * *     * *     *  *  *                        *
C*   *   *  *       * *    *   *   *******   *  *  *                        *
C*   *   *  *****   *   *  *   *      *   *  *  ***                         *
C*                                                                         *
C*   CODE FOR POISSON/HELMHOLZ SOLUTION BY LINEAR FINITE ELEMENTS          *
C*   ON TRIANGLES                                                          *
C*   RUTGERS UNIVERSITY                                                    *
C*   R.VICHNEVETSKY                                                        *
C****************************************************************************
C
C
C
C      *EQUATION SOLVED:
C      -DEL.SIG(X,Y)DEL.U(X,Y)+Q(X,Y) U(X,Y) = F(X,Y)
C      WITH BOUNDARY CONDITIONS OF FIRST AND THIRD KIND
C      *UB(NB) TO BE GIVEN IN BOUNDARY POINTS OF FIRST KIND
C      *G(N) AND H TO BE GIVEN ON BOUNDARY POINTS OF THIRD KIND WHERE
C      DU/DN=G(N)+H*U(X,Y)
C      (H = CONSTANT)
```

```
C
C       INPUTS:
C       NN = NUMBER OF NODES
C       NE = NUMBER OF ELEMENTS
C       NB = NUMBER OF BOUNDARY POINTS OF THE FIRST KIND
C       NS3 = NUMBER OF SIDES ON BOUNDARY OF THE THIRD KIND
C       X,Y  = COORDINATES OF NODES
C       KONECT= ELEMENTS CONNECTION ARRAY ORDERED COUNTERCLOCKWIZE
C       NBI = ADDRESS OF NODES ON BOUNDARY OF FIRST KIND (U GIVEN)
C       KONB3 = SIDE CONNECTION ARRAY OF BOUNDARY SIDES OF THE THIRD KIND
C       SOURCE = LOCAL VALUES OF SOURCE
C       SIG = DISTRIBUTED CONDUCTIVITY SIGMA
C       G = COEFICIENT IN BOUNDARY CONDITIONS OF THE THIRD KIND (N DEPENDENT)

C       H = COEFICIENT IN BOUNDARY CONDITIONS OF THE THIRD KIND (CONSTANT)

C
C       Q = LINEAR COEFFICIENT,GIVEN AT EACH NODE
C
C       INTERNAL VARIABLES NAMES:
C       ELCOND(3,3) = ELEMENT CONDUCTIVITY MATRIX TO BE ADDED TO GLOBAL MATRIX

C       CONDCT(NN,40) = GLOBAL CONDUCTIVITY MATRIX
C       R1(N) = GLOBAL RIGHT HAND SIDE VECTOR
C       RHS(2) = ELEMENT RIHGT HAND SIDE VECTOR
C       SLNGTH = SIDE LENGTH
C
C
        COMMON/BLOCK1/NN,NE,NB,NS3
        COMMON/BLOCK2/X(100),Y(100),KONECT(3,100),KONB3(2,25),
       1NBI(25),R1(100),CONDCT(100,40)
        COMMON/BLOCK5/ UB(100), F(100),TF(3),SIG(100),Q(100)
C
C
        CALL FORMK
C
        CALL B3KIND
C
        CALL B1KIND
C
        CALL SOLVE
C
C       PRINT RESULTS
        WRITE(5,100)
        DO 111 M=1,NN
111     WRITE(5,110)M,X(M), Y(M),R1(M),F(M)
100     FORMAT(///,T10,'#',T23,'X',T38,'Y',T49,'SOLUTION',T68,'F')
110     FORMAT(I10,4F15.4)
        STOP
        END
```

```
C************************************************************************
C*                                                                     *
C*      DATA BLOCK - PROVIDES THE PROGRAM WITH THE APROPRIATE DATA      *
C*                                                                     *
C************************************************************************
C
      BLOCK DATA
      COMMON/BLOCK1/NN,NE,NB,NS3
      COMMON/BLOCK2/X(100),Y(100),KONECT(3 100),KONB3(2,25),
     1NBI(25),R1(100),CONDCT(100,40)
      COMMON/BLOCK3/ELCOND(3,3),A(2,3)
      COMMON/BLOCK4/B3CO(2,2),RHS(2)   ,G(100),H
      COMMON/BLOCK5/UB(100),F(100).TF(3),SIG(100),Q(100)
C
      DATA NN,NE,NB,NS3/49,72,14,0/
C
C                   NODAL POINT  DATA
C
      DATA X/0.,1.,2.,3.,4.,5.,6.,0.,1.,2.,3.,4.,5.,6.,0.,1.,2.,3.,
     a4.,5.,6.,0.,1.,2.,3.,4.,5.,6.,0.,1.,2.,3.,4. 5.,6.,
     b0.,1.,2.,3.,4.,5.,6.,;0.,1.,2.,3.,4.,5.,6.,51*0./
      DATA Y/7*6.,7*5.,7*4.,7*3.,7*2.,7*1.,7*0.,51*0./
C
C                   ELEMENT  DATA
C
      DATA KONECT/1,8,9, 1,9,2, 2,9,10, 2,10,3, 3,10,11, 3,11,4,
     a4,11,12, 4,12,5, 5,12,13, 5,13,6, 6,13,14, 6,14,7, 8,15,9,
     b9,15,16, 9,16,10, 10,16,17, 17,11,10, 17,18,11, 18,12,11,
     c18,19,12, 12,19,13, 13,19,20, 20,14,13, 20,21,14, 22,23,15,
     d15,23,16, 16,23,24, 24,17,16, 17,24,25, 25,18,17, 25,26,18,
     e18,26,19, 19,26,27, 27,20,19, 27,28,20, 20,28,21, 22,29,23,
     f23,29,30, 30,24,23, 30,31,24, 31,25,24, 31,32,25, 32,26,25,
     g32,33,26, 33,27,26, 33,34,27, 27,34,28, 28,34,35, 29,36,37,
     h37,30,29, 37,38,30, 30,38,31, 31,38,39, 39,32,31, 39,40,32,
     i32,40,33, 33,40,41, 41,34,33, 41,42,34, 35,34,42, 43,37,36,
     j43,44,37, 44,38,37, 44,45,38, 45,39,38, 45,46,39, 39,46,40,
     k46,47,40, 47,41,40, 47,48,41, 48,42,41, 48,49,42,84*0/
C
C                   SIGMA DATA
C
      DATA SIG/5*10.,6*0.01,10.,6*0.01,10,4*.01,3*10.,4*0.01,10.,
     a6*0.01,10.,6*0.01,5*10.,51*0./
C
C                   BOUNDARY DATA
C
      DATA H/0.0/
      DATA G(0),G(0),G(0)/0.,0.,0./
      DATA KONB3/50*0/
      DATA NBI/1,8,15,22,29,36,43,7,14,21,28,35,42,49,11*0/
      DATA UB(1),UB(8),UB(15),UB(22),UB(29),UB(36),UB(43),UB(7),UB(14),
     aUB(21),UB(28),UB(35),UB(42),UB(49)/7*0.,7*1./
      DATA Q/100*0.  /
      DATA F/100*0.0/
      END
```

```
      SUBROUTINE FORMK
C*****************************************************************************
C*                                                                         *
C*   THIS ROUTINE CALLS THE SUBROUTINE * ELEMNT * ONCE FOR EACH            *
C*                                                                         *
C*   ELEMENT AND ASSEMBLES THE RESULT INTO * CONDCT *                      *
C*                                                                         *
C*   MATRIX AND * R1 * VECTOR                                              *
C*                                                                         *
C*****************************************************************************
C
C
C                         FORMS CONDUCTIVITY MATRIX
C                         IN UPPER TRIANGULAR FORM
C
      COMMON/BLOCK1/NN,NE,NB,NS3
      COMMON/BLOCK2/X(100),Y(100),KONECT(3,100),KONB3(2,25),
     1NBI(25),R1(100),CONDCT(100,40)
      COMMON/BLOCK3/ELCOND(3,3),A(2,3)
      COMMON/BLOCK5/UB(100),F(100),TF(3),SIG(100),Q(100)
C
C                         SET BANDMAX AND NO OF EQUATIONS
C
C     ZERO THE SOURCE ARRAY
C
      DO 161 J=1,NN
  161 R1(J)=0.0

      NBAND=40
C
C                         ZERO CONDUCTIVITY MATRIX
C
      DO 300 N=1,NN
      DO 300 M=1,NBAND
  300 CONDCT(N,M)=0.
C
C                         SCAN ELEMENTS
C
      DO 400 N=1,NE
      CALL ELEMNT(N+0)
C
C                         RETURNS ELCOND AS CODUCTIVITY MATRIX
C
C                         STORE ELCOND IN CONDCT
C
C                         FIRST ROWS
C
      DO 350 JJ=1,3
      NROWB=KONECT(JJ,N)
C
C     ASSEMBLE RIGHT HAND SIDE R1(N)
C
      R1(NROWB)=R1(NROWB)-TF(JJ)
C
C                         THEN COLUMNS
C
      DO 330 KK=1,3
      NCOLB=KONECT(KK,N)
      NCOL=NCOLB+1-NROWB
C
C                         SKIP STORING IF BELOW BAND
C
      IF(NCOL.LE.0) GO TO 330
      CONDCT(NROWB,NCOL)=CONDCT(NROWB,NCOL)+ELCOND(JJ,KK)
  330 CONTINUE
  350 CONTINUE
  400 CONTINUE
      RETURN
      END
```

```
      SUBROUTINE ELEMNT(N)
C*******************************************************************************
C*                                                                            *
C*    THIS ROUTINE COMPUTES THE COMPONENTS OF THE CONDUCTIVITY                 *
C*    MATRIX AND THE RIGHT HAND SIDE VECTOR ONE ELEMENT AT A                   *
C*    TIME                                                                     *
C*                                                                            *
C*******************************************************************************
C
C
      COMMON/BLOCK1/NN,NE,NB,NS3
      COMMON/BLOCK2/X(100),Y(100),KONECT(3,100),KONB3(2,25),
     1NBI(25),R1(100),CONDCT(100,40)
      COMMON/BLOCK3/ELCOND(3,3),A(2,3)
      COMMON/BLOCK5/UB(100),F(100),TF(3),SIG(100),Q(100)
C
C                          DETERMINE ELEMENT CONNECTIONS
C
      I=KONECT(1,N)
      J=KONECT(2,N)
      K=KONECT(3,N)
C
C                          SET UP LOCAL COORDINATE SYSTEM
C
C
C
      A(1,1)=Y(J)-Y(K)
      A(1,2)=Y(K)-Y(I)
      A(1,3)=Y(I)-Y(J)
      A(2,1)=X(K)-X(J)
      A(2,2)=X(I)-X(K)
      A(2,3)=X(J)-X(I)
      AREA=(A(2,3)*A(1,2)-A(2,2)*A(1,3))/2.0
      IF(AREA.LE.0.) GO TO 220
      QA=(Q(I)+Q(J)+Q(K))/3.0
      SIGM=(SIG(I)+SIG(J)+SIG(K))/3.0
      TF(1)=((AREA/12.0)*(2*F(I)+F(J)+F(K)))
      TF(2)=((AREA/12.0)*(F(I)+2*F(J)+F(K)))
      TF(3)=((AREA/12.0)*(F(I)+F(J)+2*F(K)))
C
C                 ELCOND IS ELEMENT 3BY3 CONDUCTIVITY MATRIX
C
      DO 210 II=1,3

      DO 210 JJ=1,3

C
C     INSERTION OF THE LINEAR TERM
C
      IF(II.EQ.JJ) ELCOND(II,JJ)=AREA*QA/6.
      IF(II.NE.JJ) ELCOND(II,JJ)=AREA*QA/12.
C     DO 210 KK=1,2
      ELCOND(II,JJ)=ELCOND(II,JJ)+SIGM*A(KK,II)*A(KK,JJ)/(4.*AREA)

  210 CONTINUE
      RETURN
C
C                          ERROR EXIT FOR BAD CONNECTIONS
C
  220 WRITE(5,100)N
  100 FORMAT(9X,'ZERO OR NEGATIVE AREA ELEMENT N',I4,5X,'EXECUTION
     1TERMINATED  '        )
      STOP
      END
```

```
      SUBROUTINE B3KIND
C********************************************************************************
C*                                                                            *
C*    THIS ROUTINE INSERTS BOUNDARY CONDITIONS OF THE THIRD KIND              *
C*                                                                            *
C********************************************************************************
C
C
      COMMON/BLOCK1/NN,NE,NB,NS3
      COMMON/BLOCK2/X(100),Y(100),KONECT(3,100),KONB3(2,25),
     1NBI(25),R1(100),CONDCT(100,40)
      COMMON/BLOCK4/B3CO(2,2),RHS(2)   ,G(100),H
C
C     SCAN ALL SIDES OF THIRD KIND
C
      DO 400 N=1,NS3
C
      CALL SIDEB3(N+0)
C
      DO 350 JJ=1,2
      NROWB=KONB3(JJ,N)
C
C     ASSEMBLE SOURCE TERM + RHS
C
      R1(NROWB)=R1(NROWB)+RHS(JJ)
      DO 330 KK=1,2
      NCOLB=KONB3(KK,N)
      NCOL=NCOLB+1-NROWB
      IF(NCOL.LE.0) GO TO 330
      CONDCT(NROWB,NCOL)=CONDCT(NROWB,NCOL)+B3CO(JJ,KK)
  330 CONTINUE
  350 CONTINUE
  400 CONTINUE
      RETURN
      END




      SUBROUTINE SIDEB3(N)
C********************************************************************************
C*                                                                            *
C*    THIS ROUTINE COMPUTES THE ADDITIONAL COEFFICIENTS OF THE CONDUCTIVITY   *
C*    MATRIX AND THE RIGHT HAND SIDE VECTOR ONE BOUNDARY SIDE AT A TIME       *
C*                                                                            *
C********************************************************************************
C
C
      COMMON/BLOCK1/NN,NE,NB,NS3
      COMMON/BLOCK2/X(100),Y(100),KONECT(3,100),KONB3(2,25),
     1NBI(25),R1(100),CONDCT(100,40)
      COMMON/BLOCK4/B3CO(2,2),RHS(2)   ,G(100),H
C
      IF (NS3.EQ.0)RETURN
C     DETERMINE SIDE CONNECTIONS
C
      I=KONB3(1,N)
      J=KONB3(2,N)
      IF(I.EQ.0.OR.J.EQ.0) RETURN
C
```

```
C     COMPUTE SIDE LENGTH
C
      SLNGTH=SQRT((X(I)-X(J))**2+(Y(I)-Y(J))**2)
C
C     COMPUTE COEFFICIENTS OF B3CO AND RIGHT HAND SIDE
C
      B3CO(1,2)=-H*SLNGTH/6
      B3CO(1,1)=-H*SLNGTH/3
      B3CO(2,2)=B3CO(1,1)
      B3CO(2,1)=B3CO(1,2)
      RHS(1)  =SLNGTH*(2*G(I)+G(J))/6.
      RHS(2)  =SLNGTH*(G(I)+2*G(J))/6.
  100 CONTINUE
      RETURN
      END




      SUBROUTINE B1KIND
C********************************************************************************
C*                                                                            *
C*    THIS ROUTINE INSERTS BOUNDARY CONDITIONS OF FIRST KIND                  *
C*                                                                            *
C********************************************************************************
C
C
      COMMON/BLOCK1/NN,NE,NB,NS3
      COMMON/BLOCK2/X(100),Y(100),KONECT(3,100),KONB3(2,25),
     1NBI(25),R1(100),CONDCT(100,40)
      COMMON/BLOCK5/UB(100),F(100),TF(3),SIG(100),Q(100)

      NBAND=40
      DO 500 N=1,NB
      M=NBI(N)
      DO 100 I=1,NN
C
C     COMPUTE EFFECTIVE COLUMN NUMBER
C
      J=IABS(M-I)+1
      IF(J.GT.NBAND) GO TO 100
      L=I
C
C     IF BELOW DIAGONAL TAKE THE VALUES ALONG THE ROWS (SYMMETRY)
C
      IF(I.GT.M) L=M
C
C     COMPUTE RIGHT HAND SIDE
C
      R1(I)=R1(I)-CONDCT(L,J)*UB(M)
  100 CONTINUE
  500 CONTINUE
      DO 501 N=1,NB
      M=NBI(N)
      DO 101 I=1,NN
C
C     COMPUTE EFFECTIVE COLUMN NUMBER
C
      J=IABS(M-I)+1
      IF(J.GT.NBAND) GO TO 101
      L=I
```

```
C
C      IF BELOW DIAGONAL TAKE THE VALUES ALONG THE ROWS (SYMMETRY)
C
       IF(I.GT.M)  L=M
C
C      SET ELEMENTS WHICH CORRESPOND WITH BOUNDARY NODES ZERO
C
       CONDCT(L,J)=0.
  101 CONTINUE
C
C      SET DIAOGONAL ELEMENTS EQUAL TO ONE AND RIGHT HAND SIDE EQUAL TO
C      BOUNDARY VALUE
C
       CONDCT(M,1)=1.
       R1(M)=UB(M)
  501 CONTINUE
       RETURN
       END


       SUBROUTINE SOLVE
C*********************************************************************
C*                                                                  *
C*      THIS ROUTINE SOLVES THE SYSTEM OF EQUATIONS AU=B            *
C*      INPUT(B) AND OUTPUT(U) ARE BOTH STORED IN R1(100)           *
C*                                                                  *
C*********************************************************************
C
       COMMON/BLOCK1/NN,NE,NB,NS3
       COMMON/BLOCK2/X(100),Y(100),KONECT(3,100),KONB3(2,25),
      1NBI(25),R1(100),CONDCT(100,40)
C
       NBAND=40
C
C                             REDUCE MATRIX
C
       DO 300 N=1,NN
       I=N
       DO 290 L=2,NBAND
       I=I+1
       IF(CONDCT(N,L))240,290,240
  240 C=CONDCT(N,L)/CONDCT(N,1)
       J=0
       DO 270 K=L,NBAND
       J=J+1
       IF(CONDCT(N,K))260,270,260
  260 CONDCT(I,J)=CONDCT(I,J)-C*CONDCT(N,K)
  270 CONTINUE
  280 CONDCT(N,L)=C
C
C                             AND SOURCE VECTOR
C                             FOR EACH EQUATION
C
       R1(I)=R1(I)-C*R1(N)
  290 CONTINUE
  300 R1(N)=R1(N)/CONDCT(N,1)
C
```

```
C                    BACK-SUBSTITUTION
C
      N=NN
  350 N=N-1
      IF(N)500,500,360
  360 L=N
      DO 400 K=2,NBAND
      L=L+1
      IF(CONDCT(N,K))370,400,370
  370 R1(N)=R1(N)-CONDCT(N,K)*R1(L)
  400 CONTINUE
      GO TO 350
C
  500 RETURN
      END
```

10.6 BILINEAR ELEMENTS ON RECTANGLES

Another type of simple element suitable for the solution of Poisson's equation will be presented here. The shape of the elements in which the domain D is subdivided is assumed to consist in *regular, adjacent rectangles*. As a prelude to the mathematics that follow, it is useful to note that in practice, the restriction of regular rectangular shapes can be (and generally is) removed by local transformations of the geometry. These are known as *isoparametric transformations* generating any straight-edged quadrilateral as a (relatively simple) transformation of the rectangles shown here (see Section 12.3).

Consider a rectangular division of the (x, y) plane by grids of lines parallel to the x and y axes, respectively, as shown in Figure 10.6.1. The simplest finite-element representation of a function $U(x, y)$ over this mesh

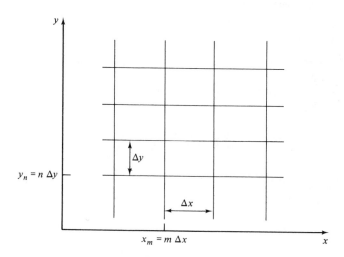

Figure 10.6.1

is given by elements which are, in each elementary rectangle,

$$x \in (x_m, x_{m+1}); \qquad y \in (y_n, y_{n+1}) \tag{10.6.1}$$

of the form

$$u(x, y) = a_0 + a_1 x + a_2 y + a_3 xy \tag{10.6.2}$$

The coefficients $\{a_0, a_1, a_2, a_3\}$ are uniquely defined by the collocation conditions of (10.6.2) with the nodal values of U at the four vertices of (10.6.1):

$$
\begin{aligned}
u(x_{m'}, y_{n'}) = u_{m', n'} \qquad & \text{for } (m' = m; \ n' = n) \\
& (m' = m + 1; \ n' = n) \\
& (m' = m; \ n' = n + 1) \\
& (m' = m + 1; \ n' = n + 1)
\end{aligned}
\tag{10.6.3}
$$

Such an element is shown in Figure 10.6.2.

We may observe that (10.6.2) varies linearly when either x or y is kept constant. In particular, *the intersection of $u(x, y)$ with the sides of the rectangle* (10.6.1) *are straight lines*, connecting the four vertices. We may call finite-element shapes such as (10.6.2) *bilinear*, for reasons that are self-evident.

Continuity.   Note that when a mosaic of bilinear elements is used to approximate a function over the rectangular mesh illustrated in Figure 10.6.1,

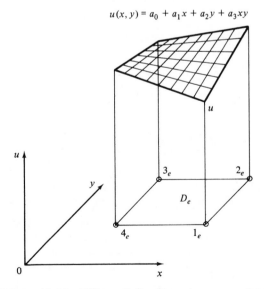

$$u(x, y) = a_0 + a_1 x + a_2 y + a_3 xy$$

Figure 10.6.2   Bilinear finite element over a rectangle.

the piecewise function $u(x, y)$ is *continuous* across interelement lines. This is a consequence of the fact that u is, on such interelement lines, simply the (unique) linear interpolant between the two adjacent nodes.

This property exists also with higher-order polynomial finite elements: the intersection of the finite-element solution with one of the sides of the element domain is the (unique) polynomial that interpolates between the nodal values located on that side (and is therefore identical for the two elements which are separated by that side).

10.6.1 BILINEAR BASIS FUNCTIONS

Consider the two-dimensional basis functions obtained by taking the product of linear chapeau functions in x and y, respectively (Figure 10.6.3):

$$\varphi_{m,n}(x, y) = \varphi_m(x)\varphi_n(y)$$

$$= \begin{cases} \left(1 - \left|\dfrac{x - x_m}{\Delta x}\right|\right)\left(1 - \left|\dfrac{y - y_n}{\Delta y}\right|\right) & \text{when } |x - x_m| \leq \Delta x \\ & \text{and } |y - y_n| \leq \Delta y \\ 0 & \text{elsewhere} \end{cases}$$

$$(10.6.4)$$

It is easily shown that the function

$$u(x, y) = \sum_{m,n} \varphi_{m,n}(x, y)u_{m,n} \qquad (10.6.5)$$

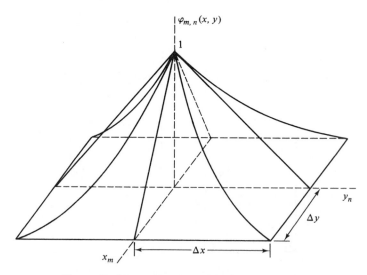

Figure 10.6.3 Bilinear basis function $\varphi_{m,n}(x, y)$.

is identical with the piecewise-bilinear finite-element representation previously defined by (10.6.2). Indeed:

1. (10.6.5) is of the form (10.6.2) in each rectangular element.

2. $u(x_m, y_n) = u_{m,n}$, as does (10.6.2); that is, both satisfy the same collocation conditions.

10.6.2 POISSON'S EQUATION WITH BILINEAR FINITE ELEMENTS

We again consider Poisson's equation,

$$-\nabla^2 U = F(x, y) \qquad \text{in } D$$
$$U = U_B(x, y) \qquad \text{on } \partial D \tag{10.6.6}$$

to be approximated over a division of D in equal rectangular finite elements of dimension $(\Delta x, \Delta y)$ (Figure 10.6.4). We shall use Galerkin's procedure (as has been noted before, the same final result would be obtained in the application of the Rayleigh–Ritz variational method). The residual orthogonality conditions are

$$0 = \int \varphi_{m,n} \left(-\sum_{m',n'} \nabla^2 \varphi_{m',n'} u_{m',n'} - \sum_{m',n'} \varphi_{m',n'} f_{m',n'} \right) dx\, dy \tag{10.6.7}$$

where integration is carried out over the rectangle of vertices

$$(x_{m-1}, y_{n-1}); \quad (x_{m-1}, y_{n+1}); \quad (x_{m+1}, y_{n+1}); \quad (x_{m+1}, y_{n-1}) \tag{10.6.8}$$

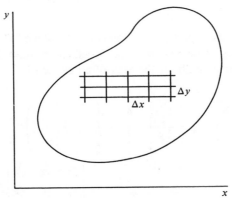

Figure 10.6.4

See Figure 10.6.5. The elementary integrals are (details are left to the reader)

$$\langle \varphi_{m,n}, \varphi_{m,n} \rangle = \tfrac{4}{9} \Delta x \, \Delta y$$

$$\langle \varphi_{m,n}, \varphi_{m,n-1} \rangle = \langle \varphi_{m,n}, \varphi_{m,n+1} \rangle$$

$$= \langle \varphi_{m,n}, \varphi_{m-1,n} \rangle = \langle \varphi_{m,n}, \varphi_{m+1,n} \rangle = \tfrac{1}{9} \Delta x \, \Delta y \qquad (10.6.9)$$

$$\langle \varphi_{m,n}, \varphi_{m+1,n+1} \rangle = \langle \varphi_{m,n}, \varphi_{m+1,n-1} \rangle$$

$$= \langle \varphi_{m,n}, \varphi_{m-1,n+1} \rangle = \langle \varphi_{m,n}, \varphi_{m-1,n-1} \rangle = \tfrac{1}{36} \Delta x \, \Delta y$$

and, using integration by parts,

$$\left\langle \varphi_{m,n}, \frac{\partial^2 \varphi_{m,n}}{\partial x^2} \right\rangle = -\frac{4}{3} \left(\frac{\Delta x \, \Delta y}{\Delta x^2} \right)$$

$$\left\langle \varphi_{m,n}, \frac{\partial^2 \varphi_{m+1,n}}{\partial x^2} \right\rangle = \left\langle \varphi_{m,n}, \frac{\partial^2 \varphi_{m-1,n}}{\partial x^2} \right\rangle$$

$$= \frac{2}{3} \left(\frac{\Delta x \, \Delta y}{\Delta x^2} \right)$$

$$\left\langle \varphi_{m,n}, \frac{\partial^2 \varphi_{m+1,n+1}}{\partial x^2} \right\rangle = \left\langle \varphi_{m,n}, \frac{\partial^2 \varphi_{m+1,n-1}}{\partial x^2} \right\rangle$$

$$= \left\langle \varphi_{m,n}, \frac{\partial^2 \varphi_{m-1,n+1}}{\partial x^2} \right\rangle = \left\langle \varphi_{m,n}, \frac{\partial^2 \varphi_{m-1,n-1}}{\partial x^2} \right\rangle \qquad (10.6.10)$$

$$= \frac{1}{6} \left(\frac{\Delta x \, \Delta y}{\Delta x^2} \right)$$

$$\left\langle \varphi_{m,n}, \frac{\partial^2 \varphi_{m,n+1}}{\partial x^2} \right\rangle = \left\langle \varphi_{m,n}, \frac{\partial^2 \varphi_{m,n-1}}{\partial x^2} \right\rangle$$

$$= -\frac{2}{6} \left(\frac{\Delta x \, \Delta y}{\Delta x^2} \right)$$

and similarly for the $\langle \varphi, \partial^2 \varphi / \partial y^2 \rangle$ integrals. In stencil-like notations, the pulling together of these results is [all coefficients have been divided by $(\Delta x \, \Delta y)$, which, incidentally, is the area of an element]

$$-\left(\frac{1}{\Delta x^2} \begin{bmatrix} \tfrac{1}{6} & -\tfrac{2}{6} & \tfrac{1}{6} \\ \tfrac{4}{6} & -\tfrac{8}{6} & \tfrac{4}{6} \\ \tfrac{1}{6} & -\tfrac{2}{6} & \tfrac{1}{6} \end{bmatrix} u_{m,n} + \frac{1}{\Delta y^2} \begin{bmatrix} \tfrac{1}{6} & \tfrac{4}{6} & \tfrac{1}{6} \\ -\tfrac{2}{6} & -\tfrac{8}{6} & -\tfrac{2}{6} \\ \tfrac{1}{6} & \tfrac{4}{6} & \tfrac{1}{6} \end{bmatrix} u_{m,n} \right)$$

$$= \begin{bmatrix} \tfrac{1}{36} & \tfrac{4}{36} & \tfrac{1}{36} \\ \tfrac{4}{36} & \tfrac{16}{36} & \tfrac{4}{36} \\ \tfrac{1}{36} & \tfrac{4}{36} & \tfrac{1}{36} \end{bmatrix} f_{m,n} \qquad (10.6.11)$$

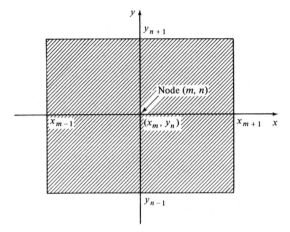

Figure 10.6.5 Domain of integration of (10.6.7). This domain is the *support* of the basis function $\varphi_{m,n}$.

Vertical position in these arrays denotes displacement in y, and horizontal position, displacement in x: for example, the first terms in (10.6.11) read

$$\frac{-1}{6\,\Delta x^2}(u_{m-1,n-1} + 4u_{m-1,n} + u_{m-1,n+1} - 2u_{m,n-1} - 8u_{m,n} - 2u_{m,n+1}$$

$$+ u_{m+1,n-1} + 4u_{m+1,n} + u_{m+1,n+1}) \qquad (10.6.12)$$

and so on.

10.6.3 MIXED ELEMENTS WITH LINEAR SIDES

Consider a division of the plane into rectangles and triangles with corner nodes only, as shown in Figure 10.6.6. The shape functions in the triangles are *linear* and those in the rectangles are *bilinear*, of the kind described above. What both types of elements have in common is that the variation of $u(x, y)$ along each side is the (unique) *linear* interpolant between the two end nodes of the side.[5] Thus, C^0 continuity of the finite-element function is preserved and this mixed-elements representation is acceptable for the approximation of second-order equations.

As we shall see later, this property will remain true for higher-order elements, and variations of u on interelement lines will be in the form of

---

[5]These are called *linear* sides.

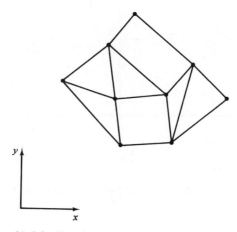

Figure 10.6.6 Simple mixed elements with linear sides.

linear, quadratic, cubic, and so on, *polynomials*, defined as interpolants between the value of u in the 2, 3, 4, ... nodes located on that side (the names linear side, quadratic side, and so on, are used to describe those cases). A finite-element approximation of Poisson's equation over such a mixed domain is obtained without complication. The calculations implied in

$$\sum_e \langle \varphi_n, \mathfrak{R} \rangle_e = 0$$

are carried out element by element. The mathematics of Section 10.2 apply to the triangular elements, and those of the present section apply to the rectangular elements (the elementary stiffness matrix for rectangular element is 4 by 4).

Theory
and Geometry

Those few things having been considered, the whole matter is reduced to pure geometry, which is the one aim of physics and mechanics.

G. W. Leibniz

11.1 FINITE ELEMENTS IN TWO DIMENSIONS

Most of the basic theoretical concepts relevant to the implementation of the finite-element method were discussed in previous chapters. We have purposely limited the analysis to one-dimensional problems and to only the simplest element shapes in two dimensions.

In practical applications, engineers often use element shapes more complex than those we have considered so far. The extension of the preceding results to such element shapes requires little additional approximation theory: *the mathematics of those extensions are mostly geometry.*

11.1.1 FINITE-ELEMENT SHAPES

We introduced in Section 10.6 bilinear elements which were possibly the simplest shapes on rectangles, and in Section 10.2 we described linear elements on triangles. Such simple element shapes were among the first to be described historically (e.g., Courant, 1943). But more complex forms have rapidly come to be used. As computers' capacity grew, so did computational sophistication. A multitude of papers and at least one book have been written on the specific subject of element geometry.[1]

It is well beyond our scope to investigate this topic in detail. Our purpose here is limited to providing a somewhat sketchy overview of the main concepts that are used to generate and classify finite-element shapes.

Fundamental to the finite-element method is the question of *representation*. The domain of the problem is divided into the subdomains of the elements, in which the solution (and other dependent variables) is represented approximately in a *piecewise fashion*, element by element. This brings to the fore the several factors that enter this process.

1.   Choice of the *form of the element subdivision* $\{D_e\}$ of the domain D.

2.   Choice of the analytic form or *shape of the dependent variables* inside each element.

3.   Choice of the *discrete parameters* that defines these shapes in each element.

It is in problems involving several spatial dimensions that the finite-element method has achieved its major success. Although more than one reason may be invoked to explain this success, it is undoubtedly true that the flexibility of choice of finite-element shapes, coupled with the ability to construct compact computer codes which derive the corresponding discrete equations, provides one of the foremost reasons.[2]

The Galerkin formulation of the finite-element method leads to the expression of integrals of the form

$$\langle \varphi_n, \Re \rangle = \int_D \varphi_n \Re \, dD = 0 \tag{11.1.1}$$

These may be evaluated element by element:

$$\sum_e \langle \varphi_n, \Re \rangle_e = \sum_e \int_{D_e} \varphi_n \Re \, dD_e \tag{11.1.2}$$

---

[1]As an example of a sophisticated analysis of element shapes, see *A Rational Finite Element Basis*, by Wachspress (1975).

[2]By contrast (except with simple geometries that can be handled by special codes), the derivation of the discrete equations with finite-difference methods must as a rule be done by hand.

where $\langle \cdot \rangle_e$ is the integral over the domain D_e of the element e only:

$$\langle \varphi_n, \mathfrak{R} \rangle_e = \int_{D_e} \varphi_n \mathfrak{R} \, dD$$

and the sum $\sum_e$ is over those elements on which φ_n is nonzero (i.e., those elements that are adjacent to the node n). See Figure 11.1.1.

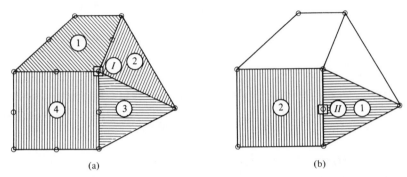

(a)   (b)

Figure 11.1.1   (a) *Four* elements that are adjacent to the node I and contribute *four* elementary integrals to the sum (11.1.2); (b) *two* elements that are adjacent to the node II and contribute *two* elementary integrals to the sum (11.1.2).

11.1.2   ASSEMBLING

If $\{n_{1_e}, n_{2_e}, \ldots, n_{n_e}\}$ are the nodes that belong to an element e, then one shall eventually need, in the evaluation of (11.1.2) for all n, the calculation of

$$\langle \varphi_{n_{1_e}}, \mathfrak{R} \rangle_e$$
$$\langle \varphi_{n_{2_e}}, \mathfrak{R} \rangle_e$$
$$\vdots$$
$$\langle \varphi_{n_{n_e}}, \mathfrak{R} \rangle_e$$

Since these calculations mainly involve factors that belong to e, it is both more convenient and more economical to evaluate all those integrals for a given element simultaneously. Computer codes for the finite-element method generally contain subroutines or blocks of calculation that perform such *element* calculations. All calculations involve knowledge of the geometry and data (material property and source terms) in D_e alone. They generate n_{n_e} relations in n_{n_e} variables (= the corresponding nodal variables). They are then summed together into a global (stiffness) matrix in the operation called assembling. This operation is illustrated in Figure 10.2.5.

11.1.3 SOLVING

After the operation of assembling has been completed and the contributions of the boundaries have been appropriately introduced, the program is ready to solve the discrete equation to produce the solution. What is involved in the solution process is essentially the same as with finite-difference methods; that is, it consists in solving a system of linear equations of the form

$$\mathbf{Ku} = \mathbf{b}$$

For elliptic equations, the stiffness matrix $\mathbf{K}$ is often symmetric, positive definite. Knowledge of this fact can be taken advantage of in reducing both the storage requirement and the computing effort.

11.1.4 CONTINUITY CONDITIONS

Although great freedom is allowed in the choice of finite-element shapes and basis functions, limitations are imposed by the mathematics to be performed.

From a practical viewpoint, we may look at the replacement of (11.1.1) by (11.1.2) as a "divide-and-conquer" step, which makes the calculations (and computer programming) more convenient. But this replacement will be permitted only if certain minimum *conditions of continuity* are satisfied by the function $u(x, y, \ldots)$ across interelement boundary lines. Since the contribution of boundary lines is implicitly left out of (11.1.2), this expression will be identical to (11.1.1) only if those contributions are zero. If the operator L of our equation is of order m, then $\Re$ contains derivatives of u of order up to m. Within the mathematics of the Galerkin method, integration by parts results in the shifting of $m/2$ of those derivatives from u in $\Re$ onto φ_n. Then both terms in the integral (11.1.1) contain partial derivatives of degree $m/2$; specifically, they contain $\nabla^{m/2}\varphi_n(x, y, \ldots)$. That these terms be finite on the interelement boundaries requires that the basis functions $\varphi_n(x, y, \ldots)$ belong to $C^{(m/2)-1}(D)$, the class of functions with continuous derivatives up to order $(m/2) - 1$ everywhere in D (in particular across boundary lines). *From the theoretical viewpoint* of functional analysis, we reach the same conclusion by considering $u(x, y, \ldots)$ as a weak solution with respect to the trial space

$$\mho \equiv \{\varphi_n\}$$

consisting of the space spanned by the basis functions. In Galerkins's method, the trial space $\mathcal{V}$ is identical to the solution space $\mathcal{U}$. Since $m/2$ derivatives can be shifted from u onto elements of $\mathcal{V}$, it is required that u belong to $C^{(m/2)-1}(D)$ only to qualify as a weak solution.

Two cases of interest are:

1. *Second-order problems* (such as Poisson's equation), which require that the u belong to $C^0(D)$ (i.e., be simply continuous across interelement boundaries).

2. *Fourth-order problems* (the biharmonic equation of structural mechanics), which require that the gradients $\nabla\varphi_n$ be continuous across interelement boundaries [i.e., that $\{\varphi_n\}$ belong to $C^1(D)$].

11.1.5 POLYNOMIALS

Polynomial basis functions are the most widely used. The simplest form consists in polynomials of varying degree over regular straight-edged elements in two or three dimensions. It is easily demonstrated that such polynomials have the required C^0 continuity. With the addition of partial derivatives as nodal variables, Hermitian elements (i.e., elements with C^1 continuity) are also easily built with polynomials.

Polynomial shape functions over elements which are no more regular nor straight-edged, but which are irregular or have curved sides, are obtained by means of polynomial transformations of regular elements. Such transformations (called *parametric* transformations) conserve continuity.

One does not cease to be surprised by the elegance and simplicity of the numerous useful properties that polynomial basis functions on polynomial element possess. As appropriately said by Wachspress (1975)[3]:

> One fascinating aspect of the analysis is the coordination of geometric and algebraic arguments to exploit the interrelationship of element geometry and basis functions. As one proceeds through the successive stages of the development to increasingly more complex elements, the geometric simplicity of the basis function construction becomes more striking, and one suspects that the theory is not an invention, but rather the discovery of a natural phenomenon.

11.1.6 NODAL AND HERMITIAN PARAMETERIZATIONS

Although parameterization of the dependent variables is most often achieved by keeping track of the *nodal values* $\{u_n\}$ *only*, there are applications where it is desirable to also use partial derivatives at these nodes. For example,

---

[3]In this work, Wachspress considers "rational" finite elements (i.e., expressed by ratios of polynomials) rather than just polynomials which are only a subset of the former.

in two dimensions (x, y), one may use the four parameters[4]

$$\begin{Bmatrix} u_n \\ \left(\dfrac{\partial u}{\partial x}\right)_n \\ \left(\dfrac{\partial u}{\partial y}\right)_n \\ \left(\dfrac{\partial^2 u}{\partial x \partial y}\right)_n \end{Bmatrix} = \begin{Bmatrix} u_n \\ u'_{x,n} \\ u'_{y,n} \\ u''_{xy,n} \end{Bmatrix}$$

to quantify for each n the dependent variable $u(x, y)$. The approximate solution $u_e(x, y)$ in a given element D_e then has to satisfy not only the collocation conditions

$$u_e(x_{n_e}, y_{n_e}) = u_{n_e} \qquad \text{for all } n_e$$

(where $\{n_e\}$ is the set of nodes that are either in or are adjacent to D_e), but must also satisfy the derivative collocation conditions at those nodes:

$$\left(\frac{\partial u_e}{\partial x}\right)_{n_e} = u'_{x,n_e}$$

$$\left(\frac{\partial u_e}{\partial y}\right)_{n_e} = u'_{y,n_e}$$

$$\left(\frac{\partial^2 u_e}{\partial x \partial y}\right)_{n_e} = u''_{xy,n_e}$$

A finite-element representation that uses as parameters nodal values $\{u_n\}$ *and* some of the derivatives of u at those nodes (as above) is called *Hermitian* (since such finite elements may be considered as generalizations of Hermite interpolation of a function). A one-dimensional example was provided by the Hermite cubic elements in one dimension described in Section 9.3.

A distinctive characteristic of Hermitian elements is that more continuity of the approximate solution is preserved at the nodes. They are therefore used principally in problems involving higher derivatives, in particular *problems in elasticity*, whose formulation is in terms of the biharmonic operator with fourth-order derivatives. They are also used with collocation methods.

11.1.7 NONCONFORMING ELEMENTS

Finite elements that satisfy the required conditions of continuity are called *conforming elements*. If there was no major difficulty, they would be the only ones to be used.

---

[4]For simplicity we use a single index n to identify each node.

But difficulties may exist. To some purists' sorrow, finite elements that do not satisfy these requirements came to be used for convenience (this is the case, in particular, in structural mechanics, since finite elements that belong to C^1 are not very easy to construct); such elements are called *nonconforming*. If the left-out contribution of the boundaries in using (11.1.2) to evaluate (11.1.1) may be shown to go to zero with element size, then the approximation is still convergent and may be used (with great care!) in practice.

11.2 FAMILIES OF ELEMENT SHAPES ON RECTANGLES

11.2.1 FINITE-ELEMENT SHAPES ON RECTANGLES

Rectangular elements provide a simple geometrical form in a plane with Cartesian coordinates (x, y). Nodes may be located (1) at the four vertices, (2) *and* along the sides, (3) *and/or* inside the element. In all cases, the collocation conditions specify that the basis function φ_{n_e} must be equal to 1 at the node n_e, and equal to zero in the other nodes in D_e. As before, $\{n_e\}$ is the set of nodes that are either in or are adjacent to the element D_e, and φ_{n_e} is the basis function (in D_e) that corresponds to the node n_e.

Corner Nodes Only. With nodes located at the four vertices only, as in Figure 11.2.1, an appropriate shape function is the simple polynomial

$$u_e = a_0 + a_1 x + a_2 y + a_3 xy \qquad (11.2.1)$$

We have shown in Section 10.6 that constructing this polynomial may be

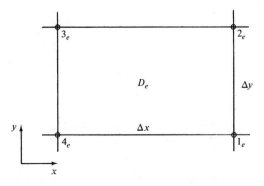

Figure 11.2.1 Four-nodes rectangular element.

achieved by using the bilinear basis function

$$\varphi_{n_e} = \left(1 - \left|\frac{x - x_{n_e}}{\Delta x}\right|\right)\left(1 - \left|\frac{y - y_{n_e}}{\Delta y}\right|\right) \quad \text{in } D_e \qquad (11.2.2)$$

The corresponding basis functions and element shapes are illustrated in Figure 11.2.2.

Moreover, the approximation is *linear* along the edges ∂D_e. Thus, the function $u(x, y)$ is *continuous* when one crosses such an edge. However, its first derivative normal to an edge is in general discontinuous. Bilinear shape functions will therefore be suitable for second-order problems but not for fourth-order problems.

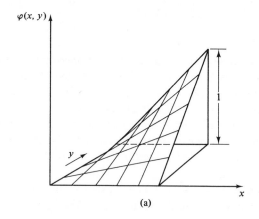

(a)

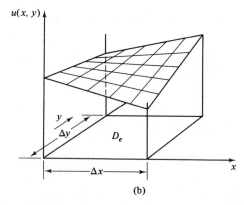

(b)

Figure 11.2.2 (*a*) Bilinear basis function on a rectangular element; (*b*) bilinear finite element on a rectangle.

Side Nodes.   Next, we may consider nodes at the middle of each side (Figure 11.2.3). A polynomial shape function with the corresponding number of degrees of freedom (eight) is

$$u_n(x, y) = a_1 + a_2 x + a_3 y + a_4 xy + a_5 x^2 + a_6 y^2$$
$$+ a_7 xy^2 + a_8 x^2 y \qquad (11.2.3)$$

The variation of $u_e(x, y)$ when x or y are kept constant is now *quadratic*. In fact, the (one-dimensional) shape of u_e along any one side is precisely the corresponding one-dimensional quadratic finite element of the kind described in Section 9.2. The corresponding basis functions are obtained by expressing that each of them is of the form (11.2.3) and must satisfy the usual collocation conditions

$$\varphi_{n_e} = \begin{cases} 1 & \text{in } (x_{n_e}, y_{n_e}) \\ 0 & \text{in the other seven nodes} \end{cases}$$

These basis functions are shown in Figure 11.2.4.

It is, of course, not necessary that the number of nodes be the same in each direction; for example, the element shown in Figure 11.2.5 is linear in x, cubic in y.

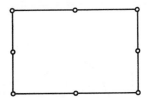

Figure 11.2.3

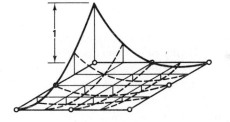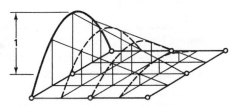

Figure 11.2.4

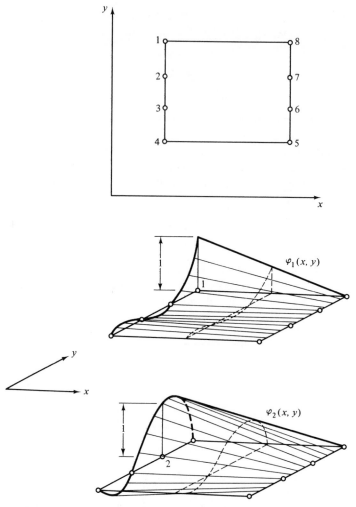

Figure 11.2.5 Rectangular finite element and the corresponding basis functions.

11.2.2 SEMANTICS

There is no particular organization in the construction of the finite-element basis functions just described. This family of elements is sometimes referred to as the "Serendipity" family[5] (see, e.g., Zienkiewicz, [1977] p. 155). By contrast, the elements to be described in the next paragraph are obtained

---

[5] After the princes of Serendip (in Horace Walpole's "The Three Princes of Serendip," 1754), who were noted for their chance discoveries.

systematically by using the Lagrange form of interpolation and the product of one-dimensional Lagrange interpolation polynomials to generate two-dimensional basis functions on rectangles. These are called the "Lagrangian" family of element shapes.

11.2.3 LAGRANGIAN ELEMENTS

A systematic way to generate higher-order finite elements in a rectangle is by using the Lagrange form of interpolation polynomials. As a reminder, the *Lagrange* form of an interpolation polynomial in one dimension between N collocation points is the expression

$$P(x) = \frac{(x - x_2)(x - x_3)\ldots(x - x_N)}{(x_1 - x_2)(x_1 - x_3)\ldots(x_1 - x_N)} u_1$$
$$+ \frac{(x - x_1)(x - x_3)\ldots(x - x_N)}{(x_2 - x_1)(x_2 - x_3)\ldots(x_2 - x_N)} u_2 \qquad (11.2.4)$$
$$\cdot$$
$$\cdot$$
$$\cdot$$
$$+ \frac{(x - x_1)(x - x_2)\ldots(x - x_{N-1})}{(x_N - x_1)(x_N - x_2)\ldots(x_N - x_{N-1})} u_N$$

which is the *unique* polynomial of degree $N - 1$ which passes through these N collocation points, that is, satisfies the N conditions

$$P(x_n) = u_n; \qquad n = 1, 2, \ldots, N$$

The expression (11.2.4) may be rewritten in the familiar form

$$P(x) = \sum_{n=1}^{N} l_n^N(x) u_n \qquad (11.2.5)$$

where each of the $l_n^N(x)$ is a polynomial of degree $N - 1$, formally expressed as

$$l_n^N(x) = \prod_{n' \neq n} \left(\frac{x - x_{n'}}{x_n - x_{n'}} \right) \qquad (11.2.6)$$

The $\{l_n^N(x)\}$ form a set of one-dimensional *basis functions* with the desired collocation properties

$$\{l_n^N(x_{n'})\} = \begin{cases} 1 & \text{if } n = n' \\ 0 & \text{if } n \neq n' \end{cases} \qquad (11.2.7)$$

We observe that the one-dimensional linear and the quadratic finite-element basis functions studied in Sections 9.1 and 9.2 may be expressed in this form. See Figure 11.2.6.

The extension to several dimensions of the Lagrange expression of interpolation polynomials generates a family of shape or basis functions which, quite naturally, is called the "Lagrange family." The Lagrange family of interpolation polynomials in a rectangle is obtained by:

1. Placing on each pair of opposing sides M and N points, respectively (usually equally spaced).

2. Expressing the shape functions in two dimensions as the product

$$\varphi_{m,n}(x, y) = l_m^M(x) l_n^N(y) \tag{11.2.8}$$

where l_m^M and l_n^N are the basic Lagrange interpolation polynomials (11.2.6). The number of nodes per element is MN and contains (except in the case where M and/or $N = 2$) internal nodes as well as nodes located on the side of the elements.

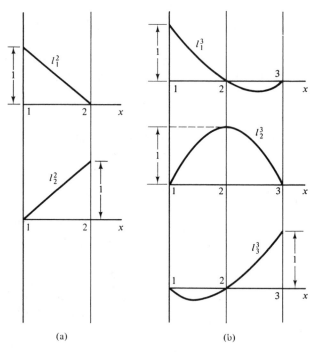

(a)   (b)

Figure 11.2.6 Fundamental Lagrange interpolation polynomials: (*a*) linear ($N = 2$); (*b*) quadratic ($N = 3$).

A simple application is that of the bilinear rectangular elements previously described, where the basis functions were products of the linear (Lagrange) interpolation functions in one dimension. Examples of higher-order Lagrangian rectangular elements and their corresponding basis functions are shown in Figure 11.2.7 and 11.2.8, respectively.

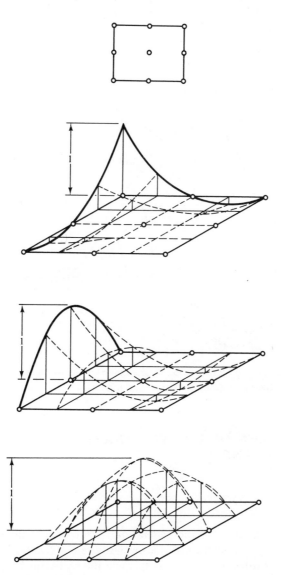

Figure 11.2.7 Quadratic Lagrangian element on a rectangle and the corresponding basis functions.

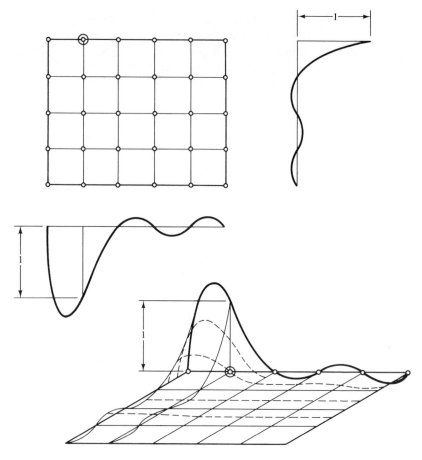

Figure 11.2.8 Higher-order Lagrangian element on a rectangle and one of the corresponding basis functions.

11.3 FAMILIES OF ELEMENT SHAPES ON TRIANGLES

11.3.1 AREA COORDINATES

An elegant tool for the generation of-polynomial shape functions on *triangles* is provided by the use of a special system of local coordinates known as *area coordinates*. Consider the triangle illustrated in Figure 11.3.1. The coordinates L_1, L_2, and L_3 are defined as

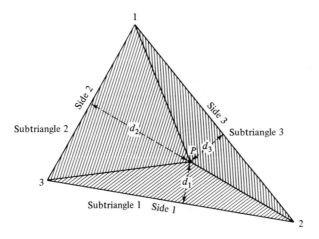

Figure 11.3.1

$$L_1 = \frac{\text{distance from } P \text{ to side 1}}{\text{distance from vertex 1 to side 1}} \Bigg\}$$

$$L_2 = \frac{\text{distance from } P \text{ to side 2}}{\text{distance from vertex 2 to side 2}} \Bigg\} \qquad (11.3.1)$$

$$L_3 = \frac{\text{distance from } P \text{ to side 3}}{\text{distance from vertex 3 to side 3}} \Bigg\}$$

It is obvious from these definitions that the $\{L_i\}$ are nonnegative numbers in [0, 1].

Since P has only *two* Cartesian coordinates (x, y), it must be the case that the *three* coordinates $\{L_i\}$ cannot be independent of one another. Their respective interdependence is easily found by noting that the definitions (11.3.1) are equivalent to the following (referring again to Figure 11.3.1):

$$L_1 = \frac{\text{area of subtriangle 1}}{\text{area of total triangle}} \Bigg\}$$

$$L_2 = \frac{\text{area of subtriangle 2}}{\text{area of total triangle}} \Bigg\} \qquad (11.3.2)$$

$$L_3 = \frac{\text{area of subtriangle 3}}{\text{area of total triangle}} \Bigg\}$$

Since the area of the total triangle is the sum of that of the three subtriangles, we have the simple relation

$$L_1 + L_2 + L_3 = 1 \qquad (11.3.3)$$

The coordinates $\{L_i\}$ are also known as *Möbius coordinates*, as *area coordinates* [the name being justified by relations (11.3.2)], and as *barycentric coordinates*.

Relation of Area Coordinates to Linear Basis Functions. It is easily verified that the area coordinates L_1, L_2, and L_3 are identical to the linear basis functions of Section 10.2 (Figure 11.3.2). That is,

$$\left.\begin{aligned} \varphi_1(x, y) &= L_1 \\ \varphi_2(x, y) &= L_2 \\ \varphi_3(x, y) &= L_3 \end{aligned}\right\} \tag{11.3.4}$$

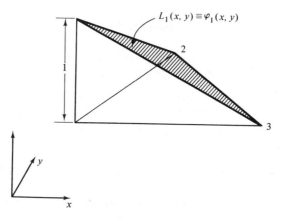

Figure 11.3.2 The area coordinates $\{L_i\}$ are identical to the linear basis functions on the triangle.

They have the useful property of varying linearly along any straight line in (x, y), and provide elementary building blocks for constructing higher-order polynomial basis functions on triangles.

It is of interest to note that the lines $L_i = $ constant are straight lines, parallel to the triangle sides. In particular, $L_i = 0$ is the equation of the side opposed to the vertex i. The relationship between the area coordinates $\{L_i\}$ and the Cartesian coordinates (x, y) may also be expressed as

$$\begin{bmatrix} 1 \\ x \\ y \end{bmatrix} = \begin{bmatrix} 1 & 1 & 1 \\ x_1 & x_2 & x_3 \\ y_1 & y_2 & y_3 \end{bmatrix} \begin{bmatrix} L_1 \\ L_2 \\ L_3 \end{bmatrix} \tag{11.3.5}$$

11.3.2 HIGHER-ORDER POLYNOMIAL SHAPE FUNCTIONS ON TRIANGLES

In a manner similar to the generation of Lagrangian polynomial basis functions on rectangles (Section 11.2), one may generate Lagrangian basis functions on triangles in terms of the area coordinates instead of Cartesian coordinates. Since the $\{L_i\}$ are linear in x and y, shape functions that are expressed as polynomials in the $\{L_i\}$ will be polynomials of the same degree in the global coordinates (x, y).

As an example, consider the triangle with *quadratic* sides of Figure 11.3.3. Basis functions corresponding to the three vertices may be expressed

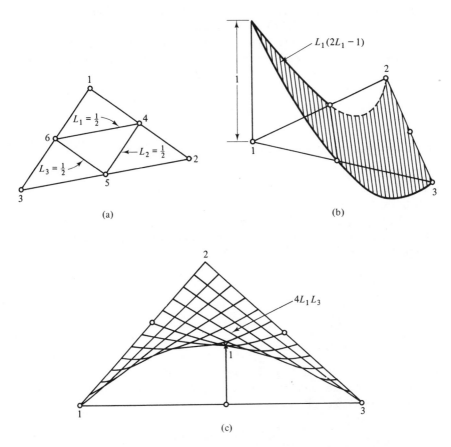

(a)

(b)

(c)

Figure 11.3.3 (a) Triangular element with quadratic sides; (b) quadratic basis function associated to a vertex node; (c) quadratic basis function associated to a midside node.

as

$$\left.\begin{array}{l}\varphi_1 = L_1(2L_1 - 1) \\ \varphi_2 = L_2(2L_2 - 1) \\ \varphi_3 = L_3(2L_3 - 1)\end{array}\right\} \qquad (11.3.6a)$$

Those corresponding to the side midpoints are, respectively,

$$\left.\begin{array}{l}\varphi_4 = 4L_1L_2 \\ \varphi_5 = 4L_2L_3 \\ \varphi_6 = 4L_3L_1\end{array}\right\} \qquad (11.3.6b)$$

These functions are illustrated in Figure 11.3.3. It may be observed that, as expected, (11.3.5) and (11.3.6) are quadratic polynomials along any straight line. In particular, they are quadratic polynomials along the sides $L_i = 0$, uniquely defined by the three nodes located on the corresponding side. Finite-element C^0 continuity across interelement quadratic boundaries is thus preserved. (C^0 continuity shall be preserved across interelement boundaries of triangular elements of higher order as well.)

11.4 HERMITIAN ELEMENTS IN RECTANGLES

Hermitian elements in rectangles are easily obtained by using as basis functions the products of the one-dimensional Hermite basis functions described in Chapter 9. Over the square element D_e in (x, y) (Figure 11.4.1), the one-dimensional fundamental Hermite basis functions associated with the node $n = 1$ are (from Section 9.3)

$$\begin{array}{l}\varphi_1(x) = 1 - 3x^2 + 2x^3 = (2x + 1)(x - 1)^2 \\ \psi_1(x) = x(x - 1)^2 \\ \varphi_1(y) = 1 - 3y^2 + 2y^3 = (2y + 1)(y - 1)^2 \\ \psi_1(y) = y(y - 1)^2\end{array} \qquad (11.4.1)$$

and similarly for the other nodes. From these, two-dimensional Hermitian basis functions are obtained by cross products. For example, the basis functions associated with node 1 are

$$\begin{array}{l}H_1^1(x, y) = \varphi_1(x)\varphi_1(y) \\ H_1^2(x, y) = \psi_1(x)\varphi_1(y) \\ H_1^3(x, y) = \varphi_1(x)\psi_1(y) \\ H_1^4(x, y) = \psi_1(x)\psi_1(y)\end{array} \qquad (11.4.2)$$

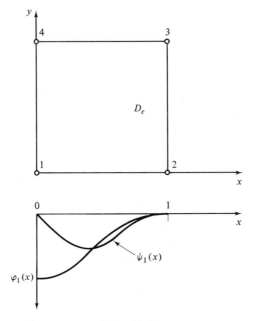

Figure 11.4.1

and similarly for the other three nodes, by symmetry. Some of those basis functions in two dimensions are illustrated in Figure 11.4.2. The variables corresponding to the four basis functions (11.4.2) may be shown to be

$$u, \quad \frac{\partial u}{\partial x}, \quad \frac{\partial u}{\partial y}, \quad \frac{\partial^2 u}{\partial x \, \partial y} \qquad (11.4.3)$$

at the corresponding node. That is, the expression of the finite-element function $u(x, y)$ in D_e is

$$u = \sum_{n=1}^{4} H_n^1 u_n + H_n^2 \left(\frac{\partial u}{\partial x}\right)_n + H_n^3 \left(\frac{\partial u}{\partial y}\right)_n + H_n^4 \left(\frac{\partial^2 u}{\partial x \, \partial y}\right)_n \qquad (11.4.4)$$

where u_n, $(\partial u/\partial x)_n$, $(\partial u/\partial y)_n$, and $(\partial^2 u/\partial x \, \partial y)_n$ are the components of the numerical solution at the nodes.

A valid question is whether one need to carry the fourth degree of freedom,

$$\frac{\partial^2 u}{\partial x \, \partial y}$$

and the corresponding basis function, H_n^4, in each node. In other words, what

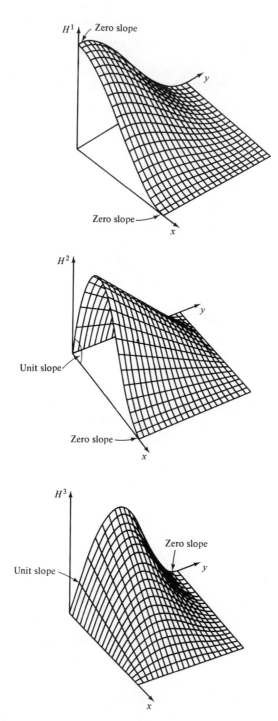

Figure 11.4.2 First three Hermite basis functions.

would happen to the representation (11.4.4) if we were to limit the sum to its first three terms?

The answer is simply:

> The finite-element representation (11.4.4) is C^1-continuous when all four degrees of freedom (11.4.3) and corresponding basis functions (11.4.2) are used, and is C^0-continuous only when the degrees of freedom $(\partial^2 u/\partial x\, \partial y)$ are deleted.

To prove this, consider a Hermitian representation using u, $\partial u/\partial x$, and $\partial u/\partial y$ only at the nodes of the square element of Figure 11.4.1. Along side 1–2 it may be observed that the displacement u varies as a cubic. Since a cubic is uniquely defined by four constants, the four end values of shapes and displacements

$$u_1, \quad \left(\frac{\partial u}{\partial x}\right)_1, \quad u_2, \quad \left(\frac{\partial u}{\partial x}\right)_2$$

will therefore uniquely define the displacements along the boundary. Because such end values are common to adjacent elements, C^0 continuity of u is verified along any interelement boundary.

It will be observed that along any line $y = $ constant (in particular, the line 1–2), the normal derivative $(\partial u/\partial y)$ also varies in a cubic way with x. With only two values of $(\partial u/\partial y)$ specified on (1, 2), the cubic is not defined uniquely by these values, and discontinuity in the normal derivative across the boundary line will occur. With the addition of the fourth nodal variable $(\partial^2 u/\partial x\, \partial y)$ at each node, the four constants

$$\left(\frac{\partial u}{\partial y}\right)_1, \quad \left(\frac{\partial^2 u}{\partial x\, \partial y}\right)_1, \quad \left(\frac{\partial u}{\partial y}\right)_2, \quad \left(\frac{\partial^2 u}{\partial x\, \partial y}\right)_2$$

are now available to define uniquely $\partial u/\partial y$ on the boundary, thus ensuring C^1 continuity.

11.5   COLLOCATION ON FINITE ELEMENTS

Collocation on finite elements, is a method in which collocation is used instead of integrals over the element subdomains (see also Section 8.4). The steps involved in this approach to approximate solutions of the equation $LU = F$ are:

1. The numerical solution is expressed in a usual finite-element formulation:

$$U \simeq u = \sum_n \varphi_n u_n \qquad (11.5.1)$$

   where the $\{u_n\}$ are nodal values.

2. The equation residual $\Re$ is formed:

$$\Re \equiv Lu - F \qquad (11.5.2)$$

3. The residual $\Re$ is equated to zero in selected collocation points $\{n_c\}$ (normally located inside the elements), to obtain the number of equations needed to determine the unknown nodal values.

$$\Re_{n_c} \equiv (Lu)_{n_c} - F_{n_c} = 0; \qquad n_c = 1, 2, \ldots \qquad (11.5.3)$$

The collocation points are in general chosen different from the nodes. The final discrete equations to be solved may be written either in terms of the solution at the nodal points *or* in terms of the solution at the collocation point, since equation (11.5.1) gives a simple, geometric relationship between those two equivalent sets of values inside of each element. At no time is integration by parts invoked, and the trial solution u must thus belong to the space of functions that have $m - 1$ derivatives (i.e., must belong to C^{m-1}; see Section 8.3) if m is the order of L. With second-order (elliptic) equations, u must be in C^1; that is, the finite-element shape functions must be of the Hermite type (first derivative continuous across boundary lines).

To illustrate the concept of collocation on finite elements, consider Poisson's equation:

$$-\nabla^2 U \equiv -\left(\frac{\partial^2 U}{\partial x^2} + \frac{\partial^2 U}{\partial y^2}\right) = F(x, y) \qquad \text{in } D \qquad (11.5.4)$$

A finite-element representation of an approximate solution is obtained by covering D with a regular grid:

$$x_m = m \, \Delta x$$
$$y_n = n \, \Delta y$$

over which polynomial finite-element shapes will be used. One of the ways in which *C^1 continuity which is required with collocation methods* can be achieved is with Hermite cubic polynomials. The solution is expressed as

$$u = \sum_n \left[H_n^1 u_n + H_n^2 \left(\frac{\partial u}{\partial x}\right)_n + H_n^3 \left(\frac{\partial u}{\partial y}\right)_n + H_n^4 \left(\frac{\partial^2 u}{\partial x \, \partial y}\right)_n \right] \qquad (11.5.5)$$

where the H_n^i are the Hermite basis functions described in Section 11.4.

To find the number of collocation points needed per element we note that:

1. *Four* unknowns exist per node: u, $(\partial u/\partial x)$, $(\partial u/\partial y)$, $(\partial^2 u/\partial x \, \partial y)$.

2. The ratio of the number of nodes to the number of elements is *one*. Thus, four collocation points are needed in each element. This is illustrated in Figure 11.5.1.

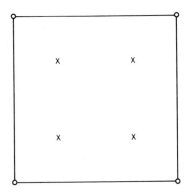

Figure 11.5.1   Square collocation finite element for Poisson's equation. The nodal values are u, u_x, u_y, and u_{xy} in each vertex and the finite-element solution is the Hermite interpolant. Four collocation points per element are needed. Although they could in principle be located anywhere as illustrated here, placing them at the Gaussian points (resulting in what is called *orthogonal* collocation) results in improved accuracy. Circles represent nodes; crosses represent collocation points.

11.5.1   ORTHOGONAL COLLOCATION[6]

The word "orthogonal" is added in "*orthogonal* collocation on finite elements" when the collocation points are placed at the Gaussian points on the elements. To understand why this may be superior to placing the collocation points anywhere else, one must look at the theory of Gaussian quadrature. This theory is given in some detail later (in Section 12.4).

The main result is as follows. When an integral

$$I = \int_a^b f(x)\, dx \tag{11.5.6}$$

is approximated by numerical quadrature with m sample points,

$$I \simeq \sum_{i=1}^{m} w_i f(x_i) \tag{11.5.7}$$

[where the fixed weights w_i are chosen so as to make the approximation exact when $f(x)$ is a polynomial of as high as possible a degree], then, in general,

---

[6]See Finlayson (1972), Carey and Finlayson (1975), and Finlayson (1980).

the highest degree for which the approximation is exact is equal to $m - 1$. But *if the m sample points x_i are chosen as the roots of the Legendre polynomial of degree m shifted on the interval $[a, b]$ (these are called Gaussian points in this context)*, then the highest-degree polynomial for which (11.5.7) is exact becomes $2m - 1$, i.e., more than twice as much, see Figure 11.5.2. (In several dimensions, the Gaussian points are simply the roots of shifted Legendre polynomials corresponding to each coordinate.) It may thus be observed that placing the collocation points at the Gaussian points is in a sense equivalent to making the residual orthogonal to polynomials of as high a degree as possible.

A detailed theoretical analysis of the convergence of polynomial approximations to the solution of differential equations using collocation at Gaussian points may be found in de Boor and Swartz (1973).

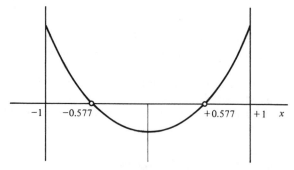

Figure 11.5.2 Legendre polynomial of degree 2 and corresponding two Gaussian points on the interval $[-1, +1]$.

12

Geometric Transformations and Isoparametric Elements

Gabriel Lamé (1795–1870), a French mathematician and engineer concerned primarily with the heat equation, introduced the use of curvilinear coordinate systems, which could also be used for many types of equations. Lamé pointed out that the heat equation had been solved only for conducting bodies whose surfaces are normal to the coordinate planes $x =$ const., $y =$ const., and $z =$ const. Lamé's idea was to introduce new systems of coordinates and the corresponding coordinate surfaces. To a very limited extent this had been done by Euler and Laplace, both of whom used spherical coordinates ρ, θ, and ϕ, in which case the coordinate surfaces $\rho =$ const., $\theta =$ const., and $\phi =$ const. are spheres, planes, and cones respectively. Knowing the equations that transform from rectangular to spherical coordinates, one can, as Euler and Laplace did, transform the potential equation from rectangular to spherical coordinates.

M. KLINE

12.1 CHANGE OF COORDINATES

We have seen how general families of finite elements, limited to straight-edged triangular or rectangular shapes, could be obtained. It is the case, however, that many real rather than academic problems have geometries that could benefit from a division into finite elements of more general shape.

Moreover, boundaries are often curved, and curved elements become desirable in approximating a problem near such a boundary. The techniques used to generate such more general element shapes fall into two basic categories:

1. Use of a *global* system of coordinates which may be distorted and/or curved.

2. Use of a *local* transformation of coordinates, different for each element, in particular by what is known as *isoparametric transformations*.

The theory that we are about to enter permits the extension of the finite-element method to divisions of space into irregular subdomains, with curved sides of a rather general nature. It will be found that, save for a reasonable extra amount of rather trivial calculations to be executed by the computer, there is no difficulty of principle in this extension to general curvilinear element subdomains. Such an extension *is not easy* with finite-difference methods, and this contrast between the two methods is one of the strong points that has contributed to the success of finite elements.

12.1.1 SOME ALGEBRA OF COORDINATE CHANGES

Changing coordinates affects the expression of the fundamental integrals that appear in the finite-element method. We shall describe those changes with reference to second-order elliptic problems in two-dimensional space. Results carry over to three-dimensional spaces and fourth-order problems without much complication.

As we have seen in preceding chapters, the fundamental integrals that appear in applications of the finite-element method to second-order problems are of the kind

$$\int \varphi_n \varphi_m \, dD \tag{12.1.1}$$

and for second-order problems,

$$\int \nabla \varphi_n^T \nabla \varphi_m \, dD \tag{12.1.2}$$

The change in coordinates affects the expression of the element of area dD and of the gradient vectors $\nabla \varphi_n$.

Area.   In Cartesian coordinates, the element of area is simply

$$dD = dx \, dy$$

which is also the cross-vector product

$$dD = d\mathbf{x} \times d\mathbf{y} \tag{12.1.3}$$

where $d\mathbf{x}$ and $d\mathbf{y}$ are the elementary displacement *vectors* in the Cartesian directions x and y, respectively (Figure 12.1.1). Consider the transformation to new coordinates (ξ, η) by algebraic relations of the form

$$\begin{aligned} x &= x(\xi, \eta) \\ y &= y(\xi, \eta) \end{aligned} \tag{12.1.4}$$

Except when these are linear expressions, the lines $\xi = \text{constant}$ and $\eta = \text{constant}$ will be curves in the (x, y) plane.

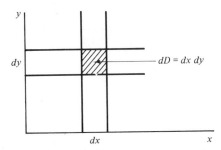

Figure 12.1.1 Element of area in Cartesian coordinates.

Next, consider in the (x, y) plane the area defined by the parallelogram on two elementary vectors:

$$d\boldsymbol{\xi} = \begin{bmatrix} d\xi_x \\ d\xi_y \end{bmatrix} \quad \text{and} \quad d\boldsymbol{\eta} = \begin{bmatrix} d\eta_x \\ d\eta_y \end{bmatrix} \tag{12.1.5}$$

corresponding to displacements in the directions $\eta = \text{constant}$ and $\xi = \text{constant}$, respectively, as illustrated in Figure 12.1.2. The corresponding elementary area dD expressed in coordinates (ξ, η) is that illustrated by the shaded parallelogram. Expressing this area is a classical exercise in vector algebra; it is equal to the cross product

$$\begin{aligned} dD &= d(\text{area}) \\ &= d\boldsymbol{\xi} \times d\boldsymbol{\eta} \\ &= d\xi_x\, d\eta_y - d\xi_y\, d\eta_x \\ &= \det \begin{bmatrix} d\xi_x & d\eta_x \\ d\xi_y & d\eta_y \end{bmatrix} \end{aligned} \tag{12.1.6}$$

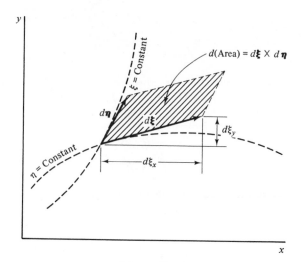

Figure 12.1.2

We now express the vectors $d\boldsymbol{\xi}$ and $d\boldsymbol{\eta}$ in terms of the corresponding scalars $d\xi$ and $d\eta$:

$$d\boldsymbol{\xi} = \begin{bmatrix} \dfrac{\partial x}{\partial \xi} \\ \dfrac{\partial y}{\partial \xi} \end{bmatrix} d\xi; \qquad d\boldsymbol{\eta} = \begin{bmatrix} \dfrac{\partial x}{\partial \eta} \\ \dfrac{\partial y}{\partial \eta} \end{bmatrix} d\eta$$

whence

$$dD = \det \begin{bmatrix} \dfrac{\partial x}{\partial \xi} & \dfrac{\partial x}{\partial \eta} \\ \dfrac{\partial y}{\partial \xi} & \dfrac{\partial y}{\partial \eta} \end{bmatrix} d\xi\, d\eta$$

$$= \det \mathbf{J}\, d\xi\, d\eta \qquad\qquad (12.1.7)$$

which is the expression to be used to compute integrals in terms of the variables ξ and η.

The discussion above introduces an important entity in coordinate changes, the Jacobian matrix:

$$\mathbf{J} \equiv \begin{bmatrix} \dfrac{\partial x}{\partial \xi} & \dfrac{\partial x}{\partial \eta} \\ \dfrac{\partial y}{\partial \xi} & \dfrac{\partial y}{\partial \eta} \end{bmatrix} \qquad\qquad (12.1.8)$$

and its determinant det $\mathbf{J}$. This determinant (a scalar) is often called simply the *Jacobian* of the transformation. We may also at times want to express the transformation in the reverse direction as

$$\xi = \xi(x, y)$$
$$\eta = \eta(x, y) \tag{12.1.9}$$

The corresponding Jacobian matrix,

$$\mathbf{J}' = \begin{bmatrix} \dfrac{\partial \xi}{\partial x} & \dfrac{\partial \xi}{\partial y} \\[2mm] \dfrac{\partial \eta}{\partial x} & \dfrac{\partial \eta}{\partial y} \end{bmatrix} \tag{12.1.10}$$

is related to $\mathbf{J}$ by

$$\mathbf{J}' = \mathbf{J}^{-1} \tag{12.1.11}$$

and their determinants obey

$$\det \mathbf{J}' = \frac{1}{\det \mathbf{J}} \tag{12.1.12}$$

Gradients.   When calculations are done in a new set of coordinates, the gradient of functions of (x, y), say

$$\nabla \varphi_n$$

needs to be expressed in terms of (ξ, η). We have the identities derived from the chain rule:

$$\frac{\partial \varphi}{\partial \xi} = \frac{\partial \varphi}{\partial x}\left(\frac{\partial x}{\partial \xi}\right) + \frac{\partial \varphi}{\partial y}\left(\frac{\partial y}{\partial \xi}\right)$$
$$\frac{\partial \varphi}{\partial \eta} = \frac{\partial \varphi}{\partial x}\left(\frac{\partial x}{\partial \eta}\right) + \frac{\partial \varphi}{\partial y}\left(\frac{\partial y}{\partial \eta}\right) \tag{12.1.13}$$

which may be expressed with the transpose of the Jacobian matrix as:

$$\nabla_{\xi,\eta}\varphi \equiv \begin{bmatrix} \dfrac{\partial \varphi}{\partial \xi} \\[2mm] \dfrac{\partial \varphi}{\partial \eta} \end{bmatrix} = \mathbf{J}^T \begin{bmatrix} \dfrac{\partial \varphi}{\partial x} \\[2mm] \dfrac{\partial \varphi}{\partial y} \end{bmatrix} = \mathbf{J}^T \nabla_{x,y}\varphi \tag{12.1.14}$$

To evaluate expressions of the form (12.1.2), we need

$$\nabla_{x,y}\varphi = (\mathbf{J}^T)^{-1} \nabla_{\xi,\eta}\varphi \tag{12.1.15}$$

which in two dimensions is

$$\nabla_{x,y}\varphi = \frac{1}{\det \mathbf{J}} \begin{bmatrix} \dfrac{\partial y}{\partial \eta} & -\dfrac{\partial x}{\partial \eta} \\[2ex] -\dfrac{\partial y}{\partial \xi} & \dfrac{\partial x}{\partial \xi} \end{bmatrix} \nabla_{\xi,\eta}\varphi \qquad (12.1.16)$$

Occasionally, we shall need to express $\nabla_{x,y}u$ in terms of (ξ, η). The preceding expression holds: namely,

$$\nabla_{x,y}u \equiv \begin{bmatrix} \dfrac{\partial u}{\partial x} \\[2ex] \dfrac{\partial u}{\partial y} \end{bmatrix} = \frac{1}{\det \mathbf{J}} \begin{bmatrix} \dfrac{\partial y}{\partial \eta} & -\dfrac{\partial x}{\partial \eta} \\[2ex] -\dfrac{\partial y}{\partial \xi} & \dfrac{\partial x}{\partial \xi} \end{bmatrix} \nabla_{\xi,\eta}u \qquad (12.1.17)$$

Expression of the Fundamental Integrals. The fundamental integrals (12.1.1) and (12.1.2) may now be expressed as

$$\int \varphi_n \varphi_m \, dD = \int \varphi_n \varphi_m \det \mathbf{J} \, d\xi \, d\eta \qquad (12.1.18)$$

and

$$\int \nabla_{x,y}\varphi_n^T \nabla_{x,y}\varphi_m \, dD$$

$$= \int \nabla_{\xi,\eta}\varphi_n^T \mathbf{J}^{-1}(\mathbf{J}^T)^{-1} \nabla_{\xi,\eta}\varphi_m \det \mathbf{J} \, d\xi \, d\eta$$

$$= \int \nabla_{\xi,\eta}\varphi_n^T \begin{bmatrix} \left(\dfrac{\partial y}{\partial \xi}\right)^2 + \left(\dfrac{\partial y}{\partial \eta}\right)^2 & -\left(\dfrac{\partial x}{\partial \xi}\dfrac{\partial y}{\partial \xi} + \dfrac{\partial x}{\partial \eta}\dfrac{\partial y}{\partial \eta}\right) \\[3ex] -\left(\dfrac{\partial x}{\partial \xi}\dfrac{\partial y}{\partial \xi} + \dfrac{\partial x}{\partial \eta}\dfrac{\partial y}{\partial \eta}\right) & \left(\dfrac{\partial x}{\partial \xi}\right)^2 + \left(\dfrac{\partial x}{\partial \eta}\right)^2 \end{bmatrix} \nabla_{\xi,\eta}\varphi_m \frac{d\xi \, d\eta}{\det \mathbf{J}}$$

$$(12.1.19)$$

12.2 FINITE ELEMENTS IN GLOBAL CURVILINEAR COORDINATES

Suppose that we are able to choose a set of coordinates (ξ, η) related to (x, y) by a transformation of the form

$$x = x(\xi, \eta)$$
$$y = y(\xi, \eta)$$

such that the lines $\xi = $ constant and $\eta = $ constant generate elements with a desirable shape in the (x, y) plane. The expression of an elementary area dD in the new coordinates (ξ, η) is given by (12.1.7):

$$dD = \det \mathbf{J} \, d\xi \, d\eta$$

where $\mathbf{J}$ is the Jacobian matrix

$$\mathbf{J} = \begin{bmatrix} \dfrac{\partial x}{\partial \xi} & \dfrac{\partial x}{\partial \eta} \\ \dfrac{\partial y}{\partial \xi} & \dfrac{\partial y}{\partial \eta} \end{bmatrix}$$

For a set of curvilinear coordinates to be acceptable, it is required that the Jacobian determinant be everywhere nonzero.

If we define rectangular elements in (ξ, η) they will become general quadrilateral, possibly curved elements in (x, y). Similarly, rectilinear triangular elements in (ξ, η) may become curved triangular elements in (x, y).

12.2.1 CIRCULAR COORDINATES

In engineering applications, curvilinear coordinates may often be suggested by the geometry of the system being analyzed. For instance, a round membrane shall be naturally described in *circular coordinates* (r, α), and rectilinear elements in (r, α) generate the curvilinear elements shown in Figure 12.2.1. The transformation is:

$$\begin{aligned} x &= r \cos \alpha \\ y &= r \sin \alpha \end{aligned} \tag{12.2.1}$$

The Jacobian matrix

$$\mathbf{J} = \begin{bmatrix} \dfrac{\partial x}{\partial r} & \dfrac{\partial x}{\partial \alpha} \\ \dfrac{\partial y}{\partial r} & \dfrac{\partial y}{\partial \alpha} \end{bmatrix} = \begin{bmatrix} \cos \alpha & -r \sin \alpha \\ \sin \alpha & r \cos \alpha \end{bmatrix} \tag{12.2.2}$$

has the determinant

$$\det \mathbf{J} = r(\cos^2 \alpha + \sin^2 \alpha) = r \tag{12.2.3}$$

By the application of (12.1.7) we find

$$dD = r \, dr \, d\alpha \tag{12.2.4}$$

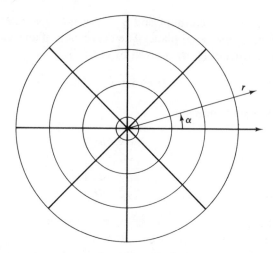

Figure 12.2.1 "Natural" division of a circular plate into curved elements formed by the lines r = constant and α = constant. The curvilinear coordinate system used here is global (i.e., the same over the entire domain D).

and, by (12.1.16),

$$\nabla_{x,y}\varphi = (\mathbf{J}^T)^{-1}\,\nabla_{\xi,\eta}\varphi = \frac{1}{r}\begin{bmatrix} r\cos\alpha & -\sin\alpha \\ r\sin\alpha & \cos\alpha \end{bmatrix}\begin{bmatrix} \dfrac{\partial\varphi}{\partial r} \\ \dfrac{\partial\varphi}{\partial\alpha} \end{bmatrix} \qquad (12.2.5)$$

The dot products in (12.1.2) become

$$\nabla_{x,y}\varphi_n^T\,\nabla_{x,y}\varphi_m = \nabla_{\xi,\eta}\varphi_n^T\mathbf{J}^{-1}(\mathbf{J}^T)^{-1}\,\nabla_{\xi,\eta}\varphi_m$$

$$= \frac{\partial\varphi_n}{\partial r}\frac{\partial\varphi_m}{\partial r} + \frac{1}{r^2}\frac{\partial\varphi_n}{\partial\alpha}\frac{\partial\varphi_m}{\partial\alpha} \qquad (12.2.6)$$

EXAMPLE

Consider again our simple problem of Section 10.2 (Poisson's equation)

$$\begin{aligned} -\nabla^2 U = F = \text{constant} &\qquad \text{in } D \\ U = 0 &\qquad \text{on } \partial D \end{aligned} \qquad (12.2.7)$$

where D is the disk of radius R. Instead of using rectilinear triangular elements as before, we divide D into the curved triangular elements illustrated in Figure 12.2.2. As basis functions we choose the triangular linear functions in (r, α). For the center node (Figure 12.2.3)

$$\varphi_1(r, \alpha) = 1 - \frac{r}{R}$$

Referring to equation (10.1.13), we must evaluate for $n = 1 =$ center node:

$$\frac{\partial J(u)}{\partial u_1} = \int_{D_e} (\nabla \varphi_1^T \nabla u_e - \varphi_1 F)\, dD_e$$

$$= \int_{D_e} \left[\frac{\partial \varphi_1}{\partial r} \frac{\partial u_e}{\partial r} + \frac{1}{r^2} \frac{\partial \varphi_1}{\partial \alpha} \frac{\partial u_e}{\partial \alpha} + \varphi_1 F \right] r\, dr\, d\alpha$$

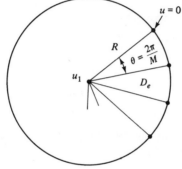

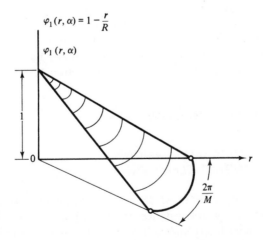

Figure 12.2.2 Division of D in curved triangular elements.

Figure 12.2.3 Curved triangular basis function corresponding to the center node.

We have

$$u_e = \varphi_1 u_1$$

$$\nabla \varphi_1 = \begin{bmatrix} -\dfrac{1}{R} \\ 0 \end{bmatrix}; \qquad \nabla u_e = u_1 \begin{bmatrix} -\dfrac{1}{R} \\ 0 \end{bmatrix}$$

Upon evaluation of the integrals

$$\int_{D_e} \nabla \varphi_1^T \nabla u_e \, dD_e = S_e \frac{u_1}{R^2}; \qquad S_e = \frac{\pi R^2}{M}$$

$$\int_{D_e} \varphi_1 F \, dD_e = \int_0^R \left(1 - \frac{r}{R}\right) F \theta r \, dr = S_e \frac{F}{3}$$

we find

$$\frac{u_1}{R^2} - \frac{F}{3} = 0$$

or

$$u_1 = \frac{FR^2}{3}$$

which incidentally is independent of θ and (as expected) is identical to [(10.2.30b), Section 10.2] for $M \to \infty$.

12.2.2 NUMERICAL INTEGRATION

In many choices of elements, in particular those defined in curvilinear coordinates, the elementary integrals may become hard (or impossible) to evaluate analytically. One may then evaluate them by numerical integration. More comments on these points are given in Section 12.4.

12.3 ISOPARAMETRIC ELEMENTS

12.3.1 LOCAL TRANSFORMATIONS AND ISOPARAMETRIC ELEMENTS

There are many cases in which one may want the freedom to specify curved elements *locally*: for instance, to match points on the domain boundary ∂D or to have interelement lines pass through interior nodal points that have been chosen unevenly in the (x, y) plane. The concepts of the preceding section, which were predicated on the use of a *global*, unique system of curvilinear coordinates, no longer apply. What remains applicable is that one may generate distorted elements by a transformation of the general form

$$\xi_e = \xi_e(x, y)$$
$$\eta_e = \eta_e(x, y)$$

(12.3.1)

But now this transformation will, in principle, be different from element to element (see Figure 12.3.1).

The manner in which these transformations are defined is by:

1. First choosing the nodal points in the global space (x, y).

2. Choosing the division of D into finite elements D_e defined by interelement lines interconnecting those nodes.[1]

3. Implying that in each element there exists a transformation of the form (12.3.1) such that D_e is a simple rectilinear polygon in (ξ_e, η_e).

In two dimensions, we may limit our attention to three- and four-sided elements, which are to be obtained by transformations of triangles and rectangles. These are by far the most commonly used element shapes. Similar elements in three dimensions are obtained by transforming tetrahedrons and cubes.

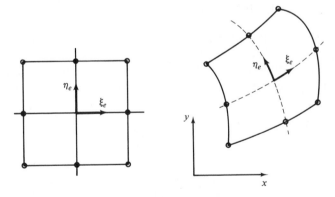

Figure 12.3.1   Generation of a curved element by a local transformation of coordinates

$$\xi_e = \xi_e(x, y)$$
$$\eta_e = \eta_e(x, y)$$

This transformation is different for each element D_e in D.

---

[1]When more than two nodes define such an interelement line, the line will, in general, be curved.

12.3.2 CONTINUITY REQUIREMENTS

As we have seen in Section 11.1, it is required, when dealing with problems involving second derivatives in space, that the finite-element solution $u(x, y)$ be *continuous* across interelement boundaries. (More precisely, we require that integrals of the form

$$\langle \nabla \varphi_n, \nabla \varphi_m \rangle = \int \nabla \varphi_n \nabla \varphi_m \, dD \tag{12.3.2}$$

exist, which is the case if u is continous.) Specifically, we want the following conditions to be satisfied (Figure 12.3.2).

Condition 1. *Preservation of the interelement boundary.* We want the interelement boundary lines generated by the transformation of the common boundary of two adjacent elements (with their own two coordinate transformations) to be the same line in the global coordinates (x, y).

Condition 2. *Continuity of the finite-element solution.* We want the function $u = \sum_n \varphi_n u_n$ to be continuous across interelement boundaries: that is,

$$\sum_{n_1} \varphi_{n_1}(\xi_1, \eta_1) u_{n_1} = \sum_{n_2} \varphi_{n_2}(\xi_2, \eta_2) u_{n_2} \tag{12.3.3}$$

for every point on the boundary.

An ingenious and simple way to satisfy these conditions is known as the *isoparametric method.* Isoparametric elements hold an extremely important place in practical applications. The basic idea may be found in a paper by Taig (1961), but the real development and generalization of the method

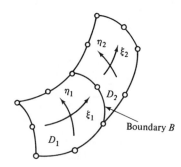

Figure 12.3.2 Isoparametric transformations in the two adjacent elements D_1 and D_2 are different. Continuity conditions for second-order problems require that (1) the distorted boundary B be the same in the two transformations; (2) the finite-element solution $u = \sum \varphi_{n_e} u_{n_e}$ be the same on B for the two elements.

took place in the late 1960s and was the doing of the well-known British group including Irons, Zienkiewicz, and others.[2]

As with other methods invoking changes of coordinates, the objective of isoparametric transformations is to obtain odd-shaped elements which have *simple* shapes in their own coordinate system, so that the mechanics of evaluating integrals in the new coordinates is relatively simple due to the simplicity of the domain (evaluation over an odd-shaped subdomain is almost always very complicated).

We start our discussion of the isoparametric method with the observation that a convenient and ingenious way to generate classes of transformations of simple rectilinear elements into irregular and/or curvilinear ones is by using *basis functions*, in much the same manner as one represents a dependent variable u on (x, y) in the finite-element method.

To be specific, consider, for example, the eight-node element D_e, which is a square in a set of local coordinates (ξ, η), as illustrated in Figure 12.3.3. In D_e, the function $u(x, y)$ is expressed in the form

$$u_e = \sum_{n_e} \varphi_{n_e}(\xi, \eta) u_{n_e} \tag{12.3.4}$$

where $\{n_e\}$ is the set of nodal points that belong to D_e and $\{\varphi_{n_e}\}$ are the simple polynomial basis functions on a square defined in Chapter 11.

Consider next the eight points in the global coordinates (x, y) to which we want D_e to be fitted (Figure 12.3.4). We may choose for such a transformation the expression

$$x = \sum_{n_e=1}^{8} \varphi_{n_e}(\xi, \eta) x_{n_e}$$
$$y = \sum_{n_e=1}^{8} \varphi_{n_e}(\xi, \eta) y_{n_e} \tag{12.3.5}$$

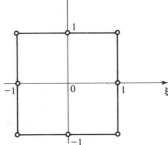

Figure 12.3.3 Representation of D_e in local coordinates (ξ, η).

[2]See, e.g., Ergatoudis (1966), Ergatoudis, Irons, and Zienkiewicz (1968a, 1968b), Felippa and Clough (1969).

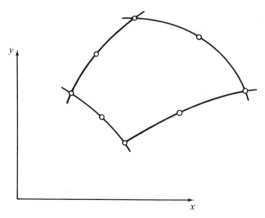

Figure 12.3.4

where the basis function φ_{n_e} are the same as those used in (12.3.4). This is called an *isoparametric transformation*. The distorted elements so obtained are called *isoparametric elements*.

The crucial importance of this technique is that it leads to distorted interelement boundary lines *and* finite-element functions $u(x, y)$ on those interelement boundary lines that are, as required, *identical* through the two *different* transformations which belong to the two elements lying on either side of that line:

> **Proof.** That the preceding statement is true is a direct consequence of the fact that when polynomials are used as basis functions on polygons in the manner illustrated in Chapter 11, then the function $u(x, y)$ on an element boundary B is equal to
>
> $$\sum_{n_B} \varphi_{n_B}(x_B, y_B)u_{n_B} \tag{12.3.6}$$
>
> where $\{n_B\}$ *are the points located on the straight line B alone* [all other $\varphi_n(x_B, y_B)$ are equal to zero!]. Moreover, (12.3.6) is the unique polynomial of degree $(N_B - 1)$ (if N_B nodal points are located on that interelement boundary) in the independent variable s, which expresses linear displacement along B. *The function* (12.3.6) *on B is independent of both the shape of the element and of nodal values in points not on B.* Thus, if we use the *same polynomial basis functions to generate local transformations of coordinates* of the form (12.3.5), then the values of x_B and y_B on the element boundary line B corresponding to a linear variation in (ξ, η) will be the *same* for the transformations (12.3.5) which originate from the two elements that lie on either side of B [thus, B will be unique in the (x, y) plane]. Similarly, any function

expressed in finite-element form (12.3.4) will be continous across the (curved) interelement line B in (x, y). $\square$

12.3.3   BILINEAR ISOPARAMETRIC ELEMENTS

A simple application of isoparametric elements is obtained by the transformation of the bilinear elements of Section 10.6. The fundamental element shape in (ξ, η) may be taken to be a square with corner nodes at the points $(1, 1), (-1, +1), (-1, -1), (1, -1)$ (Figure 12.3.5). Using $n = 1, 2, 3$, and 4 as a local numbering of these nodes, the corresponding bilinear basis functions are

$$\varphi_1(\xi, \eta) = \frac{(1 + \xi)(1 - \eta)}{4}$$

$$\varphi_2(\xi, \eta) = \frac{(1 - \xi)(1 + \eta)}{4}$$

$$\varphi_3(\xi, \eta) = \frac{(1 - \xi)(1 - \eta)}{4} \qquad (12.3.7)$$

$$\varphi_4(\xi, \eta) = \frac{(1 + \xi)(1 - \eta)}{4}$$

This fundamental element is to be mapped into an arbitrarily shaped quadrilateral in the (x, y) plane,[3] defined by its four corner nodes (Figure 12.3.6),

$$(x_1, y_1), \quad (x_2, y_2), \quad (x_3, y_3), \quad (x_4, y_4) \qquad (12.3.8)$$

Following the isoparametric concept, the transformation is

$$x = \sum_{n=1}^{4} \varphi_n(\xi, \eta) x_n$$

$$y = \sum_{n=1}^{4} \varphi_n(\xi, \eta) y_n \qquad (12.3.9)$$

We observe that when ξ or η is varied along one of the sides of the element of Figure 12.3.5, the variation of both x and y as expressed by (12.3.9) is linear in both ξ and η. *Thus, each side of the transformed element is a segment of straight line in the (x, y) plane;* the transformed element is a rectilinear quadrilateral.

---

[3] The (x, y) coordinates are the physical spatial coordinates in which the given equation to be solved is formulated. They are global, in the sense that they are consistently the same for all the elements covering D.

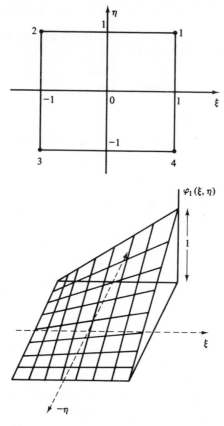

Figure 12.3.5 Bilinear basis function on a square element in (ξ, η).

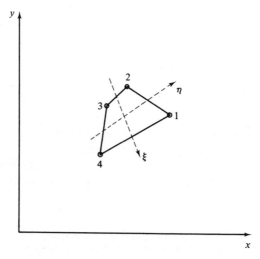

Figure 12.3.6 Mapping of the square element of Figure 12.3.5 into a quadrilateral element in the global coordinates (x, y).

The dependent variable u is expressed locally as

$$u(\xi, \eta) = \sum_{n=1}^{4} \varphi_n(\xi, \eta)u_n \qquad (12.3.10)$$

and varies linearly along the element sides in the (ξ, η) local coordinates. Thus, *variations of u along the transformed element edges in (x, y) remain linear* (Figure 12.3.7). This property was expected. As previously stated, it ensures that $u(x, y)$ as represented by bilinear shape functions over adjacent isoparametric quadrilateral elements remains continuous across interelement boundaries.

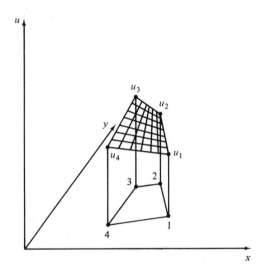

Figure 12.3.7   Transformed bilinear element in (x, y).

12.3.4 HIGHER-ORDER ISOPARAMETIC ELEMENTS

In the preceding example a square element with linear sides in (ξ, η) is transformed by isoparametric transformations into quadrilaterals in (x, y) which also have linear sides; that is, the variation of interpolated functions is linear along each side, and the sides are straight lines in (x, y). If an isoparametric transformation is applied to a square element in (ξ, η) with *quadratic* sides, then the variation of x and y expressing the shape of the sides in the (x, y) plane will be as quadratic polynomials in ξ and η, and the resulting element shape in (x, y) will consist of segments of quadratic curves (Figure 12.3.8).

Similarly, for higher-order elements the sides of an element in (x, y) corresponding to a square in (ξ, η) are polynomial curves of a degree equal

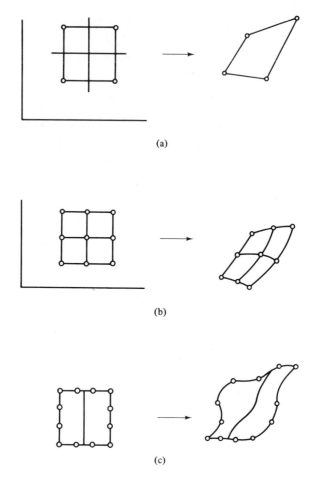

Figure 12.3.8 (*a*) Linear; (*b*) quadratic; (*c*) cubic.

to the degree of the corresponding side in the original element in (ξ, η). Thus, higher-order isoparametric tranformations give the means tc generate finite elements with increasingly twisted shapes.

12.3.5 SUB- AND SUPERPARAMETRIC TRANSFORMATIONS

One may, as in Figure 12.3.9, use corner nodes only to obtain a linear parametric transformation of a quadratic element (the transformed sides are straight lines; the variation of u on those sides is quadratic). Conversely, one may use a quadratic parametric transformation on an element that is otherwise linear [the transformed sides are quadratic curves in (x, y), but

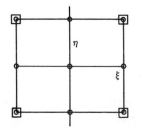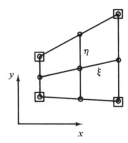

Figure 12.3.9 Linear transformation of a biquadratic Lagrangian element from a square in (ξ, η) into a general quadrilateral in (x, y). Circles represent nine nodes used to describe u; squares represent four nodes used to define the local transformation from (ξ, η) to (x, y).

the variation of u on these sides remains linear]. These are examples of what is called sub- and superparametric transformations (see, e.g., Zienkiewicz, 1977, p. 188).

12.4 NUMERICAL INTEGRATION

Other than for relatively simple cases (such as linear elements in triangles or bilinear elements in rectangles for second-order problems), the integrals that appear in the formulation of the finite-element method cannot be evaluated or expressed in simple analytic form. Those integrals can then be computed (or approximated) by numerical quadrature, and numerical quadrature algorithms are thus to be included inside the "forming" part of finite-element codes or programs.

It is, of course, with higher-order, odd-shaped finite elements in two and three spatial dimensions that the difficulty of analytic integration makes numerical integration almost mandatory. The applicability of standard numerical integration formulas in those cases is predicated on the use of coordinate changes and of isoparametric transformations. Each element becomes a regular polygon or polyhedron (square, cube, triangle, tetrahedron, etc.) in its own coordinate system, in which numerical integration formulas are applicable in their standard, tabulated form.

Numerical quadrature formulas in one dimension are of the form

$$\int_a^b f(x)\,dx \simeq \sum_{i=1}^N w_i f(x_i) \tag{12.4.1}$$

where the x_i are N fixed abcissae in the interval of integration $[a, b]$ and the w_i are fixed weights (similar expressions apply for two and three dimensions).

The classical derivation of the weights w_i consists in choosing them so as to make (12.4.1) exact for

$$f(x) = 1, x, x^2, \ldots, x^{N-1} \tag{12.4.2}$$

that is, exact for any polynomial of degree up to $(N - 1)$.

Two basic families of numerical quadrature formulas are used:

1. *Newton–Cotes formulas.* The sample points x_i are equally spaced in $[a, b]$.
2. *Gauss formulas.* The sample points x_i are placed at selected points in $[a, b]$ so as to give a higher accuracy than Newton–Cotes formulas for the same number of samples.

Both will be described below. The straightforward extension to regular squares and regular, straight-edged triangles will also be outlined.

12.4.1 NEWTON–COTES INTEGRATION

The class of numerical quadrature formulas in one dimension expressed as a weighted sum of samples evaluated at equally spaced points

$$x_i = ih; \qquad i = 1, 2, \ldots, N - 1 \tag{12.4.3}$$

is known as the Newton–Cotes formulas (see Figure 12.4.1). These formulas are derived by integrating analytically the Lagrangian interpolation polynomial of order $(N - 1)$ fit to the N values of $f(x)$. Newton–Cotes formulas with N sample points are thus exact for polynomials up to degree $(N - 1)$ (one actually gets one more degree of accuracy when N is odd). They are

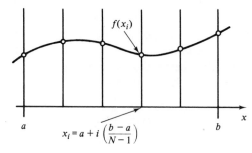

Figure 12.4.1 Location of sample points for Newton–Cotes (closed) numerical quadrature formulae.

tabulated in a number of classical texts (e.g., Abramowitz and Stegun, 1964). The first eight members of this family are summarized in Table 12.4.1.

Table 12.4.1   Newton–Cotes numerical integration (closed) formulas

---

$n = 2$:

$$\int_{x_0}^{x_1} f(x)\, dx = \frac{h}{2}(f_0 + f_1) - \frac{h^3}{12} f^{(2)}(\xi)$$

$n = 3$:

$$\int_{x_0}^{x_1} f(x)\, dx = \frac{h}{3}(f_0 + 4f_1 + f_2) - \frac{h^5}{90} f^{(4)}(\xi)$$

$n = 4$:

$$\int_{x_0}^{x_3} f(x)\, dx = \frac{3h}{8}(f_0 + 3f_1 + 3f_2 + f_3) - \frac{3h^5}{80} f^{(4)}(\xi)$$

$n = 5$:

$$\int_{x_0}^{x_4} f(x)\, dx = \frac{2h}{45}(7f_0 + 32f_1 + 12f_2 + 32f_3 + 7f_4) - \frac{8h^7}{945} f^{(6)}(\xi)$$

$n = 6$:

$$\int_{x_0}^{x_5} f(x)\, dx = \frac{5h}{288}(19f_0 + 75f_1 + 50f_2 + 50f_3 + 75f_4 + 19f_5) - \frac{275h^7}{12,096} f^{(6)}(\xi)$$

$n = 7$:

$$\int_{x_0}^{x_6} f(x)\, dx = \frac{h}{140}(41f_0 + 216f_1 + 27f_2 + 272f_3 + 27f_4 + 216f_5 + 41f_6)$$
$$- \frac{9h^9}{1400} f^{(8)}(\xi)$$

$n = 8$:

$$\int_{x_0}^{x_7} f(x)\, dx = \frac{7h}{17.280}(751f_0 + 3577f_1 + 1323f_2 + 2989f_3 + 2989f_4$$
$$+ 1323f_5 + 3577f_6 + 751f_7) - \frac{8183h^9}{518,400} f^{(8)}(\xi)$$

$n = 9$:

$$\int_{x_0}^{x_8} f(x)\, dx = \frac{4h}{14,175}(989f_0 + 5888f_1 - 928f_2 + 10,496f_3 - 4540f_4$$
$$+ 10,496f_5 - 928f_6 + 5888f_7 + 989f_8) - \frac{2368h^{11}}{467,775} f^{(10)}(\xi)$$

EXAMPLE:

Consider the integral

$$I = \int_{-h}^{h} f(x)\, dx \tag{12.4.4}$$

where $f(x)$ is known to be a polynomial of second degree:

$$f(x) = a_0 + a_1 x + a_2 x^2 \tag{12.4.5}$$

The $N = 3$ case of Newton–Cotes integration formulas (also known as Simpson's rule) is

$$I \simeq \frac{h}{3}[f(-h) + 4f(0) + f(h)] \tag{12.4.6}$$

and is *exact* for polynomials up to degree 3. Because $f(x)$ is a polynomial of degree 2, (12.4.6) is exact. Indeed, analytical integration gives the result

$$I = \left[a_0 x + a_1 \frac{x^2}{2} + a_2 \frac{x^3}{3} \right]_{-h}^{h} \tag{12.4.7}$$

which may be verified to be, as predicted, strictly equal to (12.4.6).

12.4.2 GAUSS QUADRATURE

Numerical integration with a small number of sample points over a finite interval is particularly well suited to Gauss quadrature. Newton–Cotes numerical integration formulas (i.e., formulas based on equally spaced sample points) are exact for polynomials up to order $(N - 1)$,

By contrast, Gauss numerical integration formulas are exact for polynomials up to degree $(2N - 1)$ for the same number of sample points. This is achieved by placing the samples at selected points in the interval of integration (a, b) called *Gaussian points*. It turns out that these points are precisely the zeros of the shifted Legendre polynomial of degree N over the same interval.

Let $(-1, +1)$ be the (normalized) interval over which we want to evaluate the integral

$$I = \int_{-1}^{+1} f(x)\, dx \tag{12.4.8}$$

LEGENDRE[4] (ORTHOGONAL) POLYNOMIALS

The Legendre polynomials are one of the classical families of orthogonal functions. Specifically, they satisfy

---

[4]Adrien Marie Legendre (1752–1833), French mathematician.

$$\int_{-1}^{+1} P_n(x)P_m(x)\, dx = \begin{cases} 0 & \text{if } m \neq n \\ \dfrac{2}{2n+1} & \text{if } m = n \end{cases} \qquad (12.4.9)$$

They also satisfy the simple recurrence relation

$$(n+1)P_{n+1}(x) - (2n+1)xP_n(x) + nP_{n-1}(x) = 0 \qquad (12.4.10)$$

This relation can be used to generate the Legendre polynomials. Starting with

$$\begin{aligned} P_0 &= 1 \\ P_1 &= x \end{aligned} \qquad (12.4.11)$$

we find, by application of (12.4.10) (see Figure 12.4.2),

$$\left. \begin{aligned} P_2 &= \tfrac{1}{2}(3x^2 - 1) \\ P_3 &= \tfrac{1}{2}(5x^3 - 3x) \\ P_4 &= \tfrac{1}{8}(35x^4 - 30x^2 + 3) \\ &\vdots \end{aligned} \right\} \qquad (12.4.12)$$

Another property is that $P_N(x)$ has N real zeros in $(-1, +1)$.

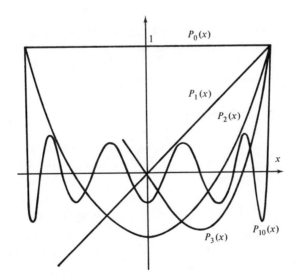

Figure 12.4.2 Legendre (orthogonal) polynomials on the interval $(-1, +1)$. Their zeros are the Gaussian points for numerical quadrature.

Suppose now that we wish to integrate a polynomial of degree $2N - 1$ [say, $p_{2N-1}(x)$] by numerical quadrature on the interval $(-1, +1)$. Prior to this integration, let us divide this polynomial by the Nth Legendre polynomial

$$p_{2N-1}(x) = q(x)P_N(x) + r(x) \qquad (12.4.13)$$

where q and r, respectively the quotient and remainder of this division, are each polynomials of degree at most $(N - 1)$. We may rewrite

$$\int_{-1}^{+1} p_{2N-1}(x)\, dx \simeq \sum_{i=1}^{N} w_i f(x_i) \qquad (12.4.14)$$

as

$$\int_{-1}^{+1} q(x)P_N(x)\, dx + \int_{-1}^{+1} r(x)\, dx \simeq \sum_i w_i q(x_i)P_N(x_i) + \sum_i w_i r(x_i) \quad (12.4.15)$$

The first term of the left-hand side is equal to zero since $P_N(x)$ is orthogonal over $(-1, +1)$ to all polynomials of degree up to $(N - 1)$. Moreover, if we choose the abcissae x_i at the zeros of $P_N(x)$, then the first term on the right-hand side also vanishes and (12.4.14) becomes identical to

$$\int_{-1}^{+1} r(x)\, dx \simeq \sum_{i=1}^{N} w_i r(x_i) \qquad (12.4.16)$$

If the N weights w_i are chosen to make this expression exact for polynomials $r(x)$ up to degree $(N - 1)$, then automatically (12.4.14) will be exact for polynomials up to degree $(2N - 1)$. See Table 12.4.2 for abscissae and weight coefficients of the quadrature formulas.

12.4.3 GAUSS INTEGRATION

Squares. The application of Gauss numerical integration to squares is straightforward (see Figure 12.4.3). It is a simple matter to show that the corresponding expression is[5]

$$\int_{-1}^{+1} \int_{-1}^{+1} f(x, y)\, dx\, dy = \sum_i \sum_j w_i w_j f(x_i, x_j) \qquad (12.4.17)$$

where $\{x_i\}$ and $\{y_j\}$ are the Gaussian points in the x and y directions, respectively, and $\{w_i\}$ and $\{w_j\}$ are the corresponding weights. Note that the degree

---

[5]This is also true for Newton–Cotes formulas, of course.

Table 12.4.2 Abscissae and weight cofficients of the Gaussian quadrature formula

$$\int_{-1}^{1} f(x)\, dx = \sum_{i=1}^{n} w_i f(x_i)$$

$\pm x_i$	w_i
$n = 2$	
0.57735 02691 89626	1.00000 00000 00000
$n = 3$	
0.77459 66692 41483	0.55555 55555 55556
0.00000 00000 00000	0.88888 88888 88889
$n = 4$	
0.86113 63115 94053	0.34785 48451 37454
0.33998 10435 84856	0.65214 51548 62546
$n = 5$	
0.90617 98459 38664	0.23692 68850 56189
0.53846 93101 05683	0.47862 86704 99366
0.00000 00000 00000	0.56888 88888 88889
$n = 6$	
0.93246 95142 03152	0.17132 44923 79170
0.66120 93864 66265	0.36076 15730 48139
0.23861 91860 83197	0.46791 39345 72691
$n = 7$	
0.94910 79123 42759	0.12948 49661 68870
0.74153 11855 99394	0.27970 53914 89277
0.40584 51513 77397	0.38183 00505 05119
0.00000 00000 00000	0.41795 91836 73469
$n = 8$	
0.96028 98564 97536	0.10122 85362 90376
0.79666 64774 13627	0.22238 10344 53374
0.52553 24099 16329	0.31370 66458 77887
0.18343 46424 95650	0.36268 37833 78362
$n = 9$	
0.96816 02395 07626	0.08127 43883 61574
0.83603 11073 26636	0.18064 81606 94857
0.61337 14327 00590	0.26061 06964 02935
0.32425 34234 03809	0.31234 70770 40003
0.00000 00000 00000	0.33023 93550 01260
$n = 10$	
0.97390 65285 17172	0.06667 13443 08688
0.86506 33666 88985	0.14945 13491 50581
0.67940 95682 99024	0.21908 63625 15982
0.43339 53941 29247	0.26926 67193 09996
0.14887 43389 81631	0.29552 42247 14753

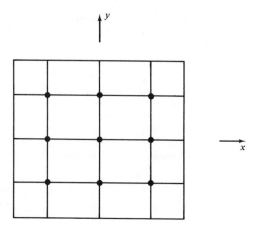

Figure 12.4.3 Gaussian points in a square, giving exact numerical integration for polynomial functions that are of degree up to 5 in both x and y.

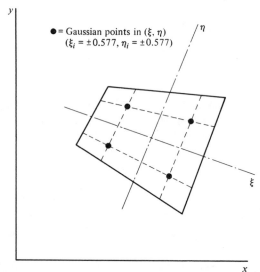

• = Gaussian points in (ξ, η)
$(\xi_i = \pm 0.577, \eta_i = \pm 0.577)$

Figure 12.4.4 Four Gaussian points in a bilinear element obtained by an isoparametric transformation of a square. All integrals over the element are approximated by a weighted sum of the integrand evaluated in those Gaussian points. The approximation is exact for all functions that are polynomials of degree not greater than 3 in ξ and η.

need not be the same in each direction. Order of accuracy is preserved; that is, if M and N points are used in the x and y directions, then the two-dimensional quadrature formula (12.4.17) is exact for functions $f(x, y)$ which are polynomials of degree $(2M - 1)$ and $(2N - 1)$ in each respective direction. When used with isoparametric elements, Gaussian quadrature is applied to the evaluation of integrals in the (ξ, η) plane, where domains have regular, well-behaved shapes. This leads to an evaluation of the integrand in points in (x, y) which are obtained by isoparametric transformation of the Gaussian points in (ξ, η). An example is given in Figure 12.4.4.

Triangles. Gaussian integration over triangles can be obtained by using the area coordinates (Section 11.3) in which shape functions are expressed as polynomials. The appropriate reference for these formulas is Cowper (1973), from which Table 12.4.3 is taken.

Table 12.4.3 Tabulated truncated values of Gauss points and weighting factors for triangles*

$$(1/A) \iint\limits_A f \, dA = \sum_{j=1}^{N} w_j f(L_{1j}, L_{2j}, L_{3j})$$

Number of Gauss Points	Coordinate			Weighting Factors w_j
	L_1	L_2	L_3	
3	0.66666667	0.16666667	0.16666667	0.33333333
	0.16666667	0.66666667	0.16666667	0.33333333
	0.16666667	0.16666667	0.66666667	0.33333333
4	0.60000000	0.20000000	0.20000000	0.52083333
	0.20000000	0.60000000	0.20000000	0.52083333
	0.20000000	0.20000000	0.60000000	0.52083333
	0.33333333	0.33333333	0.33333333	−0.56250000
6	0.81684757	0.09157621	0.09157621	0.10995174
	0.09157621	0.81684757	0.09157621	0.10995174
	0.09157621	0.09157621	0.81684757	0.10995174
	0.10810302	0.44594849	0.44594849	0.22338159
	0.44594849	0.10810302	0.44594849	0.22338159
	0.44594849	0.44594849	0.10810302	0.22338159

*From Cowper (1973).

Bibliography

Abramowitz, M., and I. A. Stegun (1964), *Handbook of Mathematical Functions*, Publ. No. AMS 55. National Bureau of Standards, Washington, D.C.

Ahlberg, J. H., E. N. Nilson, and J. L. Walsh (1967), *The Theory of Splines and Their Applications*, Academic Press, Inc., New York.

Ahmad, S., B. M. Irons, and O. C. Zienkiewicz (1969), "Curved thick shell and membrane elements with particular references to axisymmetric problems," *Proc. 2nd Conf. Matrix Methods Struct. Mech.*, AFFDL-TR-68-150 (Oct. 15–17, 1968), Wright-Patterson AFB, Dayton, Ohio, pp. 539–572.

d'Alembert, J. le Rond (1785), "Mathematics," in *Encyclopédie méthodique*, Vol. 2, Part II, Paris.

Allen, D. N. de G. (1954), *Relaxation Methods*, McGraw-Hill Book Company, New York.

Allen, R. C., Jr., G. M. Wing, and M. R. Scott (1969), "Solution of a certain class of non-linear two-point boundary value problems," *J. Comp. Phys.*, Vol. 4, pp. 250–257.

Ames, W. F. (1965), *Nonlinear Partial Differential Equations in Engineering*, Academic Press, Inc., New York.

Ames, W. F. (1969), *Numerical Methods for Partial Differential Equations*, Barnes & Noble Books, New York.

Andre, H. (1970), "Extensions to the CSDT methods for the hybrid solution of partial differential equations," *Proc. AICA-IFIP Congr. Hybrid Computation*, Munich, September 1970. Pub. European Academic Press, Brussels.

Angel, E. and R. Bellman (1972), *Dynamic Programming and Partial Differential Equations*, Academic Press, Inc., New York.

Argyris, J. H. (1954), "Energy theorems and structural analysis," *Aircr. Eng.*, Vol. 26, pp. 347–356 (Oct.), 383–387, 394 (Nov.)

Argyris, J. H. (1955), "Energy theorems and structural analysis," *Aircr. Eng.*, Vol. 27, pp. 42–58 (Feb.), 80–94 (Mar.), 125–134 (Apr.), 145–158 (May)

Argyris, J. H. (1956), "The Matrix Analysis of Structures with Cut-outs and Modifications," *Proc. 9th Int. Congr. Appl. Mech., Sec. II: Mech. Solids.*

Argyris, J. H. (1957), "The matrix theory of statics," *Ing. Arch.*, Vol. 25, pp. 174–192.

Argyris, J. H. (1958), "On the analysis of complex structures," *Appl. Mech. Rev.*, Vol. 11, pp. 331–338.

Argyris, J. H. (1959), "Recent developments of matrix theory of structures," *10th Meet. AGARD Struct. Mater. Panel*, Aachen, Germany.

Argyris, J. H. (1965a), "Matrix analysis of three-dimensional elastic media, small and large deflections," *AIAA J.*, Vol. 3, pp. 45–51.

Argyris, J. H. (1965b), "Tetrahedron elements with linearly varying strain for the matrix displacement method," *J. R. Aeronaut. Soc.*, Vol. 69, pp. 877–880.

Argyris, J. H. (1965c), "Triangular elements with linearly varying strain for the matrix displacement method," *J. R. Aeronaut. Soc.*, Vol. 69, pp. 711–713.

Argyris, J. H., and S. Kelsey (1956), "Structural analysis by the matrix force method with applications to aircraft wings," *Wiss. Ges. Luftfahrt: Jahrb.*

Argyris, J. H., and S. Kelsey (1959), "The analysis of fuselages of arbitrary cross-section and taper," *Aircr. Eng.*, Vol. 31, pp. 62–74 (March), 101–112 (April), 133–143 (May), 169–180 (June), 192–203 (July), 224–256 (Aug.), 272–283 (Sept.).

Argyris, J. H., and S. Kelsey (1960), *Energy Theorems and Structural Analysis*, Butterworth & Co. (Publishers) Ltd., London.

Aziz, A. K., and B. E. Hubbard (1964), "Bounds on the truncation error by finite differences for the Goursat problem," *Math. Comput.*, Vol. 18, pp. 19–35.

Babuska, I. (1970), "Finite element method for domains with corners," *Computing*, Vol. 6, pp. 264–273.

Babuska, I. (1971a), "The rate of convergence for the finite element method," *SIAM J. Numer. Anal.*, Vol. 8, No. 2, pp. 304–315.

Babuska, I. (1971b), "Error bounds for the finite element method," *Numer. Math.*, Vol. 16, pp. 322–333.

Babuska, I. (1973), "The finite element method with Lagrangian multipliers," *Numer. Math.*, Vol. 20, pp. 179–192.

Babuska, I., M. Prager, and E. Vitasek (1966), *Numerical Processes in Differential Equations*, Interscience Publishers, New York.

Bailey, P., L. F. Shampine, and P. E. Waltman (1968), *Nonlinear Two-Point Boundary Value Problems*, Academic Press, Inc., New York.

Baker, V. A., ed. (1976), *Sparse Matrix Techniques*, Notes from a course held at the Technical University of Denmark in Aug. 1976, Lecture Notes in Mathematics, No. 572, Springer-Verlag, New York.

Beckenbach, E. F., ed. (1956), *Modern Mathematics for the Engineer*, McGraw-Hill Book Company, New York.

Bellman, R., H. Kagiwada, and R. Kalaba (1967), "Invariant imbedding and the numerical integration of boundary value problems for unstable linear systems of ordinary differential equations," *Commun. ACM*, Vol. 10, No. 2, pp. 100–102.

Bergland, G. D. (1969), "A guided tour of the fast Fourier transform," *IEEE Spectrum*, July, pp. 41–52.

Bernoulli, D. (1753a), "Réflexions et éclaircissemens sur les nouvelles vibrations des cordes . . . ," *Mem. Acad. Sci. Berlin*, 9 (1753; publ. 1755), pp. 147–172.

Bernoulli, D. (1753b), "Sur le mélange de plusieurs espèces de vibrations simples isochrones . . . ," *Mem. Acad. Sci. Berlin*, 9 (1753; publ. 1755), pp. 173–195.

Besseling, J. F. (1963), "The complete analogy between the matrix equations and the continuous field equations of structural analysis," *Int. Symp. Analogue Digital Tech. Appl. Aeronaut.*, Liège, Belgium.

Beuken, C. L. (1936), *Compt. Rend. Int. Congr. Electr. Heating*, Scheveningen (Holland).

Bickley, W. G. (1941), "Experiments in approximating to the solution of a partial differential equation," *Phil. Mag.* (7), Vol. 32, pp. 50–56.

Bickley, W. G. (1948), "Finite-difference formulae for the square lattice," *Quart. J. Mech. Appl. Math.* ,Vol. 1, pp. 35–42.

Bickley, W. G., S. Michaelson, and M. R. Osborne (1961), "On finite difference methods for the numerical solution of boundary-value problems," *Proc. R. Soc. Lond.*, Ser. *A*, Vol. 262, pp. 219–236.

Bienzeno, C. B. (1923–1924), "Over een vereenvoudiging en over een uitbreiding van de methode van Ritz," *Christiaan Huygens*, Vol. 3, p. 69.

Bienzeno, C. B. (1924), "Graphical and numerical methods for solving stress problems," in *Proc. 1st Int. Congr. Appl. Mech.*, *Delft*, pp. 3–17.

Bienzeno, C. B., and J. J. Koch (1923), "Over een nieuwene methode ter berekening van vlokke platen met toepassing op enkele voor de techniek belangrijke belastingsgevallen," *Ing. Grav.*, Vol. 38, pp. 25–36.

Birdsall, C. K., and D. Fuss (1969), "Clouds-in-clouds, clouds-in-cells physic for many-body plasma simulation," *J. Comp. Phys.*, Vol. 3, pp. 494–411.

Birkhoff, G. (1972a), "Piecewise analytic interpolation and approximation in triangular polygons," in A. K. Aziz, ed., *The Mathematical Foundation of the FEM with Applications to PDE's*, Academic Press, Inc., New York.

Birkhoff, G. (1972b), *The Numerical Solution of Elliptic Equations*, SIAM/CBMS Regional Conference Series in Applied Mathematics, No. 1, Society for Industrial and Applied Mathematics, Philadelphia.

Birkhoff, G., and V. A. Dougalis (1975), "Numerical solution of hydrodynamic problems," in R. Vichnevetsky, ed., *Advances in Computer Methods for Partial Differential Equations*, AICA, Rutgers University, New Brunswick, N.J.

Birkhoff, G., and S. Gulati (1974), "Optimal few-point discretizations of linear source problems," *SIAM J. Numer. Anal.*, September, Vol. 11, No. 4, pp. 700–728.

Birkhoff, G., and R. S. Varga (1959), "Implicit alternating direction methods," *Trans. Am. Math. Soc.*, Vol. 92, pp. 13–24.

Birkhoff, G., M. H. Schultz, and R. Varga (1968), "Piecewise Hermite interpolation in one and two variables with applications to partial differential equations," *Numer. Math*, Vol. 11, pp. 232–256.

Bogner, F. K., R. L. Fox, and L. A. Schmidt (1965), "The generation of interelement compatible stiffness and mass matrices by use of interpolation formulae," *Proc. 1st Conf. Matrix Methods Struct. Mech.*, AFFDL-TR66-80 (Oct. 26–28, 1965), pp. 397–443, Wright-Patterson AFB, Dayton, Ohio.

Boole, G. D. (1860), *A Treatise of the Calculus of Finite Differences*, Constable & Co. Ltd., London. 2nd ed. (1875) reprinted by Dover Publications, Inc., New York.

Boyer, C. B. (1968), *A History of Mathematics*, John Wiley & Sons, Inc., New York.

Bramble, J. H., and M. Zlamal (1970), "Triangular elements in the finite element method," *Math. Comput.*, Vol. 24, pp. 809–821.

Brandt, A. (1972), "Multi level adaptive technique (MLAT) for fast numerical solution to boundary value problems," *Proceedings of the Third International Conference on Numerical Methods in Fluid Mechanics (Paris 1972)*, Lecture Notes in Physics, No. 18, Springer-Verlag, Berlin.

Bradt, A. (1976), "Multi level adaptive techniques (MLAT)," *IBM Res. Rep. RC 6026.*

Brigham, E. O. (1974), *The Fast Fourier Transform*, Prentice-Hall, Inc., Englewood Cliffs, N.J.

Brown, R. R. (1962), "Numerical solution of boundary value problems using non-uniform grids," *J. Soc. Ind. Appl. Math.*, Vol. 10, pp. 475–495.

Bucy, R. S. (1967), "TPBVP of linear Hamiltonian systems," *SIAM J. Appl. Math.*, Vol. 15, pp. 1385–1389.

Bunch, J. R., and D. J. Rose, eds. (1976), *Sparse Matrix Computations*, Academic Press, Inc., New York.

Buneman, O. (1969), "A compact non-iterative Poisson solver," SUIPR Report No. 294, Institute of Plasma Research, Stanford University, Palo Alto, Calif.

Buzbee, B. L., G. H. Golub, and C. W. Nielson (1970), "On direct methods for solving Poisson's equation," *SIAM J. Numer. Anal.*, Vol. 7, pp. 627–656.

Carey, G. F. and B. A. Finlayson, (1975), "Orthogonal collocation on finite elements," *Chem. Eng. Sci.*, Vol. 30, pp. 587–596, Pergamon Press.

Carslaw, H. S. (1930), *Introduction to the Theory of Fourier's Series and Integrals*, 3rd, ed., Macmillan International Ltd., London.

Carslaw, H. S., and J. C. Jaeger (1947), *Conduction of Heat in Solids*, Clarendon Press, Oxford.

Caussade B. H. and G. Renard (1977), "Hybrid simulation of two-dimensional infiltration from parallel open-channels" in R. Vichnevetsky (ed.), *Advances in Computer Methods for Partial Differential Equations II*, IMACS, New Brunswick, New Jersey, pp. 245–249.

Cayles, M. A. (1961), "Solution of systems of ordinary and partial differential equations by quasi diagonal matrices," *Comput. J.*, Vol. 4, pp. 54–61.

Chang, P. W., and B. A. Finlayson (1978), "Orthogonal collocation on finite elements for elliptic equations," *Math. Comput. Simul.*, Vol. 20, no. 1, pp. 83–92.

Churchill, R. V. (1941), *Fourier Series and Boundary Value Problems*, McGraw-Hill Book Company, New York. Rev. ed., 1963.

Clough, R. W. (1960), "The finite element method in plane stress analysis," *Proc. 2nd ASCE Conf. Electron. Comput.*, Pittsburgh, Pa., Sept. 8–9.

Collatz, L. (1933), *Z. Angew. Math. Mech.*, Vol. 13, pp. 56–57.

Collatz, L. (1950), "Uber die Konvergenzkriterien bei Iterationsverfahren für lineare Gleichungssysteme," *Math. Z.*, Vol. 53, pp. 149–161.

Collatz, L. (1951), "Zur Stabilität des Differenzverfahrens bei der Stabschwin-guhsgleichung," *Z. Angew. Math. Mech.*, Vol. 31, pp. 392–393.

Collatz, L. (1960), *The Numerical Treatment of Differential Equations*, Springer-Verlag, Berlin.

Concus, P., and G. H. Golub (1973), "Use of fast direct methods for the efficient numerical solution of nonseparable elliptic equations," *SIAM J. Numer. Anal.*, Vol. 10, p. 1103.

Conte, S. D. (1966), "The numerical solution of linear boundary value problems," *SIAM Rev.*, Vol. 8, pp. 309–321.

Conte, S. D., and R. T. Dames (1958), "An alternating direction method for solving the biharmonic equation," *Math. Tables other Aids Comput.*, Vol. 12, pp. 198–205.

Conte, S. D., and R. T. Dames (1960), "On an alternating direction method for solving the plate problem with mixed boundary conditions," *J. ACM*, Vol. 7, pp. 264–273.

Cooley, J. W. (1967), "Applications of the fast Fourier transform method," *Proc. IBM Sci. Comput. Symp. Digital Simul. Contin. Syst.*, IBM, p. 83.

Cooley, J. W., and J. W. Tukey (1965), "An algorithm for the machine calculation of complex Fourier series," *Math. Comput.*, Vol. 19, pp. 297–301.

Coolery, J. W. P. W. Lewis, and P. D. Welch (1967), "Historical notes on the fast Fourier transform," *IEEE Trans. Audio Electroacoust.*, Vol. AU-15, No. 2, pp. 76–79.

Cooley, J. W., P. A. W. Lewis, and P. D. Welch (1970), "The fast Fourier transform algorithm; programming considerations in sine, cosine and Laplace transforms," *J. Sound Vib.*, Vol. 12, No. 3, pp. 315–337.

Cornock, A. F. (1954), "The numerical solution of Poisson's and the biharmonic equations by matrices," *Proc. Camb. Philos. Soc.*, Vol. 50, pp. 524–535.

Courant, R. (1924), Remark on "weighted averages of the residual" in discussion following Biezeno's paper in *Proc. 1st Int. Congr. Appl. Mech.*, Delft, p. 17.

Courant, R. (1943), "Variational methods for the solution of problems of equilibrium and vibrations," *Bull. Am. Math. Sol.*, Vol. 49, No. 1, pp. 1–23.

Courant, R., and D. Hilbert (1953), *Methods of Mathematical Physics*, Vol. I, Interscience Publishers, New York.

Courant, R., and D. Hilbert (1962), *Methods of Mathematical Physics*, Vol. II, Interscience Publishers, New York.

Courant, R., and P. Lax (1949), "On non-linear partial differential equations with two independent variables," *Commun. Pure Appl. Math.*, Vol. 2, No. 2, pp. 255–273.

Courant, R., K. Friedrichs, and H. Lewy (1928), "Uber die partiellen Differenzengleichungen der mathematischen Physik," *Math. Ann.*, Vol. 100, pp. 32–74. (A translation by Phyllis Fox has been multilithed under the title "On the partial difference equations of mathematical physics," *Rep. NYO-7689*, Institute of Mathematical Sciences, New York University, 1956.)

Cowper, G. R. (1973), "Gaussian quadrature formulas for triangles," *Int. J. Numer. Meth. Eng.*, Vol. 1, No. 3, pp. 405–408.

Crandall, S. H. (1951), "Iterative procedures related to relaxation methods for eigenvalue problems," *Proc. R. Soc. Lond.*, Ser. A., Vol. 207, pp. 416–423.

Crandall, S. H. (1956), *Engineering Analysis, A Survey of Numerical Procedures*, McGraw-Hill Book Company. New York.

Crank, J. (1948), *The Differential Analyser*, Longmans, Green and Co., London.

Cross, H. (1932), "Analysis of continuous frames by distributing fixed-end moments," Paper No. 1793, *Trans. ASCE*, Vol. 96, pp. 1–10.

Czerwinska, A. (1979), "Application of hybrid computing system APH-600 for partial differential equations," in R. Vichnevetsky and R. Stepleman, eds., *Advances in Computer Methods for Partial Differential Equations*, IMACS, Rutgers University, New Brunswick, N.J.

Dahlquist, G., and A. Bjork (1974), *Numerical Methods*, Prentice-Hall, Inc., Englewood Cliffs, N.J. Translated from the 1969 Swedish edition.

d'Alembert, J. le Rond (1747), "Recherches sur la courbe que forme une corde tendue mise en vibration," *Mem. Acad. Sci. Berlin*, 3 (1747, publ. 1749), pp. 214–219, 220–253.

Danielson, G. C., and C. Lanczos (1942), "Improvements in functional Fourier analysis and their application to X-ray scattering from liquids," *J. Franklin Inst.*, Vol. 233, pp. 365–380, 435–452.

de Boor, C., and B. Swartz (1973), "Collocation at Gaussian points," *Siam J. Numer. Anal.*, Vol. 10, No. 4, pp. 582–606.

Desai, C. S. (1979), *Elementary Finite Element Method*, Prentice-Hall, Inc., Englewood Cliffs, N.J.

Desai, C. S., and J. F. Abel (1972), *Introduction to the Finite Element Method*, Van Nostrand Reinhold Company, New York, pp. 144–145.

Diaz, J. B., and R. C. Roberts (1952), "On the numerical solution of the Dirichlet problem for Laplace's equation," *Quart. Appl. Math.*, Vol. 9, pp. 355–360.

Diaz, J. B., and R. C. Roberts (1952a), "Upper and lower bounds for the numerical solution of the Dirichlet difference boundary value problem," *J. Math. Phys.*, Vol. 31, pp. 184–191.

Di Guglielmo, F. (1969), "Construction d'approximations des espaces de Sobolev sur des réseaux en simplexes," *Calcola*, Vol. 6, pp. 279–331.

Dorr, F. W. (1970), "The direct solution of the discrete Poisson equation on a rectangle," *SIAM Rev.*, Vol. 12, pp. 248–263.

Douglas, J., Jr. (1962), "Alternating direction iteration for mildly non-linear elliptic differential equations," *Numer. Math.*, Vol. 4, pp. 41–63.

Douglas, J., and T. Dupont (1971), "Alternating direction Galerkin methods on rectangles," in B. Hubbard, ed., *Proceedings of Symposium on Numerical Solution of Partial Differential Equations*, Academic Press, Inc., New York, pp. 133–214.

Douglas, J., Jr., and T. Dupont (1973), "A finite-element collocation method for quassilinear parabolic equations," *Math. Comput.*, Vol. 27, p. 17.

Duncan, W. J. (1937), "Galerkin's method in mechanics and differential equations." *G. B. Aeronaut. Res. Coun. Rep. Memo 1798.* Reprinted in *G. B. Air Ministry Aeronaut. Res. Commun. Tech. Rep. 1*, pp. 484–516.

Duncan, W. J. (1938), "The principles of Galerkin's method," *G. B. Aeronaut. Res. Coun. Rep. Memo. 1848.* Reprinted in *G. B. Air Ministry Aeronaut. Res. Commun. Tech. Rep. 2*, pp. 589–612.

Dupont, T., R. Kendall, and H. H. Rachford (1968), "An approximate factorization procedure for solving self adjoint elliptic difference equations," *SIAM J. Numer. Anal.*, Vol. 5, pp. 559–573.

Ergatoudis, J. (1966), "Quadrilateral elements in plane analysis," M. S. thesis, University of Wales.

Ergatoudis, J., B. M. Irons, and O. C. Zienkiewicz (1968a), "Curved isoparametric, 'quadrilateral' elements for finite element analysis," *Int. J. Solids Struct.*, Vol. 4, pp. 31–42.

Ergatoudis, J., B. M. Irons, and O. C. Zienkiewicz (1968b), "Three dimensional analysis of arch dams and their foundations," *Proc. Symp. Arch. Dams*, pp. 21–34.

Eringen, A. C. (1954), "The finite Sturm Liouville transform," *Quart. J. Math.*, *Oxford Ser.*, Vol. 2, pp. 120–124.

Eringen, A. C. (1955), "A transform technique for boundary-value problems in fourth-order partial differential equations," *Quart. J. Math.*, *Oxford Ser.*, Vol. 6, No. 34, pp. 241–249.

Euler, L. (1752), "Principles of the motion of fluids," *Novi Comm. Acad. Sci. Petrop. 6 in Opera* (2), Vol. 12, pp. 133–68.

Evans, D. J., and C. V. D. Farrington (1963), "Note on the solution of certain tridiagonal systems of equations," *Comput. J.*, Vol. 5, pp. 327–328.

Eve, J., and H. I. Scions (1956), "A note on the approximate solution of Poisson and Laplace by finite difference methods," *Quart. J. Math.* (2), Vol. 7, pp. 217–223.

Faddeeva, V. N. (1959), *Computational Methods of Linear Algebra*. English translation published by Dover Publications, Inc., New York.

Felippa, C. A. and R. W. Clough (1969), "The finite element method in solid mechanics," in G. Birkhoff and R. S. Varga, eds., *Numerical solution of field problems in continuum mechanics*, SIAM-AMS proceedings 2—AMS, Providence.

Finlayson, B. A. (1972), *The Method of Weighted Residuals and Variational Principles*, Academic Press, Inc., New York.

Finlayson, B. A. (1979), "Orthogonal collocation on finite elements—progress and potential," in R. Vichnevetsky and R. Stepleman, eds., *Advances in Computer Methods for Partial Differential Equations*, IMACS, Rutgers University, New Brunswick, N.J. Reprinted in *Mathematics and Computers in Simulation*, Volume 22, no. 1, North-Holland Publishing Co., pp. 11–17 (1980).

Finlayson, B. A., and L. E. Scriven (1965), "The method of weighted residuals and its relation to certain variational principles for the anlysis of transport processes," *Chem. Eng. Sci.*, Vol. 20, pp. 395–404.

Finlayson, B. A., and L. E. Scriven (1966), "The method of weighted residuals—a review," *Appl. Mech. Rev.*, Vol. 19, No. 9, pp. 735–748.

Finlayson, B. A., and L. E. Scriven (1967), "On the search for variational principles," *Int. J. Heat Mass Transfer*, Vol. 10, pp. 799–821.

Fix, G., and G. Strang (1969), "Fourier analysis of the finite element method in Ritz-Galerkin theory," *Stud. Appl. Math.*, Vol. 48, pp. 265–273.

Forsythe, G. E. (1953), "Solving linear algebraic equations can be interesting," *Bull. Am. Math. Soc.*, Vol. 59, pp. 299–329.

Forsythe, G. E. (1956), "What are relaxation methods?" in E. F. Beckenbach, ed., *Modern Mathematics for the Engineer*, McGraw-Hill Book Company, New York.

Forsythe, George E. (1965), "Difference methods on a digital computer for Laplacian boundary value and eigenvalue problems," *Commun. Pure Appl. Math.*, Vol. 9, pp. 426–434.

Forsythe, G., and C. B. Moler (1967), *Computer Solution of Linear Algebraic Systems*, Prentice-Hall, Inc., Englewood Cliffs, N.J.

Forsythe, G. E., and W. R. Wasow (1960), *Finite Difference Methods for Partial Differential Equations*, John Wiley & Sons, Inc., New York.

Fourier, J. B. J. (1807), "Sur la propagation de la chaleur," manuscript presented to the "Academy," Paris. Reprinted in I. Grattan-Guinness, *Fourier*, The MIT Press, Cambridge, Mass., 1972.

Fourier, J. B. J. (1822), *La theorie analytique de la chaleur*, Paris. Translated by A. Freeman, Cambridge, 1878. Reprinted by Dover Publications, Inc., New York, 1955.

Fox, L. (1948), "A short account of relaxation methods," *Quart. J. Mech. Appl. Math.*, Vol. 1, pp. 253–280.

Fox, L. (1950), "The numerical solution of elliptic differential equations when the boundary conditions involve a derivative," *Philos. Trans. R. Soc. Lond.*, *Ser. A*, Vol. 242, pp. 345–378.

Fox, L. (1957), *The Numerical Solution of Two-Point Boundary Value Problems*, Oxford University Press, New York.

Fox, L. (1960), "Some numerical experiments with eigenvalue problems in ordinary differential equations," in R. E. Langer, ed., *Boundary problems in differential equations.*, University of Wisconsin Press, Madison, pp. 243–255.

Fox, L. (1962), *Numerical Solution of Ordinary and Partial Differential Equations*, Pergamon Press, Inc., Elmsford, N.Y.

Fraeijs de Veubeke, B. M., ed. (1964a), *Matrix Methods of Structural Analysis*, Pergamon Press Ltd., Oxford.

Fraeijs de Veubeke, B. M. (1964b), "Upper and lower bounds in matrix structural analysis," in B. M. Fraeijs de Veubeke, ed., *Matrix Methods of Structural Analysis*, Pergamon Press Ltd., Oxford.

Fraeijs de Veubeke, B. M. (1965), "Displacements and equilibrium models in the finite element method," in O. C. Zienkiewicz and G. S. Holister, eds., *Stress Analysis*, John Wiley & Sons, Inc., New York, pp. 145–197.

Fraeijs de Veubeke, B. M. (1966), "Bending and stretching of plates— special models for upper and lower bounds," *Proc. 1st Conf. Matrix Methods Struct. Mech.*, AFFDI-TR-66-80 (Oct. 26–28, 1965), Wright-Patterson AFB, Dayton, Ohio, pp. 863–886.

Fraeijs de Veubeke, B. M. (1968). "A conforming finite element for plate bending," *Int. J. Solids Struct.*, Vol. 4, pp. 96–108.

Frankel, S. P. (1950), "Convergence rates of iterative treatments of partial differential equations," *Math. Tables Other Aids Comput.*, Vol. 4, pp. 65–75.

Frazer, R. A., W. P. Jones, and S. W. Skan (1937), "Approximation to functions and the solutions of differential equations," *G. B. Aeronaut. Res. Coun. Rep. Memo 1799.*

Fried, I. (1971a), "Discretization and round-off error in the finite analysis of elliptic boundary value problems and eigenvalue problems," Doctoral dissertation, Massachusetts Institute of Technology.

Fried, I. (1971b), "Discretization and computational errors in high-order finite elements," *AIAA J.*, Vol. 9, pp. 2071–2073.

Fried, I. (1971c), "Basic computational problems in the finite element analysis of shells," *Int. J. Solids Struct.*, Vol. 7, pp. 1705–1715.

Fried, I. (1971d), "Condition of finite element matrices generated from nonuniform meshes," *AIAA J.*, Vol. 10, pp. 219–221.

Fried, I. (1971e), "Accuracy of finite element eigenproblems," *J. Sound Vib.*, Vol. 18, pp. 289–295.

Friedman, B. (1956), *Principles and Techniques of Applied Mathematics*, John Wiley & Sons, Inc., New York.

Friedrichs, K. O., and H. B. Keller (1966), "A finite difference scheme for generalized Neumann problems," in J. Bramble, ed., *Numerical Solutions of Partial Differential Equations*, Academic Press, Inc., New York.

G-AE Subcommittee (1967), "What is the fast Fourier transform?" *IEEE Trans. Audio Electroacoust.*, Vol. AU-15, No. 2, pp. 45–55.

Galerkin, B. G. (1915), *Vestn. Inzh. Tekh. Petrograd*, Vol. 19, pp. 897–908. Translation 63–18924, Clearinghouse Fed. Sci. Tech. Info.

Garabedian, P. R. (1964), *Partial Differential Equations*, John Wiley & Sons, Inc., New York.

Garebedian, P. R. (1965), "Estimation of the relaxation factor for small mesh size," *Math. Tables Other Aids Comput.*, Vol. 10, pp. 183–185.

Gaur, S. P., and A. Brandt (1977), "Numerical solution of semiconductor transport equations in two dimensions by multi grid methods," in R. Vichnevetsky, ed., *"Advances in Computer Methods for Partial Differential Equations II,"* IMACS, Rutgers University, New Brunswick, N.J.

Gauss, C. F. (1823), Brief an Gerling, 26. Dez. 1823, *Werke*, Vol. 9, pp. 278–281. Translated by G. E. Forsythe under the title "Gauss to Gerling on relaxation," *Math. Tables Other Aids Comput.*, Vol. 5, (1951), pp. 255–258.

Gentleman, W. M., and G. Sande (1966), "Fast Fourier transform—for fun and profit," *Proc. 1966 Fall Joint Comput. Conf., AFIPS*, Vol. 29, Spartan Books.

George, J. A. (1971), "Computer implementation of the finite element method," Ph.D. thesis, Stanford University.

Gerschgorin, S. (1930), "Fehlerabschätzung für das Differenzenverfahren zur Lösung partieller differentialgleichungen," *Z. Angew, Math. Mech.*, Vol. 10, pp. 373–382.

Giese, John H. (1958), "On the truncation error in a numerical solution for the Neumann problem for a rectangle," *J. Math. Phys.*, Vol. 37, pp. 169–177.

Godunov, S. K. (1956), "Finite difference method for the numerical computation of discontinuous solution of the equations of fluid dynamics," *Mathe. Sb.*, Vol. 47 (89), No. 3, p. 271.

Godunov, S. K. (1962), "The method of orthogonal sweeps for the solution of difference equations," *Zh. Vychisl. Mat. Mat. Fiz., Moscow*, Vol. 2 (5).

Godunov, S. K., and V. S. Ryabenkii (1963), "Special criteria of stability of boundary-value problems for non-self-adjoint difference equations," *Usp. Mat. Nauk*, Vol. 18.

Godunov, S. K., and V. S. Ryabenkii (1964), *Introduction to the Theory of Difference Schemes*, Interscience Publishers, New York. Also Fixmatigiz, Moscow, 1962.

Goel, J. J. (1968), "Construction of basic functions for numerical utilization of Ritz's method," *Numer. Math.*, Vol. 12, pp. 435–447.

Goldstine, Herman H., and John von Neumann (1951), "Numerical inverting of matrices of high order, II," *Proc. Am. Math. Soc.*, Vol. 2, pp. 188–202.

Goodman, T. R., and C. N. Lance (1956), "The numerical integration of two-point boundary value problems," *Math. Tables Other Aids Comput.*, Vol. 10, No. 54, pp. 82–86.

Gottlieb, D., and S. A. Orszag (1977), *Numerical Analysis of Spectral Methods*, CBMS–NSF Regional Conference Series, No. 26, Society for Industrial and Applied Mathematics, Philadelphia.

Goult, R. J., R. F. Hoskins, J. A. Milner, and M. J. Pratt (1974), *Computational Methods in Linear Algebra*, Stanley Thornes (Publishers) Ltd., London.

Grattan-Guinness, I. (1970), *The Development of the Foundations of Mathematical Analysis from Euler to Riemann*, The MIT Press, Cambridge, Mass.

Grattan-Guinness, I. (1972), *Joseph Fourier—1768–1830*, The MIT Press, Cambridge, Mass.

Greenstadt, J. (1971), "Cell discretization," in J. L. Morris, ed., *Conference on Applications of Numerical Analysis*, Lecture Notes in Mathematics, No. 228, Springer-Verlag, New York.

Guthenmaker, L. I. (1949), "Electrical Analogues," *Elektricheskie modeli*, Academy of Sciences Moscow, USSR.

Hadamard, J. (1908), "Mémoire sur le problème d'analyse relatif à l'équilibre des plaques élastiques encastrées," in *Mémoires présentés par divers savants à l'Académie des sciences*, Vol. 33, pp. 1–128.

Hadamard, J. (1923), *Lectures on Cauchy's Problem in Linear Partial Differential Equations*, Yale University Press, New Haven, Conn.

Hadamard, J. (1932), *Le Problème de Cauchy et les équations aux derivées partielles lineaires hyperboliques*, Hermann et Cie., Paris.

Hamming, R. W. (1973), *Numerical Methods for Scientists and Engineers*, 2nd ed., McGraw-Hill Book Company, New York.

Hamming, R. W. (1977), *Digital Filters*, Prentice-Hall, Inc., Englewood Cliffs, N.J.

Hedstrom, G. W. (1968), "The rate of convergence of some difference schemes," *SIAM J. Numer. Anal.*, Vol. 5, pp. 363–406.

Heideman, J. C. (1968), "Use of the method of particular solutions in nonlinear two-point boundary value problems," *J. Optimization Theory Appl.*, Vol. 2, No. 6, pp. 450–462.

Heller, D. E. (1977), "Direct and iterative methods for block tridiagonal linear systems," Ph. D. thesis, Carnegie Mellon University.

Heller, J. (1960), "Simultaneous, successive, and alternating direction iteration schemes," *J. Soc. Ind. Appl. Math.*, Vol. 8, pp. 150–173.

Hockney, R. W. (1965), "A fast direct solution of Poisson's equation using Fourier analysis," *J. ACM*, Vol. 12, No. 1, pp. 95–113.

Houstis, E. N., and T. S. Papatheodorou (1979), "A fast sixth order Helmholtz equation solver and its performance," in R. Vichnevetsky and R. Stepleman, eds., *Advances in Computer Methods for Partial Differential Equations III*, IMACS, Rutgers University, New Brunswick, N.J.

Hrenikoff, A. (1941), "Solution of problems in elasticity by the framework method," *J. Appl. Mech.*, Vol. 8, pp. 169–175.

Huang, Y. H., and S. J. Wu (1975), "Simulation of confined aquifers by three-dimensional finite elements," in R. Vichnevetsky, ed., *Advances in Computer Methods for Partial Differential Equations*, IMACS (AICA), Rutgers University, New Brunswick, N.J.

Hulme, B. L. (1968), "Interpolation by Ritz approximation," *J. Math. Mech.*, Vol. 18, pp. 337–342.

Irons, B. M. (1966), "Engineering applications of numerical integration in stiffness methods," *AIAA J.*, Vol. 4, No. 11, pp. 2035–2037.

Irons, B. M. (1968), "Roundoff criteria in direct stiffness solutions," *AIAA J.*, Vol. 6, pp. 1308–1312.

Irons, B. M. (1969), "Economical computer techniques for numerically integrated finite elements," *Int. J. Numer. Methods Eng.*, Vol. 1, pp. 201–203.

Irons, B. M. (1970), "A frontal solution program for finite element analyses," *Int. J. Numer. Methods Eng.*, Vol. 2, pp. 5–32.

Irons, B. M. (1971), "Quadrature rules for brick based finite elements," *AIAA J.*, Vol. 9, pp. 293–294.

Irons, B., and K. J. Draper (1965), "Inadequacy of nodal connections in a stiffness solution for plate bending," *AIAA J.*, Vol. 3, No. 5, p. 961.

Irons, B. M., and A. Razzaque (1971), "A new formulation for plate bending elements," manuscript.

Irons, B. M., E. A. de Oliveira, and O. C. Zienkiewicz (1970), "Comments on the paper 'Theoretical foundations of the finite element method'," *Int. J. Solids Struct.*, Vol. 6, pp. 695–697.

Isaacson, E., and H. B. Keller (1966), *Analysis of Numerical Methods*, John Wiley & Sons, Inc., New York.

Jacobi, C. G. J. (1845), "Ueber eine neue Auflösungsart bei der Methode der kleinsten Quadrate vorkommenden linearen Gleichungen," *Astrron. Nachr.*, Vol. 22, No. 523, pp. 294–306.

Janenko, N. N. (1967), *Nauka*, Sibirian Section, Novosibirsk.

Jolley, L. B. W. (1961), *Summation of Series*, Dover Publications, Inc., New York.

Jones, R. E. (1964), "A generalization of the direct stiffness method of structural analysis," *AIAA J.*, Vol. 2. pp. 821–826.

Kantorovitch., L. V. (1948), "Functional analysis and applied mathematics," *Usp. Mat. Nauk*, USSR, Vol. 3, p. 89.

Kantorovitch, L. V., and V. I. Krylov (1964), *Approximate Methods of Higher Analysis*, Interscience Publishers, New York.

Kaplan, S. (1962), "Some new methods of flux synthesis," *Nucl. Sci. Eng.*, Vol. 13, pp. 22–31.

Kaplan, S. (1963), "On the best method for choosing the weighting functions in the method of weighted residuals," *Trans. Am. Nucl. Soc.*, Vol. 6, pp. 3–4.

Karplus, W. J. (1954), "Electronic analogue solution of free surface boundary value problems—water coning," Dissertation, University of California, Los Angeles.

Karplus, W. J. (1956), "Water-coning before breakthrough—an electronic analog treatment," *Pet. Trans. Soc. Mech. Eng.*, Vol. 207, pp. 240–245.

Karplus, W. J. (1958), *Analog Simulation—Solution of Field Problems*, McGraw-Hill Book Company, New York.

Keast, P., and A. R. Mitchell (1967), "Finite difference solution of the third boundary problem in elliptic and parabolic equations," *Numer. Math.*, Vol. 10, pp. 67–75.

Keller, H. B. (1958), "On some iterative methods for solving elliptic difference equations," *Quart. Appl. Math.*, Vol. 16, pp. 209–226.

Keller, H. B. (1960), "On the pointwise convergence of the discrete ordinate method," *J. Soc. Ind. Appl. Math.*, Vol. 8, p. 560.

Keller, H. B. (1968), *Numerical Methods for Two-Point Boundary Value Problems*, Blaisdell Publishing Co., Waltham, Mass.

Kolmogoroff, A. N., and S. V. Fomin (1957), *Elements of the Theory of Functions and Functional Analysis* (English trans.), Graylock Press, Rochester, N.Y.

Kron, G. (1944a), "Tensorial analysis and equivalent circuits of elastic structures," *J. Franklin Inst.*, Vol. 238, pp. 399–442.

Kron, G. (1944b), "Equivalent circuits of the elastic field," *J. Appl. Mech.*, Vol. 66, pp. 149–167.

Laasonen, P. (1949), "Uber eine Methode zur Lösung der Wärmeleitungsgleichung," *Acta Math.*, Vol. 81, p. 309.

Laasonen, P. (1957), "On the degree of convergence of discrete approximations for the solutions of the Dirichlet problem," *Ann. Acad. Sci. Fenn.*, *Ser. A1*, Vol. 246, pp. 1–19.

Laasonen, P. (1958), "On the solution of Poisson's difference equation," *J. Assoc. Comput. Mach.*, Vol. 5, pp. 370–382.

Lagrange, J. L. (1759), "Recherches sur le nature de le propagation du son," *Miscell. Taurin*, 1. Also in *Works*, Vol. 1, pp. 39–148.

Lagrange, J. L. (1788), *Mécanique analytique*, Paris.

Lanczos, C. (1956), *Applied Analysis*, Prentice-Hall, Inc., Englewood Cliffs, N.J.

Lanczos, C. (1960), *The Variational Principles of Mechanics*, 2nd ed., University of Toronto Press, Toronto.

Langmuir, I., E. Q. Adams, and G. S. Meikle (1913), "Flow of heat through furnace walls: the shape factor," *Trans. Am. Electrochem. Soc.*, Vol. 24, pp. 53–76.

Laning, J. H., and R. H. Battin (1956), *Random Processes in Automatic Control*, McGraw-Hill Book Company, New York.

Lapidus, L. (1962), *Digital Computation for Chemical Engineers*, McGraw-Hill Book Company, New York.

Le Bail, R. C. (1972), "Use of fast Fourier transforms for solving partial differential equations in physics," *J. Comp. Phys.*, Vol. 9, 203, pp. 440–465.

Liebmann, G. (1954), "Resistance network analogues with unequal meshes or subdivided meshes," *Br. J. Appl. Phys.*, Vol. 5, pp. 362–366.

Liebmann, G. (1955), "Solution of elastic problems by the resistances network analogue," *Proc. 1st Int. Analogy Comput. Meet.*, Brussels, Sept. 1955, AICA, Rutgers University, New Brunswick, N.J.

Liebmann, G. (1956), "A new electrical method for the solution of Transient heat conduction problems," *Trans., ASME*, Vol. 78, pp. 655–665.

Liu, B. [ed.] (1975), *Digital Filters and the Fast Fourier Transform*, Halstead Press (division of John Wiley & Sons, Inc.), New York. A collection of reprints of papers.

McHenry, D. (1943), "A lattice analogy for the solution of plane stress problems," *J. Inst. Civ. Eng.*, Vol. 21, no. 2, pp. 59–82.

MacNeal, R. H. (1953), "An asymmetrical finite difference network," *Quart. Appl. Math.*, Vol. 11, pp. 295–310.

Mann, W. R., C. L. Bradshaw, and J. G. Cox (1957), "Improved approximations to differential equations by difference equations," *J. Math. Phys.*, Vol. 35, pp. 408–415.

Marchuk, G. I. (1966), *Numerical Methods for Nuclear Reactor Calculations.* Translated from the Russian by Consultants Bureau, New York.

Marchuk, G. I., and N. N. Yanenko (1965), "Applications of the method of splitting (fractional steps) to the solution of problems of mathematical physics," *IFIP Congress '65*, Spartan Books, New York.

Mayo, A. (1979), "Fast, high order accurate methods for solving elliptic equations on general regions," in R. Vichnevetsky and R. Stepleman, eds., *Advances in Computer Methods for Partial Differential Equations III*, IMACS, Rutgers University, New Brunswick, N.J.

Metcalfe, R. (1974), "Spectral methods for boundary value problems in fluid mechanics," Ph.D. thesis, Massachusetts Institute of Technology.

Meyer, G. H. (1973), *Initial Value Methods for Boundary Value Problems— Theory and Application of Invariant Imbedding*, Academic Press, Inc., New York.

Miele, A. (1968), "Method of particular solutions in non-linear, two-point boundary value problems," *J. Optimization Theory Appl.*, Vol. 2, No. 4, pp. 260–273.

Milne, W. E. (1953), *Numerical Solution of Differential Equations*, John Wiley & Sons, Inc., New York.

Milne, W. E., and L. M. Thomson (1951), *The Calculus of Finite Differences*, Macmillan International Ltd., London.

Mitchell, A. R. (1969), *Computational Methods in Partial Differential Equations*, John Wiley & Sons, Inc., New York.

Mitchell, A. R., and R. McLeod (1973), "Curved elements in the finite element method," in Lecture Notes in Mathematics, No. 363, Springer-Verlag, New York.

Mitchell, A. R., and R. Wait (1977), *The Finite Element Method in Partial Differential Equations*, John Wiley & Sons, Inc., New York.

Morrison, D. D., J. D. Riley and J. F. Zancanaro (1969), "Multiple shooting method for two-point boundary value problems," *Comm. ACM 5* (1962), pp. 613–614.

Mufti, I. H., C. K. Chow, and F. T. Stock (1969), "Solution of ill conditioned linear TPBVP by the Riccati transformation," *SIAM Rev.*, Vol. 11, No. 4, pp. 616–619.

Oden, J. T. (1972), *Finite Element of Nonlinear Continua*, McGraw-Hill Book Company, New York.

Orszag, S. A. (1971), "Numerical simulation of incompressible flows within simple boundaries: Galerkin (spectral) representations," *Studies in Appl. Math.*, Vol. 50, p. 395.

Orszag, S. A. (1972), "Comparison of pseudospectral and spectral approximation," *Stud. Appl. Math.*, Vol. 51, p. 253.

Orszag, S. A. (1979), "Spectral methods for problems in complex geometries," in R. Vichnevetsky and R. Stepleman, eds., *Advances in Computer Methods for Partial Differential Equations III*, IMACS, Rutgers University, New Brunswick, N.J.

Papoulis, A. (1962), *The Fourier Integral and Its Applications*, McGraw-Hill Book Company, New York.

Parlett, B. (1966), "Accuracy and dissipation in difference schemes," *Commun. Pure Appl. Math.*, Vol. 19, No. 1, p. 111.

Paschkis, V., and F. L. Ryder (1968), *Direct Analog Computers*, Wiley–Interscience, New York.

Peaceman, D. W., and H. H. Rachford (1955), "The numerical solution of parabolic and elliptic difference equations," *J. Soc. Ind. Appl. Math.*, Vol. 3, pp. 28–41.

Pearcy, C. (1962), "On the convergence of alternating direction procedures," *Numer. Math.*, Vol. 4, pp. 172–176.

Petrovsky, I. G. (1955), *Lectures on Partial Differential Equations*, Interscience Publishers, New York.

Phillips, H., and N. Wiener (1923), "Nets and the Dirichlet problem," *J. Math. Phys.*, Vol. 2, pp. 105–124.

Pinder, G. F., and E. O. Frind (1972), "Application of Galerkin's procedure to aquifer analysis," *Water Resour. Res.*, Vol. 8, No. 1, pp. 108–120.

Pinder, G. F., and W. G. Gray (1977), *Finite Element Simulation in Surface and Subsurface Hydrology*, Academic Press, Inc., New York.

Poincare, H. (1890), "Sur les équations aux dérivées partielles de la physique mathématique," *Am. J. Math.*, Vol. 12, pp. 211–294.

Poisson, S. D. (1813), *Nouveau Bulletin de la Société Philosophique*, Vol. 3.

Rayleigh (John William Strutt, Baron Rayleigh) (1877), *Theory of Sound*, 2 vols., The Macmillan Company, London. 2nd ed. (1894) reprinted by Dover Publications, Inc., New York, in 1945.

Reich, E. (1949), "On the convergence of the classical iterative methods of solving linear simultaneous equations," *Am. Math. Stat.*, Vol. 20, pp. 448–451.

Reid, W. T. (1967), "Generalized green matrices for two-point boundary value problems," *SIAM J. Appl. Math.*, Vol. 15, No. 4, pp. 856–870.

Rice, J. R. (1977), "ELLPACK: A research tool for elliptic PDE's software," in J. R. Rice, ed., *Mathematical Software III*, Academic Press, Inc., New York.

Richardson, L. F. (1908), "Freehand graphic way of determining stream line and equipotentials," *Proc. Phys. Soc., Lond.*, Vol. 21.

Richardson, L. F. (1910), "The approximate arithmetical solution by finite-differences of physical problems involving partial differential equations, with an application to the stresses in a masonry dam," *Philos. Trans. R. Soc. Lond., Ser. A*, Vol. 210, pp. 307–357, and *Proc. R. Soc. Lond., Ser. A.*, Vol. 83, pp. 335–336.

Ritz, W. (1908), "Uber eine neue Methode zur Lösung gewisser variations Probleme der mathematischen Physik," *Z. Reine Angew. Math.*, Vol. 135, pp. 1–61.

Roache, P. J. (1972), *Computational Fluid Dynamics*, Hermosa Publ., Albuquerque, N. Mex.

Rockey, K. C., H. R. Evans, D. W. Griffiths, and D. A. Nethercot (1975), *The Finite Element Method—A Basic Introduction*, John Wiley & Sons, Inc., New York.

Rose, D. J., and R. A. Willoughby (1972), *Sparse Matrices and Their Applications*, Proc. IBM Conf., Sept. 1970, Plenum Press, New York.

Runge, C. (1903), *Z. Math. Phys.*, Vol. 48, pp. 433.

Runge, C. (1905), *Z. Math. Phys.*, Vol. 53, pp. 117.

Schiesser, W. E. (1972), "A digital simulation system for higher-aimensional partial differential equations," *Proc. 1972 Summer Comput. Simul. Conf.*, San Diego, Calif., June, pp. 62–72.

Schoenberg, I. J. (1946), "Contributions to the problem of approximation of equidistant data by analytic functions, parts A and B," *Quart. Appl. Math.*, Vol. 4, pp. 45–99, 112–141.

Schoenberg, I. J. (1973), *Cardinal Spline Interpolation*, Regional Conference Series in Applied Mathematics, Vol. 12, Society for Industrial and Applied Mathematics, Philadelphia.

Schultz, M. H. (1973), *Spline Analysis*, Prentice-Hall, Inc., Englewood Cliffs, N.J.

Seidel, L. (1874), "Ueber ein Verfahren, die Gleichungen, auf welche die Methode der kleinsten Quadrate fuhrt, sowie lineare Gleichungen überhaupt, durch successive Annäherung aufzulösen," *Abh. Math.-Phys. Kl., Bayer. Akad. Wiss. Munchen*, Vol. 2, No. 3, pp. 81–108.

Sheldon, J. W. (1958), "Algebraic approximations for Laplace's equation in the neighborhood of interfaces," *Math. Tables Other Aids Comput.*, Vol. 12, pp. 174–186.

Shortley, G. H., and R. Weller (1938), "The numerical solution of Laplace's equation," *J. Appl. Phys.*, Vol. 9, pp. 334–348.

Smirnov, V. I. (1964), *A Course of Higher Mathematics*, Vol. 4, Pergamon Press, Inc., London. English translation published by Addison-Wesley Publishing Co., Inc., Reading, Mass.

Smith, G. D. (1965), *Numerical Solution of Partial Differential Equations*, Oxford University Press, New York.

Southwell, R. V. (1940), *Relaxation methods in Engineering Science*, Oxford University Press, London.

Southwell, R. V. (1946), *Relaxation Methods in Theoretical Physics*, Oxford University Press, London.

Stacey, W. (1967), *Modal Approximations*, Research Monograph No. 41, The MIT Press, Cambridge, Mass.

Stiefel, E. (1955), "Relaxationsmethoden bester Strategie zur Lösung lineare gleichungssysteme," *Comment. Math. Helv.*, Vol. 29, pp. 157–179.

Strang, G. (1971), "The finite-element method and approximation theory," in B. Hubbard, ed., *Numerical Solution of PDE's—II (SYNSPADE 1970)*, Academic Press, Inc., New York.

Strang, G., and G. J. Fix (1973), *An Analysis of the Finite Element Method*, Prentice-Hall, Inc., Englewood Cliffs, N.J.

Strutt, J. W. (later called 3rd Baron Rayleigh) (1873), "Some general theorems relating to vibrations," *Proc. Lond. Math. Soc.*, Vol. 4, pp. 357–368.

Swartz, B., and B. Wendroff (1969), "Generalized finite-difference schemes," *Math. Comput.*, Vol. 22, pp. 37–49.

Swartz, B., and B. Wendroff (1974), "The relation between the Galerkin and collocation methods using smooth splines," *SIAM J. Numer. Anal.*, Vol. 11, No. 5, pp. 994–996.

Swartztrauber, P. N. (1974), "The direct solution of the discrete Poisson equation on the surface of a sphere," *J. Comp. Phys.*, Vol. 15, pp. 46, 54.

Taig, I. C. (1961), "Structural analysis by the matrix displacement method," *English Electric Aviation Report, No. SO17.*

Tewarson, R. P. (1973), *Sparse Matrices*, Academic Press, Inc., New York.

Timoshenko, S. P. (1954), *History of Strength of Materials*, McGraw-Hill Book Company, New York.

Timoshenko, S., and J. N. Goodier (1951), *Theory of Elasticity*, 2nd ed., McGraw-Hill Book Company, New York.

Turner, M. J., R. W. Clough, H. C. Martin, and L. C. Topp (1956), "Stiffness and deflection analysis of complex structures," *J. Aeronaut. Sci.*, Vol. 23, No. 9, pp. 805–823.

van den Dungen, F. H. (1952), "Formules pour l'intégration numérique de l'équation des ondes, I and II," *Acad. R. Belg. Bull. Cl. Sci., 5th ser.*, Vol. 38, pp. 39–49, 669–684.

Varga, R. S. (1957), "A comparison of the successive overrelaxation method and semi-iterative methods using Chebyshev polynomials," *J. Soc. Ind. Appl. Math.*, Vol. 5, pp. 39–46.

Varga, R. S. (1959), "Orderings of the successive overrelaxation scheme," *Pac. J. Math.*, Vol. 9, pp. 925–939.

Varga, R. S. (1962), *Matrix Iterative Analysis*, Prentice-Hall, Inc., Englewood Cliffs, N.J.

Varga, R. S. (1971), *Functional Analysis and Approximation Theory in Numerical Analysis*, Regional Conference Series in Applied Mathematics, Vol. 3, Society for Industrial and Applied Mathematics, Philadelphia.

Vichnevetsky, R. (1968), "A new stable computing method for the serial hybrid computer integration of partial differential equations," *Proc. 1968 Spring Joint Computer Conference*, AFIPS, Vol. 32, Thompson Books, Co., pp. 143–150.

Vichnevetsky, R. (1969a), "Generalized finite-difference approximations for the solution of initial value problems," *Simulation*, Vol. 12, No. 5, pp. 233–237.

Vichnevetsky, R. (1969b), "Serial solution of parabolic partial differential equations: The decomposition method for non-linear and space dependent problems," *Simulation*, Vol. 13, No. 1, pp. 47–48.

Vichnevetsky, R. (1970), "The method of decomposition for space dependent boundary value problems," *Princeton Conference on Information Science and Systems*, Princeton University, March 27, 28, 1970.

Vichnevetsky, R. (1971), "The method of decomposition for unstable two-point boundary value problems," *IEEE Transactions on Computers*, Vol. C-20, No. 8, pp. 910–914.

Vichnevetsky, R. (1975a), "Efficient solution of Poisson's equation in a rectangle," *NAM-186*, Dept. of Computer Science, Rutgers University, New Brunswick, N.J.

Vichnevetsky, R., ed. (1975b), *Advances in Computer Methods for Partial Differential Equations*, AICA, Rutgers University, New Brunswick, N.J.

Vichnevetsky, R., ed. (1977), *Advances in Computer Methods for Partial Differential Equations II*, IMACS, Rutgers University, New Brunswick, N.J.

Vichnevetsky, R., and R. Stepleman, eds. (1979), *Advances in Computer Methods for Partial Differential Equations III*, IMACS, Rutgers University, New Brunswick, N.J.

Volinskii, B. A., and V. Ye. Bukhman (1965), *Analogues for the Solution of Boundary Value Problems*, Pergamon Press, Oxford. Translated from the original Russian edition (1960).

Wachspress, E. L. (1966), *Iterative Solution of Elliptic Systems*, Prentice-Hall, Inc., Englewood Cliffs, N.J.

Wachspress, E. L. (1975), *A Rational Finite Element Basis*, Academic Press, Inc., New York.

Walsh, J. L., and David Young (1953), "On the accuracy of the numerical solution of the Dirilecht problem by finite differences," *J. Res. Natl. Bur. Stand.*, Vol. 51, pp. 343–363.

Walsh, J. L., and David Young (1954), "On the degree of convergence of solutions of difference equations to the solution of the Dirilecht problem," *J. Math. Phys.*, Vol. 33, pp. 80–93.

Wasow, Wolfgang (1952), "On the truncation error in the solution of Laplace's equation by finite differences," *J. Res. Natl. Bur. Stand.*, Vol. 48, pp. 345–348.

Wasow, W. (1955), "Discrete approximation to elliptic differential equations," *Z. Angew. Math. Phys.*, Vol. 6, pp. 81–97.

Witsenhausen, H. (1961), "Hybrid solution of boundary value problems by backward and forward integration," *PCC Report No. 175*, Electronic Associates, Inc., Princeton, N.J.

Wright, K. (1964), "Chebyshev collocation methods for ordinary differential equations," *Comput. J.*, Vol. 6, pp. 358–365.

Young, D. M. (1954a), "Iterative methods for solving partial difference equations of elliptic type," *Trans. Am. Math. Soc.*, Vol. 76, pp. 92–111.

Young, D. M. (1954b), "On Richardson's methods for solving linear systems with positive definite matrices," *J. Math. Phys.*, Vol. 32, pp. 243–255.

Young, D. M. (1971), *Iterative Solution of Large Linear Systems*, Academic Press, Inc., New York.

Young, L. C. (1977), "A preliminary comparison of finite-element methods for reservoir simulation," in R. Vichnevetsky, ed., *Advances in Computer Methods for Partial Differential Equations—II*, IMACS, Rutgers University, New Brunswick, N.J.

Zenicek, A. (1970), "Interpolation polynomials on the triangle," *Numer. Math.*, Vol. 15, pp. 283–296.

Zienkiewicz, O. C. (1970), "The finite element method: from intuition to generality," *Appl. Mech. Rev.*, Vol. 23, No. 3, pp. 249–256.

Zienkiewicz, O. C. (1977), *The Finite Element Method*, 3rd ed., McGraw-Hill Book Co. (UK) Ltd., Maidenhead, England.

Zienkiewicz, O. C., and Y. K. Cheung (1967), *The Finite Element Method in Structural and Continuum Mechanics*, McGraw-Hill Book Company, New York.

Zienkiewicz, O. C., and G. S. Holister, eds. (1963), *Stress Analysis*, John Wiley & Sons, Inc., New York.

Zlamal, M. (1968), "On the finite element method," *Numer. Math.*, Vol. 12, pp. 394–409.

Index

A

Acceleration factor, 127
Advection, 259
Aliasing, 51–53, 86
Alternating directions implicit
  method, 135
Andre, H., 94
Angel, E., 94
Area coordinates, 300
Argyris, J.H., 181, 204
Assembling, 201, 243, 247–48,
  287

B

Baker, V.A., 141
Banded matrices, 116, 118, 121,
  136
Basis functions, 24, 28, 179
Beam:
  cantilever, 177
  elastic, 175
  finite element approximation,
    229
Bellman, R., 94
Bernouilli, Daniel, 179
Besseling, J.F., 204
Bessel's inequality, 26
Beuken, C.L., 161
Bickley, W.G., 103
Biezeno, C.B., 191

Biharmonic equation, 95
Biharmonic operator, 102
Bilinear basis functions, 280
Bilinear elements on rectangles,
  279
  for Poisson's equation, 281
Birkoff, G., 102, 241
Bjorck, A., 140
Block tridiagonal matrices, 138
Boundary conditions:
  of the first kind, 17, 267
  of the second kind
    (Neumann), 17, 212–14
  of the third kind (Robbin's),
    17, 268
Brandt, A., 135
Brigham, E.O., 53
Bukhman, V.Ye., 162
Buneman, O., 152

C

Calculus of variations, 26, 165
Canonical classification of
  partial differential
  equations, 8
Carey, G.F., 307
Cauchy problem, 11
Caussade, B.H., 162
Change of coordinates, 309
Chapeau basis functions, 30, 196
Characteristic lines, 9

Cholesky method for symmetric
  matrices, 117, 141
Circular coordinates, 315
Circular membrane, 249–53
Clough, R.W., 203
Collatz, L., 102, 106, 132
Collocation method, 188
Collocation on finite elements,
  305–8
  orthogonal, 307
Collocation points, 60
Complex variables, 35
Computation molecule, 98
  for the biharmonic equation,
    105
Computing effort, 53, 90, 109,
  119
Concentrated source, 210, 255
Conductivity matrix, 209
Conforming elements, 290
Consistency, 101
Continuity requirements, 288,
  320
  with collocation methods, 306
Contraction mapping theorem,
  144
Convergence, 23, 81, 101, 102
  of iterative methods, 129, 131,
    143
Convolution theorems, 41
Cooley, J.W., 53
Courant, R., 12, 23, 180, 204, 286
Cowper, G.R., 335

Crandall, S.H., 188
Cross, H., 128, 129
Curved boundaries, 106
Curvilinear elements, 315
Cylindrical coordinates, 256
Czerwinska, A., 162

D

Dahlquist, G., 140
d' Alembert, J. le Rond, 3, 4, 15, 179
de Boor, C., 308
de Moivre's theorem, 37
Differential operators:
    approximation, 58
    eigenfunctions of, 27, 191
    Hermitian, 191
    self adjoint, 191–92
Diffusion equation, 181, 205, 209, 211, 255
Dirichlet problem, 17, 199, 236
Discrete Fourier series, 47
    sine series, 50
Discrete operator notations, 68
Displacement method, 225
Displacements, definition, 227
Domain of dependence, 14
Double Fourier series method for Poisson's equation, 149
Double-sweep method:
    for Poisson's equation in circular coordinates, 155
    for Poisson's equation in rectangular coordinates, 152
    for tridiagonal systems, 77

E

Elastic beam:
    finite element approximation, 229
Energy:
    external, 172
    internal, 172
    spectral decomposition, 45
Ergatoudis, J., 321
Euler, L., 7, 165, 179
Euler–Lagrange equation, 168
Euler's formula for complex numbers, 37
Extremal curves, 168

F

Fast Fourier Transform (FFT), 53
Finite difference approximation, 17
    near boundaries, 106

Finite difference (cont.):
    of differential operators, 62, 67
    of first derivative, 63
    of Laplacian operator, 103
    of Poisson's equation, 19, 99, 104–6
    of second derivative, 64
    on a triangular mesh, 106
    of two-point boundary value problems, 76
Finite element shapes:
    on rectangles, 291–98, 302–5
    on triangles, 298–302
Finlayson, B., 307
Fix, G., 187
Flexural rigidity, 228
Forsythe, G., 126, 133
Fourier, Joseph, 5
Fourier integral–Fourier transform, 39
Fourier method for two-point boundary-value problems, 89
Fourier's equation, 4, 6, 12, 13
Fourier series, 25, 43
    discrete, 47
    sine, 46
Fox, L., 92
Fraeijs de Veubeke, B.M., 204
Frankel, S.P., 133
Friedrichs, K.O., 23, 102
Functionals: 264
    definition, 165
    higher order, 173
    in several dimensions, 177
Function expansion methods, 24, 179

G

Galerkin, B.G., 29, 180
Galerkin's method, 24, 28, 29, 180, 185, 186, 288
    with bilinear elements, 281
    for Poisson's equation, 238
Gauss, C.G., 110, 128, 180
Gaussian elimination, 110
Gauss–Seidel method, 134
George, A., 140
Gerschgorin, S., 102
Global curvilinear coordinates, 315, 316, 323
Global errors, 81, 101
Golub, G.H., 152
Gradients, 243, 313
Gulati, S., 83
Guthenmaker, L.I., 162

H

Hadamard, J., 11
Harmonic equation, harmonic function, 98

Heat equation, 4, 5, 205, 309
Helmholz's equation, 236
Hermite basis functions:
    in one dimension, 220
    in two dimensions, 302
Hermite cubic finite elements, 219
Hermite cubic polynomials, 230
Hermite interpolation, 219
Hermitian elements, 219, 289–90, 302–5
Hermitian property, 191
Hilbert, D., 12
Hockney's method, 151
Hrenikoff, A., 203, 204

I

Ill-posed problems, 12-13
Initial value methods for two-point boundary-value problems, 91
Integration by parts, 31, 187, 197
Invariant imbedding, 94
Irons, B., 321
Isaacson, E., 144
Isoparametric elements, 318
    bilinear, 323
Iterative methods:
    Alternating directions (ADI), 135
    Convergence of, 127, 129
    Fixed point, 144
    Gauss–Seidel, 124
    Jacobi, 123
    nonlinear equations, 141–46
    Overrelaxation, 127
    Picard, 144
    SOR method, 128
    Young–Frankel Theory, 133

J

Jacobi, C.G., 123, 128
Jacobian, 312
Jacobian matrix, 144, 146, 312, 315
Janenko, N.N., 135
Jones, R.E., 204

K

Kantorovich, L.V., 162
Karplus, W.J., 161
Keller, H.B., 82, 92, 144
Kirchhoff's first law, 158
Krylov, V.I., 162

L

L_2 Norm, 26
    of error, 84

L_2 Norm (cont.):
  relation to Fourier transform, 42
Lagrange, J.L., 173, 179
Lagrange interpolation, 215
Lagrangian basis functions, Lagrangian elements:
  on rectangles, 295
  on triangles, 301
Lanczos, C., 53, 188, 189
Langmuir, I., 161
Laplace, P.S., 7, 58
Laplace's equation, 5, 95, 159
Laplacian operator, 19, 102
Legendre, A.M., 330
Legendre polynomials, 308, 330
Lewy, H., 23, 102
Liebmann, G., 129, 161
Linear interelement sides, 283
Lipschitz condition, 144
Loaded string, 170
Local curvilinear elements, 318
LU decomposition, 114

M

MacNeal, R.H., 162, 204
Mayo, A., 158
Mean value theorem, 98
MERMAID computer code, 269
Meyer, G., 94
Milne, W.E., 109
Minimum distance, 168
Möbius coordinates, 300
Moler, C., 126
Morrison, D.D., 92

N

Natural boundary conditions:
  for the beam equation, 177
  for Poisson's equation, 263
Neumann Problem, 17, 236
Newton–Raphson method, 144
Nodal displacements, 227, 231
Nonconforming elements, 290
Nonhomogeneous diffusion, 205, 254
Nonlinear equations, 141
Nonlinear problems, 110
Numerical differentiation, 59
Numerical integration, 307, 327–35
  Gauss formulas, 332–33
  Newton–Cotes formulas, 329
  on triangles, 335

O

Odd-even reduction, 152
Operator decomposition method for two-point

Operator decomposition (cont.):
  boundary-value problems, 92
Order of accuracy, 80, 83, 100
Orszag, S., 188
Orthogonal collocation method, 189
Orthogonality, 185, 186, 191, 330

P

Parametric transformations, 289, 318
Parseval's relation, 41–42, 45
Particle dynamics, 147
Paschkis, V., 161
Peaceman, D.W., 135
Phase angle, 36
Phillips, H., 102
Plancherel, 43
Pointers, 141
Poisson's equation, 16, 95, 147–58
  axisymmetric case in three dimensions, 256
  generalized, 239
Poisson solvers, 147–58
Polynomial interpolation, 60
  Lagrange form, 62
Positive-definition matrix, 125
Potential equation, 5
Principle of virtual works, 226

Q

Quadratic basis functions, 125
Quadratic finite-elements, 214
Quadratic form associated with a matrix, 125
Quasilinearization, 141

R

Rachford, H.H., 135
Rayleigh, Baron (J.W. Strutt), 180
Rayleigh–Ritz method, 24, 27, 181
Reich, E., 133
Relaxation method, 128
Renard, G., 162
Residuals, 24, 185, 188
Riccati equation, 93
Richardson, L.F., 15–16, 129
Ritz, W., 180
Robbin's boundary condition, 236
Roof basis functions, 196
Rose, D.J., 141
Roundoff errors, 79
Runge, C., 53

S

Sampling, 51
Seidel, L., 128
Self-adjoint operators, 191
Serendipity elements, 294
Simpson's rule, 32, 198
Solution methods for linear systems:
  direct, 109
  iterative, 123
Solution space, 289
Southwell, R.V., 128, 129
Space-dependent diffusivity, 205, 254
Sparse matrices, 135–41
  profile storage, 140
Spectral methods, 191
Splitting techniques, 135
Stability, 101
Stacey, W., 194
Stencil notation, 19, 282
Stiefel, E., 126
Stiffness matrix, 33, 209, 248
  in displacement method, 234
Störmer-Numerov formula, 83, 88
Strains, 226, 227
Strang, G., 187
Stresses, 226, 227
Strong solution, 187
Sub-and superparametric transformations, 326
Subdomain method, 191
Subroutines, 269
Surface heat flux, 212
Swartz, B., 308
Swartztrauber, P.N., 158

T

Taig, I.C., 320
Taylor, B., 58, 179
Taylor Series, 21
  symbolic expression, 68, 103
Test functions, 187
Tewarson, R.P., 141
Thomas's method, 77
Trial space, 289
Triangular elements, 240
  curved, 317
  for Poisson's equation, 235
Triangular matrices, 114
Tridiagonal systems, 77, 137
Truncation error, 80, 99
  Fourier analysis of, 83
Tukey, J.W., 53
Turner, M.J., 204
Two-point boundary-value problems, 21
  finite-difference approximation, 21, 16–24

Two-point (cont.):
  finite-element approximation,
    205–14
  Fourier method, 89

U

Undetermined coefficients,
  method of, 66

V

Varga, R., 130

Vichnevetsky, R., 92, 93, 94, 152,
  270
Virtual work, 172, 195

W

Wachspress, E.L., 289
Wasow, W., 133
Wave equation, 3
Weak solution, 187, 289
Weighted residual method, 28,
  188

Well-posed problems, 10–14
  for elliptic equations, 16–17
Wiener, N., 102
Willoughby, R.A., 141

Y

Young, D., 131, 133
Young's elasticity modules, 175

Z

Zienkiewicz, O.C., 241, 294, 321